Artificial Intelligence-Based 6G Networking

Artificial Intelligence-Based 6G Networking focuses exclusively on the upcoming sixth-generation (6G) network and services slated for implementation by 2030. It explores the paradigm shift that is 6G. It discusses the deep integration of computing and communication, supported by artificial intelligence (AI) across network elements like cloud, edge, and terminals. It also examines how AI-native interfaces will permeate various network components, from radio access networks to application servers and databases.

Proposing a unified AI-enabled framework for optimizing networks and applications as a single integrated system, the book covers how network service providers can tailor network baselines, reduce noise, and accurately identify issues. The book delves into the potential of AI-driven networks to self-correct, predict, and rectify service degradations proactively, enhancing uptime and troubleshooting efficiency. It outlines the "Connection, Communication, Collaboration, Curation, and Community" framework to enhance network effects, aiding operators in automation, cost reduction, and providing optimal user experiences.

Covering topics from MIMO and Massive MIMO to holographic communications, cybersecurity and quantum communications, the book explores cutting-edge technologies shaping the future of 6G networks. It anticipates a future where AI, along with machine learning and deep learning, enables continuous learning, self-optimization, and predictive maintenance, even with full automation, that will be the hallmark of a new era in network connectivity and innovation.

Radhika Ranjan Roy is an Electronics Engineer (Research), United States, Engineering & Systems Integration Directorate, Command Control, Computers, Communications, Cyber, Intelligence, Surveillance and Reconnaissance (C5ISR) Center, Aberdeen Proving Ground (APG), Maryland, since 2009. Before joining to US Army Research, he worked in various capacities in CACI, SAIC, AT&T/Bell Laboratories, CSC, and PDB since his graduation. He holds a PhD in electrical engineering with major in computer communications from the City University of New York, New York, in 1984, and MS in electrical engineering from the Northeastern University, Boston, Massachusetts, in 1978. He received his BS in electrical engineering from the Bangladesh University of Engineering and Technology, Dhaka, Bangladesh, in 1967. He has published more than 50 technical papers. He is holding and/or has submitted over 35 patents. In addition, Dr. Roy has authored five books:

Networked Artificial IntelligenceAI-enabled 5G Networking, CRC Press, 2024, Handbook on Networked Multipoint Multimedia Conferencing and Multistream Immersive Telepresence using SIP, CRC Press, 2021, Handbook of SDP for Multimedia Session Negotiations: SIP and WebRTC IP Telephony, CRC Press, 2018, Handbook on Session Initiation Protocol: Networked Multimedia Communications for IP Telephony, CRC Press, 2016, and *Handbook of Mobile Ad Hoc Networks for Mobility Models*, Springer Nature Customer Service Center LLC., October 2010.

Artificial Intelligence-Based 6G Networking

Radhika Ranjan Roy

CRC Press
Taylor & Francis Group
Boca Raton London New York

CRC Press is an imprint of the
Taylor & Francis Group, an **informa** business
AN AUERBACH BOOK

First edition published 2025
by CRC Press
2385 NW Executive Center Drive, Suite 320, Boca Raton FL 33431

and by CRC Press
4 Park Square, Milton Park, Abingdon, Oxon, OX14 4RN

CRC Press is an imprint of Taylor & Francis Group, LLC

© 2025 Taylor & Francis Group, LLC

ISBN: 978-1-032-81221-2 (hbk)
ISBN: 978-1-032-81371-4 (pbk)
ISBN: 978-1-003-49948-0 (ebk)

DOI: 10.1201/9781003499480

Typeset in Times
by SPi Technologies India Pvt Ltd (Straive)

Dedication

This book is dedicated to our beloved grandson Amor.

To my Grandma for her causeless love, my parents Rakesh Chandra Roy and Sneholota Roy whose spiritual inspiration remains vividly alive within all of us, my late sisters GitaSree Roy, Anjali Roy, and Aparna Roy and their spouses and my brother Raghunath Roy and his wife Nupur for their inspiration, my daughter Elora and my son-in-law Nick, my son Ajanta, daughter-in-law Nageen and their son and our grandson Amor, and finally my beloved wife Jharna for their love.

To our dearest son, Debasri Roy, Medicinae Doctoris (MD) (January 20, 1988–October 31, 2014) who had a brilliant career (summa cum laude, undergraduate) and had so much more to contribute to this country and the world. He had been so eager to see this book published, saying, "Daddy, you are my hero." Pictured with his fiancé, "Love is forever," he wrote as his lifelong wishes: "I am ready for my life, I like to see the whole world, I will read spiritual scriptures to find the mystery of life, and I want to make a difference in the world." He loved all of us from the deepest part of his kindest heart, including his long-time fiancé who was also an MD and his classmate, and to whom he was engaged and wanted to be married on October 30, 2015, in front of relatives, friends, colleagues, and neighbors who, until his last breath, he had inspired in so many ways to love God, along with his prophetic words: "Mom, I am extremely happy that you are with me and I want to be happy in life." May God let his soul live in peace in His abode.

Contents

Preface

I have been working on networked artificial intelligence for a long time at my present position at United States, Engineering & Systems Integration Directorate, Command Control, Computers, Communications, Cyber, Intelligence, Surveillance and Reconnaissance (C5ISR) Center for artificial intelligence, machine learning, and deep learning (AI/ML/DL) for large-scale global networks, a term coined with networked AI, since 2009. We have published a separate book that deals with the fifth generation (5G) wireless technologies and networking with full treatment of networked AI first of this kind. This book solely deals with the emerging sixth generation (6G) network and services which is expected to be implemented in the year 2030. In 6G, a paradigm shift is happening with the deep convergence of computing and communication with inherent support of AI in network functional entities. Distributed AI is becoming immersive in all elements of the network, for example, cloud, edge, and terminal devices, which make AI virtually operating as a networking system.

6G networks will be using AI native interfaces from radio access networks (RANs), wireline access networks, and long-haul wide area networks including smart intelligent user equipment (UE), switches/routers, application servers, and databases. AI has achieved great success in areas including computing vision, natural language processing, multimedia production, medical and medicine applications, and even human genetics, among others. However, this book has provided a unified framework for the deep convergence of computing and communications, where the network and application/service can be jointly optimized as a single integrated system using AI. Network service providers can customize the network baseline for alerts, reducing noise and false positives while enabling information technology (IT) teams to accurately identify issues, trends, anomalies, and root causes. AI/ML/DL techniques are also used to reduce unknowns and improve the level of certainty in decision making. Networked AI/ML/DL even can enable IT systems to self-correct for maximum uptime and provide prescriptive actions as to how to fix problems that occur. In addition, AI-driven networks can capture and save data prior to a network event or outage, helping to speed troubleshooting. Connection, Communication, Collaboration, Curation, and Community are the five steps that help to boost network effects. Network operators gain insights through analytics and AI/ML/DL that guide more trusted automation processes that lower the cost of network operations and provide users with an optimal connected experience. One important thing is that AI/ML/DL will increasingly enable networks to continually learn, self-optimize, and even predict and rectify service degradations before they occur even with full automation.

This book has 18 chapters as follows:

1. MIMO
2. Massive MIMO
3. Ultra-Massive MIMO

4. Reconfigurable Intelligent Surface
5. Cell-Free Communications
6. Non-Terrestrial Communications
7. Massive MIMO Radar
8. Device-to-Device Communications with Mesh Networking
9. Vehicle-to-Vehicle and Vehicle-to-Everything Communications
10. Robotics and Autonomous Communications
11. Virtual, Augmented, Mixed Reality and Extended Reality
12. Unmanned Aerial Vehicles Communication Systems
13. Unmanned Robot Network with Swarm Capabilities
14. Digital Twins
15. Holographic Communications
16. Ultra-Dense Network
17. Quantum Communications
18. Cybersecurity

The descriptions of all these technologies in detail is a huge undertaken. The book has covered each of the above topics succinctly as far as possible. Each of these sections and subsections are being covered with the texts of from the latest renown published research and development papers. I am deeply indebted to these researchers who have published these papers. I have provided a list of references in each chapter with the names and date of publication. In addition, I have authored five books: Networked Artificial Intelligence: AI-enabled 5G Network, CRC Press 2024, Handbook on Networked Multipoint Multimedia Conferencing and Multistream Immersive Telepresence using SIP, CRC Press, 2021, Handbook of SDP for Multimedia Session Negotiations: SIP and WebRTC IP Telephony, CRC Press, 2018, Handbook on Session Initiation Protocol: Networked Multimedia Communications for IP Telephony, CRC Press, 2016, and Handbook of Mobile Ad Hoc Networks for Mobility Models, Springer Nature Customer Service Center LLC., October 2010.

1 Overview

The future sixth-generation (6G) technology is going to revolutionize the entire spectrum of wireless communications in an unprecedented way. This technology is still in the research phase in both private and government arenas. Many academics, researchers, industries, and industry alliances, for example Next G Alliance (NGA) [1], Sixth Generation Partnership Project (6GPP) [2], and Institute of Electrical and Electronics Engineers (IEEE) International Network Generations Roadmap (INGR) [3], and other open industry fora and governments throughout the world have been studying the 6G technologies, assuming that the 6G network could be rolled out around the year 2030. The Federal Communications Commission (FCC) is expected to allocate spectrum 96 Gigahertz (GHz) to 3 Terahertz (THz). In 6G, AI will be used not only for computing but also for communications over the 6G network, where each network functional entity will be enabled. In addition, we are expecting legacy protocols from physical layer to link layer to network layer to transport layer to middleware layer to application layer will be artificial intelligence (AI) enabled as the legacy networking protocols of these layers are inefficient to meet the key performance indicator (KPI). However, new AI standard-based networking protocols of different layers need to be standardized by the standard bodies like international standard organization (ISO), internet engineering task force (IETF), international telecommunications union-telecommunication (ITU-T), and/or other fora appropriate for interoperability in a multivendor environment offering economies of scale. The following AI standard-based networked technologies will drive the 6G network with full automation:

- Pervasive AI standard-based immerse communications and applications, spatio and temporal services, critical services (e.g., smart manufacturing), extreme networking (e.g., autonomous driving), internet of nano things, edge and cloud computing, and global broadband
- AI standard-based cell-free communications with reconfigurable 6G radio front ends (RFE) for dynamic spectrum access
- AI standard-based distributed networked sensing and communications that include sensors that are tightly integrated with communications to support autonomous systems
- AI standard-based network-enabled robotics and autonomous systems that can perceive their surroundings using sensors, not limited to, such as global positioning system (GPS), light detection and ranging (LiDAR), sonar, radar, and odometry
- AI standard-based network-enabled multisensory extended virtual and augmented reality, and mixed (EVR/EAR/EMR) reality, which will usher immersive technologies including the things like virtual and augmented Reality (VR/AR)

DOI: 10.1201/9781003499480-1

- AI standard-based personalized fully automated user experiences will be enabled in real time, and secure personalization of devices, networks, products, and services based on a user's personal profile and context information (e.g., user's preferences, trends, biometrics, and contextual information)
- AI standard-based joint communications and sensing (JCAS)
- AI standard-based massive multiple-input multiple output (MIMO) communications with ultra-antenna arrays and beam management overcoming increased absorption and pathloss for both cell-based and cell-free communications
- AI standard-based reconfigurable intelligence surface (RIS); also known as intelligent reflecting surface (IRS)
- AI standard-based unmanned robot network (URN) with swarm capabilities
- AI standard-based unmanned air vehicle (UAV) with swarm capabilities
- AI standard-based vehicle-to-vehicle (V2V)/device-to-device (D2D) mesh communications networking
- AI standard-based start-of message (SOS) Morse procedural messaging
- AI standard-based digital twins
- AI standard-based Massive MIMO radar
- AI standard-based integrated communications and sensing based on RIS and Massive MIMO AI standard-based applications-enabled and -enhanced by RIS-aided Massive MIMO communications
- AI and blockchain standard-based secure communications and sensing
- Quantum communications
- Quantum communications-based encryption

We are at the start of the 6G research era, where both academia and industries are actively investigating new, promising use cases and technologies intended for 6G. The Key Performance Indicators (KPIs) are either evolved from previous-generation KPIs or newly derived from 3GPPs, "Beyond 5G/6G" use cases, and 6G feature/capability/network targets and other industry documentations. Table 1.1 provides the overall summary of expected KPI for 6G networks.

TABLE 1.1: 6G KPI

KPI	5G	6G
Spectrum band	1–66 GHz	0.1–3 THz
Peak data rate	20 Gbps	1 Tbps
User-experienced data rate	100 Mbps	1 Gbps
Peak spectral efficiency	0.3b/s/Hz	3 b/s/Hz
End-to-end latency	10 milliseconds	1 millisecond
Radio-only latency	1 millisecond	100 microseconds
Block error rate (BLER)	10^{-5}	10^{-9}
Connection density	10^{+6} devices/Km^2	10^{+7} devices/km^2
Network energy efficiency	Not specified	1pJ/b
Mobility	500 km/h	1,000 km/h

TABLE 1.2: 6G Technology Vision

Spectrum of 6th Generation (6G) Wireless Technology (Research Stage)	Remarks
~less than 1 microsecond	
100 GHz to 3 THz (millimeter wave to THz)	6G network routers/switches will be
FCC (USA): 96 GHz to 3 THz	inherently AI-enabled
Holographic communications	
Pervasive AI/ML/DL-based communications and applications	
Quantum communications	
Internet-of nano-things (IoNT)	
Immersed communication	
Computer AI services	
Global broadband	
Omnipresence IoT	
Spatio-temporal services	
Critical services	

6G is also likely to feature entirely new capability dimensions including extreme performance and coverage, integrated- and cognitive computing functions within the network, and even functionality beyond communication such as spatial and timing data, as well as joint communication and sensing. We envision the 6G as a union of physical space, cyberspace, and connectivity with intelligence. Also, interactivity is a critical component to provide users with a truly immersive experience. Table 1.2 depicts some of the performance parameters and spectrum.

6G is not only about moving bits. It will become a framework of services, including communication services. In 6G, all user-specific computation and intelligence may move to edge cloud. Integration of sensing, imaging, and highly accurate positioning capabilities with mobility open a myriad of new applications in 6G. The next alliance has published the 6G application-specific KPI as depicted, in Table 1.3, as well as application-specific robustness KPI shown in Table 1.4.

6G ARTIFICIAL INTELLIGENCE-NATIVE ARCHITECTURE FRAMEWORK

AI native is the concept of having intrinsic trustworthy AI capabilities, where AI is a natural part of the functionality, in terms of design, deployment, operation, and maintenance. An AI native implementation leverages a data-driven and knowledge-based ecosystem, where data/knowledge is consumed and produced to realize new AI-based functionality or augment and replace static, rule-based mechanisms with

TABLE 1.3

6G Application-Specific KPI

Examples of Sensing Use Case Families	Example Applications	Maximum Range (m)	Maximum Velocity (m/sec)	Example KPIs Range Resolution (m/sec)	Doppler Resolution (m/sec)	Required Bandwidth (GHz)
Smart manufacturing and IoT	Digital twin, mapping, industrial environment	~100–200	± 9	0.01 level	0.5	~3–5
Traffic maintenance and smart transportation	Advanced driver assistance system (ADAS)	250–300	± 70	0.07–0.3	0.1–0.6	~0.5
	Road traffic monitoring	~80–200	~± 20–40	0.5–1	~1–3	~0.15–0.3
	Parked vehicle detection	~5	0	0.5 (< vehicle dimension)	Not applicable	~0.3
	Pedestrian detection	~5	~3 to 3	10s of 0.01	Very low	~0.2–0.4
	Around the corner vehicle	~100	~15	1 (< vehicle dimension)	High	~0.15
Environmental monitoring	Weather prediction	~200–500	Not applicable	~5m	Not applicable	~0.03
	Pollution monitoring	~200–500	Not applicable	~1–5	Not applicable	~0.03–0.15
Human activity/ presence detection	Gesture recognition	0.01 to ≤1	~10 to +10	0.01-level	~0.3	~3–5
	Human presence detection	~20	~2	~0.3	~0.1	~0.2–0.4
	Body proximity detection	0.001 to <0.2	~2	0.01-level	High	~3.5
	Obstacle proximity	~2	~3	0.01-level	~0.1	~3.5
	Human proximity detection	~20	~4	0.03	~0.1	~0.2–0.4
	Fall detection	~10	~3	0.03	~0.1	~0.2–0.4
Remote sensing	Drone monitoring, detection, and/or management	~500	~30–40	1	~5	~0.15

TABLE 1.4

6G Application-Specific Robustness Parameters

Use Case	In-Vehicle	In-Robot, In-Production Module	Intra-Body
Application Examples	Engine control, electric power steering, ABS, electric parking brakes, suspensions, and advanced driver assistance system (ADAS)	Motion, torque, force control	Heartbeat control, vital sign monitoring, insulin pumping, muscle haptic control
Number of devices	~50–100	~20	<20
Data rate	<10 Mbps (control) <10 Gbps (ADAS)	<10 Mbps	<20 Mbps
Service reliability and availability (%)	99.9999–99.999999	99.9999–99.999999	99.9999999
Latency (microseconds)	~54	~100	~20

learning and adaptive AI when needed. Autonomous networks also use AI, but they are not AI-native networks. Autonomous networks and AI-native networks are both types of networks that use AI to improve their performance, but there are differences as follows (also see Table 1.5):

- Autonomous networks are being used to improve the performance of 5G networks. They are also being used to optimize the performance of cloud networks.
- AI-native networks are being used to develop new applications for 6G networks. They are also being used to develop new security solutions for networks.

As AI technology continues to evolve, we can expect to see even more applications for autonomous networks and AI-native networks. These networks have the potential to revolutionize the way we use and manage networks.

TABLE 1.5

Autonomous vs. AI-Native Network

Feature	Autonomous Network	AI-Native Network
Level of automation	Minimum human intervention (zero touch)	Built from the ground-up with AI in mind
AI applications	Monitoring and automation	Machine learning and deep learning
Performance	Improved performance and reliability	Potential for even greater performance improvements
Adaptability	Adaptable to new and emerging applications	More adaptable to new and emerging applications

6G AI-native network is to realize AI-related work such as data acquisition, data preprocessing, model training, model reasoning, model evaluation, etc. This network deeply integrates the computing force, data, algorithm, connection required by AI with 6G network functions (NFs), protocols, processes, and other functionalities [4]. 6G AI native solves a series of problems such as the generation of AI-native use case, QoS research and definition of AI native, AI life cycle management, self-generated data, and services. These problems should be considered from whole system angle, including the core network and access network. Thus, we are describing to design a wireless access network with native AI, i.e., intelligence is embedded as a native and deeply integrated part into the CP and the UP. A high-level AI-native architecture framework for 6G is depicted in Figure 1.1.

In the coming 6G era, motivated by the upcoming new services, new scenarios, and new technologies, the 6G network characteristics can be summarized, in addition to the way it was described earlier, as: on-demand fulfillment of networking, smart and lite network, soft network, AI native, and native security [1–4]. For such a 6G network, the user-centric network is a promising and competitive RAN architecture. The traditional cellular network is BS-centric network. It has different transmission

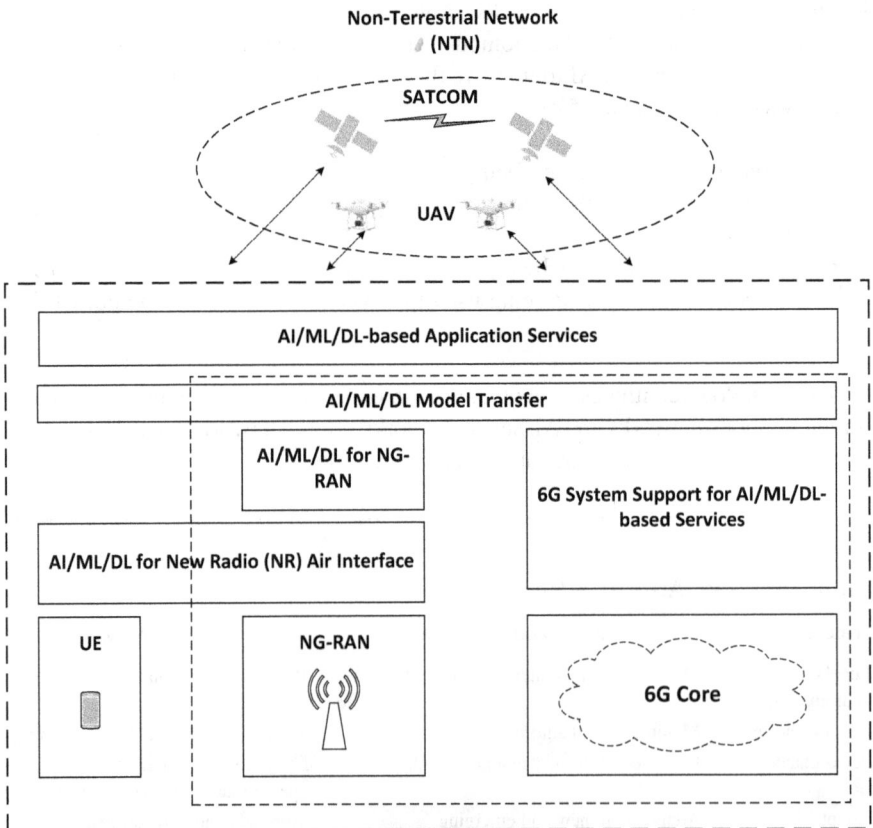

FIGURE 1.1 AI-native architecture framework for 6G.

performance between the cell center and the cell edge. The reduction of transmission latency and service interruption during handover procedure are important issues which must be resolved, especially in the high-frequency band. In 6G network, from user equipment's (UE's) aspect, it is always located in the center of a "cell" (we name it "flexible cell"). Network allocated one or more suitable access points (APs) to form this "flexible cell" to serve the UE by considering following factors: the wireless channel condition, UE location, and service requirements, and others.

It becomes even more important with a 6G multi-connectivity (MC) solution with the ability to have efficient spectrum usage and be able to aggregate resources between the current frequency bands and the new sub-THz spectrum bands. This calls for a new improved MC solution. The new 6G MC solution should replace the current stand-alone dual-connectivity (DC) and carrier aggregation (CA) solutions by combining the best features to be able to handle both extreme reliability and excellent flexibility. Note that CA is a feature of long-term evolution (LTE)-advanced, which allows mobile operators to combine two or more LTE carriers into a single data channel to increase the capacity of the network and data rates by exploiting fragmented spectrum allocations. The MC solution should support decoupled downlink (DL) and uplink (UL) and the ability to quickly add inactive connections. A general disadvantage with the DC solutions for new radio (NR) is the implementation complexity of the 3GPP specification, for example the numerous architecture options for DC between LTE and NR and the message exchange over the Xn interface between the next generation 6G base stations (gNBs) to reduce complexity.

With the combination of terrestrial networks (TNs) and non-terrestrial networks (NTNs), it will be possible to achieve 100% global coverage. NTNs can likely provide a lower capacity per square kilometer than TNs, but at a reasonable cost. Thus, an NTN is suitable for rural areas, including oceans, with low or very low population density. For urban areas, there will always be a need for TNs. There are two types of architecture options for NTN: transparent and regenerative payload architecture. Transparent is the simplest type, where the NTN basically serves as a relay of the signal between the UE and the base station (BS) on the ground. The regenerative architecture is equivalent to having the BS (RAN) functions onboard the satellite. The main research question for 6G is how the NTN and TN mobility will be solved. Since the low Earth orbit (LEO) and medium Earth orbit (MEO) satellites move, it may be necessary to find solutions that minimize the number of handovers and the signaling needs for mobility robustness. Another important research topic for 6G NTN is the actual architecture solution, e.g., if regenerative or transparent or a hybrid split should be used as suggested by 3GGP.

One important enabler for 6G architecture is function elasticity and in particular 6G-RAN-CN function elasticity, which is achieved by co-locating some of the common 6G-CN NFs with the 6G RAN-CP in the cloud environment (see Figure 1.2). Signaling procedures that benefit from being in the regional edge cloud comprise 6G mobility management and 6G session management. As a result of placing critical signaling processing together with 6G-RAN-CP in the regional edge cloud, signaling performance is improved, thus reducing latency. This approach can be applied for 6G-UE associated services since the 6G-UE context handling would remain within the control of the 6G mobility management without creating new or additional dependencies.

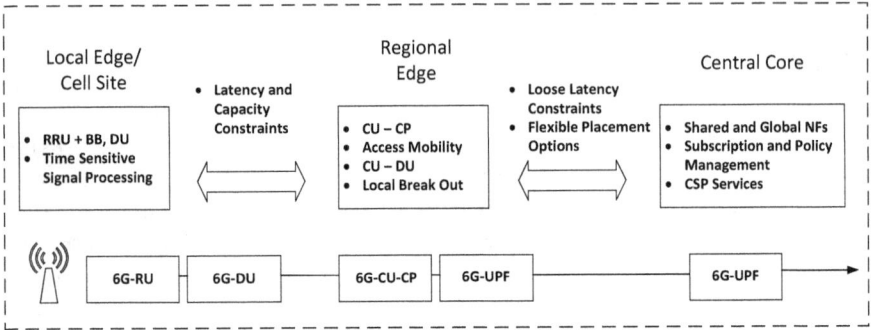

FIGURE 1.2   Functionality distribution between RAN and CN clouds.

Another way to improve the 6G architecture is to enhance the possibilities for signaling directly between NFs, that is, to remove potential bottlenecks.

6G AI-NATIVE VS. 6G AI-STANDARD-BASED NETWORK

6G AI-native (or native AI) architecture uses AI as an overlay network, whereas 6G AI-standard-based architecture makes each of the network elements inherently AI-enabled standards for both hardware and software design. In AI-standard-based network, all networking communication protocols (e.g., physical, link, network, transport, middleware, and application layer) will be AI enabled, which are supposed to be standardized by the ISO, IETF, ITU-T, and/or other standard organizations. We have published a separate book on "Networked Artificial Intelligence" (CRC Press, 2024), with detail suggestions of how these AI-based networking protocols need to be standardized by standards fora based on recent published research and development (R&D) papers. We have not discussed this here for the sake of brevity.

SEMICONDUCTOR TECHNOLOGY

Research is going on up to 30 terahertz technologies throughout the world, and industry needs terahertz technologies for many applications as follows:

- Industrial control less than 5 THz
- Homeland security less than 3 THz
- Pharmaceutical and medical up to 30 THz
- Ozone monitoring between 1.5 and 6 THz
- Hardware cybersecurity up to 30 THz
- 6G Communications up to 10 THz
- Cancer detection between 0.2 and 2.9 THz

Although 6G can go up to 30 THz, the present 6G network that is expected to be deployed in the year 2030 is looking to 1 THz matured technologies. Advances in silicon germanium (SiGe) and bipolar complementary metal-oxide semiconductor (BiCMOS) have the potential to increase silicon transistor performance while leveraging the cost and scale of mature high-volume silicon manufacturing for 6G sub-THz. Compound semiconductor technologies such as indium phosphide (InP) and gallium nitride (GaN) have the best front-end performance at sub-THz. Advances in heterogeneous and monolithic integration with silicon are needed to address the cost and scale concerns of high-frequency InP and GaN. A comprehensive approach to reliability is essential to extract maximum performance without sacrificing reliability. The large bandwidth available in the sub-THz 100–300 GHz spectrum has the potential to enable the order of magnitude improvement in network data rate, latency, and location sensing that 6G targets.

Applications of terahertz (THz) and sub-THz technology require portable, inexpensive, efficient, and reliable THz electronics. Operation at sub-THz creates significant challenges for the cellular radio, including higher losses, lower semiconductor device performance, and a smaller per element physical area that constrains circuit size, integration, power, and thermal management. The concept of ultra-massive multiple-input multiple-output (UM-MIMO) has emerged in recent years, which takes advantage of plasmonic materials for building antennas and transceivers to achieve the capacity of THz band. Materials such as graphene and metamaterials can be used to build nano antennas and transceivers. These nano antennas and transceivers can operate in the THz band [1–4]. This stimulated the search for novel semiconductor materials for THz applications. Many published research reports show that there is tremendous progress in the development of 6G THz transceiver and antennas and beyond.

RADIO TECHNOLOGIES

Advancements in semiconductor research for communications in 30 THz-band have ushered 6G radio technologies: 6G with newly available extremely high frequencies such as sub-THz/THz bands, as well as technologies enhancing the current use of millimeter wave bands. Earlier, we have discussed about 6G KPIs that deal with spectral efficiency, throughput, and latency. The basic tenant of 6G networking is to connect the high-speed user (radio) interface (UE) to the massively high bandwidth networks. MIMO, distributed MIMO, RIS, holographic beamforming, operation, administration, and management (OA&M), and other applications will be used over 6G communications networks. Unlike legacy cellular network that assumes networks are fixed, 6G radio networks will be using nonconventional advanced topology and networking such as UE cooperative communication, NTNs, and mesh networking, which will be continuously evolving network topology to adaptively meet the varying traffic demand. Figure 1.3 depicts a high-level view of 6G radio communications networks.

6G topology and networking are poised to support extreme industrial and commercial use cases opening the door for new business opportunities. The key challenge is to design the 6G native AI-based air interface that will enable distributed computing and intelligence, which are recent areas of radio technology and considered to play an important role in 6G. The 6G technologies are expected to offer

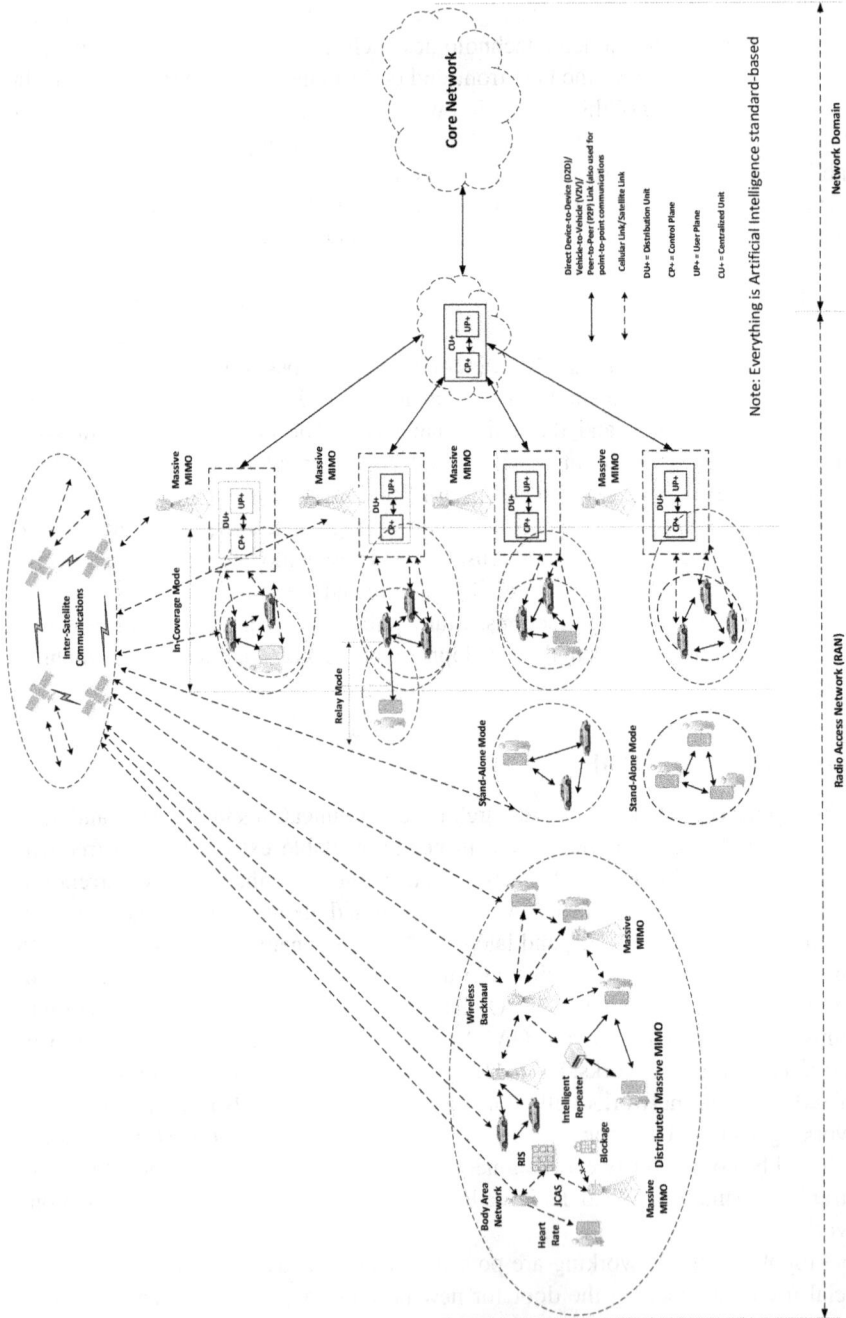

FIGURE 1.3 High-level view of 6G radio communications networking.

end-to-end system optimization, seamless automation, and true convergence of communication and computing.

6G radio technologies will enable sensing for situational awareness, that is, one of the important use cases will be used for many applications such as autonomous driving, virtual reality, augmented reality, mixed reality, and interactive gaming. It implies that a very wide sensing bandwidth is required for high-resolution applications. Very high-frequency sensing bands, such as Millimeter Wave/sub-THz/THz, are required. Note that these high-frequency bands are also considered for communication, so by integrating sensing functions into the communication systems making the system more efficient and cost-effective integrated solutions for 6G.

Seamless Mobility

The existing legacy low-bandwidth wireless networks and devices have many limitations because of their high ranges. Wireless devices are fully dependent upon the network for any cellular operation, including mobility control. For example, these cellular mobility operations consist of multiple procedures, including handovers, cell selection, cell re-selection, beam management, and Channel State Information (CSI) reporting. Extensive signaling information exchanges between the network and device are needed for these operations, which cause extensive signaling information exchanges between the network and the device.

6G will employ high-frequency spectrum like millimeter wave, sub-THz, and THz bands, which will result in a reduced range of communication and the use of narrow beams, further increasing the challenges for networks and devices during mobility. Radio link failures and handover service interruptions also impact UEs heavily. So, by introducing native AI/ML/DL-based enhancements and simplified, minimized control signaling, cellular operations could be made more power and performance efficient. In this way, 6G device users can benefit from more reliable communication in mobility scenarios enabling interruption-free and robust data transmission and reception across radio technologies, including optimized interworking with other entities of the 6G ecosystem (e.g., terrestrial and non-terrestrial links). It is expected that 6G wireless networks will benefit from improved KPIs in terms of reduced handover failure rates, reduced radio link failure rates, and reduced signaling and processing overhead.

We have explained the future 6G technologies and their performances (e.g., KPIs) that are the target for R&D robust methods for interruption-free handover between cells of same technology and across technologies while simplifying and minimizing the amount of control signaling, signaling delays, procedure delays, and processing efforts. This goal includes optimization of device power consumption under mobility scenarios and for mobility-related procedures. A key driver of these goals is the native use of AI/ML/DL technologies for, not limited to, UE, radio nodes, and networks.

SPECTRUM SHARING

6G has no precursor of any wireless networks for spectrum sharing. It comes with a clean-slate opportunity to consider designs for all deployment scenarios, including exclusively licensed, non-exclusively licensed, and unlicensed spectrum uses. Technology that efficiently shares spectrum can provide economical and societal benefits, such as reduced cost, and new use cases and deployment scenarios. The main challenges for shared-spectrum deployments include, but are not limited to, the following:

- Management of KPIs
- Predictability of available resources in meeting performances
- Efficient system performance measurements
- Interference detection and mitigation techniques
- Real-time spectrum sensing in complex radio frequency environments such as mixed radio frequency signals and heterogeneous systems
- Management of mutual interferences between networks of the same technology or across different technologies
- Alleviation of bandwidth availability constraints for public and private networks
- Deployment of AI/ML/DL technologies for spectrum sharing

Figure 1.4 depicts some example scenarios that might need to be considered for developing 6G technologies for spectrum sharing.

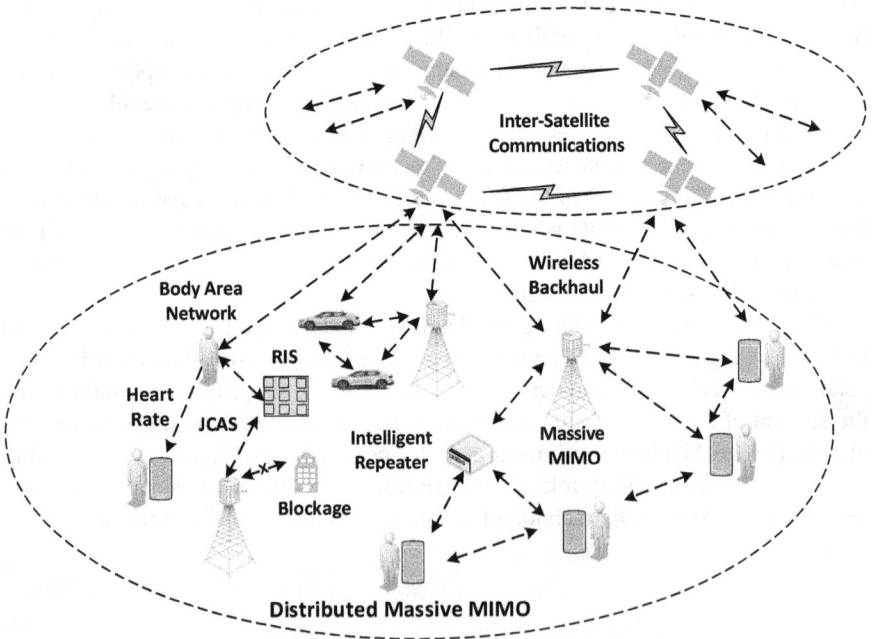

FIGURE 1.4   Example scenarios for 6G spectrum sharing.

It might be that the 6G spectrum may be exclusively licensed, shared in coexistence scenarios that focus on sharing between previous generations and 6G, or locally licensed where spectrum is shared between previous generations and 6G, or 6G only. The unlicensed spectrum is another scenario of interest, which may cover cases of license-assisted access and fully unlicensed access schemes. Research needs to be done for spectrum sharing of superhigh frequency (SHF) and extremely high frequency (EHF) bands. AI/ML/DL-based new channel-access procedures, with or without channel sensing, may be needed to enable efficient sharing with a high degree of predictability of resources among multiple authorized co-primary users or sharing among secondary services and primary users in meeting the required KPIs.

MILLIMETER WAVES

Generally, a frequency below 6 GHz is used for cellular communication, and frequency above that is mostly used for other services like medical imaging, microwave remote sensing, amateur radio, terahertz computing, and radio astronomy. The massive increase in data traffic has made the radio frequency spectrum congested. The result is that there is limited bandwidth for a user, causing a slower and unreliable connection. One way to solve this problem is by using frequencies above 6 GHz for wireless communication. Frequencies above 6 GHz has never been used for wireless communication. There has been a lot of research going on with broadcasting millimeter waves. Millimeter waves lie within the frequencies ranging between 30 and 300 GHz, and are called millimeter waves because the length varies from 1 to 10 mm compared to the radio waves that are used in the current mobile communication system, which measure tens of centimeters in length. Many aspects of millimeter waves have been published in the past few years discussing the potentials and challenges in the millimeter-wave technology. Millimeter waves can provide bandwidth ten times more than that of the entire 4G cellular band. These high-frequency waves are used in some satellite application, but never for mobile broadband. Since millimeter waves have a lower wavelength, they are not suitable for long-range applications. Another problem with millimeter waves is that they cannot penetrate buildings and obstacles, and they tend to get absorbed by rain.

SUB-MILLIMETER OR TERAHERTZ BAND

With globalization, the current wireless market is expanding rapidly. With talk of 6G networks, the demand for a higher spectrum is imminent soon. The frequency higher than the millimeter-wave band (30–300 GHz) could be used for wireless communication. The frequency band between 300 GHz and 3 THz is known as the Terahertz band. Although this idea is relatively new, research in this area can be worthwhile for the wireless communication industry. Other than just a higher spectrum, there are many advantages of THz band, such as interference friendly deployment, scalability, enhanced security, availability of greenfield spectrum, low power consumption, a front-haul boost for the wireless network, small antennas size, and focused beams. THz technology would be beneficial for applications such as imaging, spectroscopy, holographic telepresence, industry 4.0, and massive-scale communications. There are several challenges and new areas of research in THz band deployments such as complex antenna design to support higher antenna gain, access point specification and deployment, complex circuit design, high propagation loss, and complex mobility management. The millimeter-wave and terahertz wave bands are shown in Figure 1.4. The concept of ultra-massive MIMO (UM-MIMO) has emerged in recent years, which takes advantage of plasmonic materials for building antennas and transceivers to achieve the capacity of THz band. Materials such as graphene and metamaterials can be used to build nano antennas and transceivers. These nano antennas and transceivers can operate in the THz band. UM-MIMO can take advantage of these miniature antennas and transceivers to provide higher spatial multiplexing and sensors with beamforming. Thus, the data rates and communication range can be improved with the help of spatial multiplexing and beamforming, respectively. A lot of investigation is needed to realize THz UM-MIMO for 5G Networks and beyond. Some of the challenges are the fabrication of plasmonic nano array antennas, channel estimation, precoding, signal detection, beamforming, and beam steering.

JOINT COMMUNICATIONS AND SENSING

Sensing capabilities as an integral part of the communications network, termed as Joint Communications and Sensing (JCAS) in 6G, have been identified as a novel feature (see Figure 1.4). Directional sensing between the network node and the device is already deployed in wireless networks where communications and sensing are not integrated together, although infrastructure is already there. The radar-centric schemes try to use radar platforms to achieve communication functions. The communication-centric schemes try to extract target information from communication signals. The spatial sensing offers new opportunities outside the network. The principles of spatial sensing as in radar rely on detecting the reflected signal and analyzing it. Radar reflections provide more information than just delay and Doppler effect. Characteristics of the reflecting object like size, shape, material, or possible micro-movements all have an impact on the channel response.

Moreover, the time intervals between the different periodic sensing beam allocations determine the burst frequency or beam revisit rate. The total time-limited contiguous length of the sensing allocation is denoted as burst duration or beam dwell time. For communication purposes, how soon transmit opportunities are available immediately upon packet arrival at the BS or a UE in a certain beam direction determine the achievable communication latency. For a JCAS system, this might motivate a time domain comb design in combination with beam sweeping with the least-possible interruption to data transmission opportunities. To achieve balanced communication and sensing performance, devising a novel dual-function waveform is needed. An integrated communication and sensing system have incomparable benefits of low-cost, low-power consumption, and compact volume. In addition, distributed sensing needs to be supported. However, in the case of JCAS, a common hardware and waveform design that caters to sensing and communication requirements would be important. Full-duplex (FDX) transceiver operation may be needed for communication and/or sensing. In this respect, the following technical capabilities are needed for 6G meeting KPIs requirements:

- Novel self-interference measurement and cancellation techniques need to be developed at these frequencies
- Beamforming-based cancellers for joint self-interference and multiuser interference
- MIMO and Massive MIMO enhancements:
  - MIMO beamforming architectures
  - Single-user MIMO (SU-MIMO)
  - Multi-User MIMO (MU-MIMO)
  - Distributed MIMO
  - Line-of-sight MIMO (LOS-MIMO)
  - Orbital angular momentum (OAM)-MIMO
  - Other wavefronts beyond OAM-carrying Gaussian beams, such as self-healing Bessel beams and discrete Fourier transform spread-orthogonal frequency division multiplexing (DFT-OFDM)
- Hardware acceleration of transceiver signal processing for using AI/ML/DL algorithms
- Higher layer protocol enhancements to realize low latency and robust design for leveraging benefits at sub-THz/THz frequencies could require
- Multiband sensing
- Fusion of sensors information
- Native AI/ML/DL support for the full duplex (FDX) transceiver

AI/ML/DL needs to be foundational and natively designed into 6G systems along with suitable hardware components for algorithms processing for advanced and complicated requirements of JCAS. For example, deep learning (DL) has been used for a variety of sensing use cases such as static object classification, radar-based fall-motion detection, semantic segmentation of radar point clouds, detection and localization of multiple objects from various classes in a single complex frame, objects detection under both LOS and non-LOS (NLOS) conditions, and unsupervised learning for classification of radar clutter using k-means algorithm. Similarly, all other signal processing tasks need to be performed using AI/ML/DL meeting 6G performance requirements. Methods for training the neural network models will be required. End-to-end learning is needed for communication systems to both learn the transmitter and the receiver of communication systems without knowledge of a channel model using supervised and reinforcement learning. Similar concepts of learning can be used for both waveform and modulation design for JCAS systems.

The multiband sensing combining both the low-band and high-band measurements is very appealing to obtain ultra-fine measurement resolution. The combination of multiple radio sensing modalities, which is termed as "sensor fusion," will be very powerful in constructing high-resolution maps required for digital twin representation of roads, lamp posts, surroundings, and others obtaining the signals from cameras, unmanned aerial vehicles (UAVs). Sensing and localization

TABLE 1.6
MIMO Enabled Communication and Sensing

System	Metric	Role of MIMO	Function	Incompatibility
Communication	Single-user rate	Coherent beamforming	Improve directionality and SNR	May cause high PAPR and high sidelobe, and destroy correlation properties
	Outage probability	Space-time coding	Combat multipath fading, improve reliability	
	System capacity	Spatial precoding	Improve SNR	
Sensing	Detection probability	Coherent beamforming	Construct spatial independent channels, lower down the interference	Influence the spatial separability, may cause symbol distortion, and influence the data-carrying capability
	Detection reliability	Waveform diversity	Lower down the target missing probability	
	Estimation resolution	Waveform shaping	Lower down the sidelobe, ensure correlation properties, constant envelop, etc.	

can be directly applied to enhancing communication itself, for example predicting blocking of devices in high-reliability communication scenarios. A further example is to exploit geolocation information for beamforming. All these functions need to support AI/ML/DL natively. Table 1.6 explains the technology challenges for JCAS.

JCAS FOR MASSIVE MIMO

The use of many antenna elements, known as Massive MIMO, is seen as a key enabling technology in the 5G and beyond wireless ecosystem. The intelligent use of a multitude of antenna elements unleashes unprecedented flexibility and control on the physical channel of the wireless medium. Through Massive MIMO and other techniques, it is envisioned that the 5G and beyond wireless system will be able to support high throughput, high reliability, that is, low bit error rate (BER), high energy efficiency, low latency, and an internet-scale massive number of connected devices. Massive MIMO and related technologies will be deployed in the mid-band (sub 6 GHz) for coverage, all the way to mm-Wave bands to support large channel bandwidths.

It is envisioned that Massive MIMO will be deployed in different environments: frequency division duplex (FDD), time division duplex (TDD), indoor/outdoor, small cell, macro cell, and other heterogeneous network (HetNet) configurations.

Accurate and useful channel estimation remains a challenge in the efficient adoption of Massive MIMO techniques, and different performance-complexity trade-offs may be supported by different Massive MIMO architectures such as digital, analog, and/or digital/analog hybrid. Carrier frequency offset (CFO), which arises due to the relative motion between the transmitter and receiver, is another important topic. Recently, maximum likelihood (ML) methods of CFO estimation have been proposed, which achieve very low root mean square (RMS) estimation errors, with a large scope for parallel processing and well suited for application with turbo codes.

Massive MIMO opens a whole new dimension of parameters where the wireless applications or other network layers may control or influence the operation and performance of the physical wireless channel. To fully reap the benefits of such flexibility, the latest advances in artificial intelligence (AI) and machine learning (ML) techniques will be leveraged to monitor and optimize the Massive MIMO subsystem. As such, a cross-layer open interface can facilitate exposing the programmability of Massive MIMO through techniques such as network slicing (NS) and network function virtualization (NFV). Finally, security needs to be integrated into the design of the system so the new functionality and performance of Massive MIMO can be utilized in a reliable manner.

REFERENCES

1. 6G Next G Alliance Report: 6G Technologies, June 2022.
2. https://www.3gpp.org/news-events/3gpp-news/partner-pr-6g
3. Wu, J. et al., "Toward Native Artificial Intelligence in 6G Networks: System Design, Architectures, and Paradigms," arXiv:2103.02823v1 [cs.NI] 4 Mar 2021.
4. Letaief, K. B., et al., "The Roadmap to 6G – AI Empowered Wireless Networks," arXiv:1904.11686v2 [cs.NI] 19 Jul 2019 and https://abnormalsecurity.com/blog/future-cybersecurity-ai-native-observations-leading-gtm

2 MIMO

MIMO antenna design has great degrees-of-freedom (DoFs) because of utilization of both spatial multiplexing and spatial diversity using multiple antennas simultaneously. The space–time codes, termed as Vertical Bell Laboratories Layered Space-Time/Diagonal Bell Laboratories Layered Space-Time (V-BLAST/D-BLAST) and Alamouti code have revolutionized the antenna systems. Figure 2.1 provides a high-level view of the MIMO transceiver system.

Both the network and the connected mobile devices in MIMO systems must be tightly coordinated. Complex algorithms use spatial information obtained from a channel state information reference signal (CSI-RS), to enable the base station (BS) to communicate with multiple devices concurrently and independently. The CSI-RS is a type of pilot signal sent out by the BS to the UE, which enables the UE to calculate the channel state information (CSI) and report it back to the BS.

The CSI describes how the signal propagates from transmitter to receiver and includes information on how that signal suffers from effects such as scatter, fade, and power decay over distance. To recover the transmitted data stream at the receiver, the MIMO system decoder must perform a considerable amount of signal processing, using the CSI to represent the channel transfer function in matrix form (see Figure 2.1):

$$H = \begin{bmatrix} h_{1,1} & h_{1,2} & \dots h_{1,n_R} \\ h_{2,1} & h_{2,2} & \dots h_{2,n_R} \\ \dots & \dots & \dots \dots \\ h_{n_T,1} & h_{n_T,2} & \dots h_{n_T,n_R} \end{bmatrix} \equiv \begin{bmatrix} h \end{bmatrix}_{n_T,n_R}$$

The channel transfer matrix is defined as follows:

$$[R] = [H] * [T]$$

where $[R]$ is the series of signals received at the various antennae in the MIMO array, $[H]$ represents the properties of each signal path (i.e., fading between the transmission and reception), and $[T]$ the various data streams being transmitted across the network. The decoder constructs the channel transfer matrix by estimating the individual channel properties, h_{11}, h_{12}, etc. from the CSI. The individual data streams are then reconstructed by multiplying the received signal by the inverse of the transfer matrix:

$$[T] = [H]^{-1} * [R]$$

Estimating the individual channel properties and computing the inverse channel matrix is computationally intensive and can add significant overhead to the network,

DOI: 10.1201/9781003499480-2

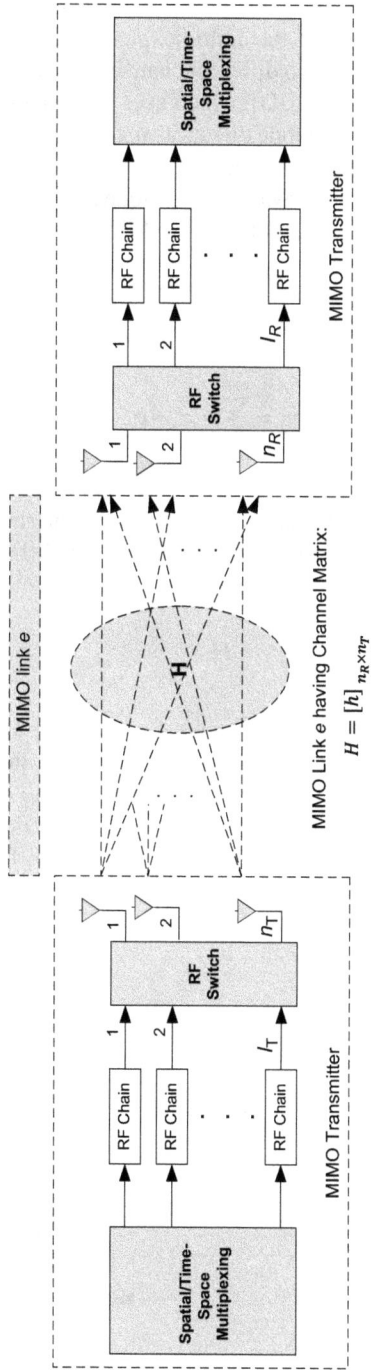

FIGURE 2.1 Basic MIMO transceiver system.

particularly as the number of antennas grows. This description is somewhat simplified as there are actually various techniques for acquiring and calculating the CSI, which depend upon factors such as the multiplexing techniques used, frequency division duplex (FDD), which uses two different channels for uplink (UL) and downlink (DL), and time division duplex (TDD), which uses the same frequency band for UL and DL but only communicates in one direction at a time, the signal frequencies, and the amount of movement of the user equipment (UE). This area is the subject of much ongoing research into how advanced techniques such as neural networking can enhance the reliability and accuracy of Massive MIMO (mMIMO).

Now, let x_i be the input data and y_j be the output data. It is assumed that noise can affect the transmission and, therefore, b_j be the White noise of zero mean. The output y_j of the system is provided as follows:

$$y_j = \sum_{i=0}^{n_T-1} h_{ji} x_i + b_j$$

h_{ji} represents the link fading between i transmitting antenna and a j receiving antenna. Then one can consider the characteristics matrix H of the MIMO channel dimension $n_T * n_R$ shown earlier. We can now generalize the result as follows:

$$\bar{y} = H.\bar{x} + \bar{b}$$

In addition, MIMO system employs signal processing algorithms with spatial and/or space–time multiplexing as indicated above. Note that MIMO has offered great benefits to wireless communications compared to those of conventional single-input single-output (SISO) systems that have much lower channel capacity. The striking capacity improvement over single antenna systems has thus kicked off the era of MIMO communications. In another example, we have shown (Figure 2.2) the typical delay diversity MIMO system.

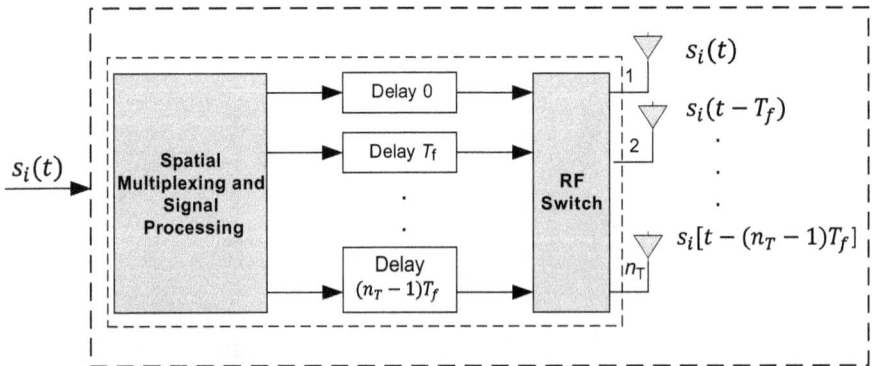

FIGURE 2.2   MIMO delay diversity scheme.

In delay diversity scheme, multiple copies of the same symbol are transmitted through multiple antennas in different time slots. Delay diversity allows a MIMO antenna system to reshape the signals using time delay diversity and even can use security algorithms making the transmission and reception more robust against cyberattacks. The high-level MIMO system capacity can be estimated as follows:

$$C_{\mathrm{MIMO}} \approx \min\{n_T, n_R\} * B * \log_2\left(1 + \frac{S}{N}\right)$$

where, C = link capacity in bits/second, B = link bandwidth in Hertz (Hz), n_T = number of transmitting antennas, n_R = number of receiving antennas, S/N = signal-to-noise ratio (SNR). Under the spatial multiplexing technique, the capacity of MIMO systems scales linearly with min $\{n_T, n_R\}$. In the same token, we can consider that the capacity of the SISO system will be as follows:

$$C_{\mathrm{SISO}} \approx B * \log_2\left(1 + \frac{S}{N}\right)$$

The single-user MIMO (SU-MIMO) has been extended to the multi-user MIMO (MU-MIMO), as depicted in Figure 2.3. In addition to using spatial DoFs to obtain the processing gains, MU-MIMO supports the spatial sharing among different

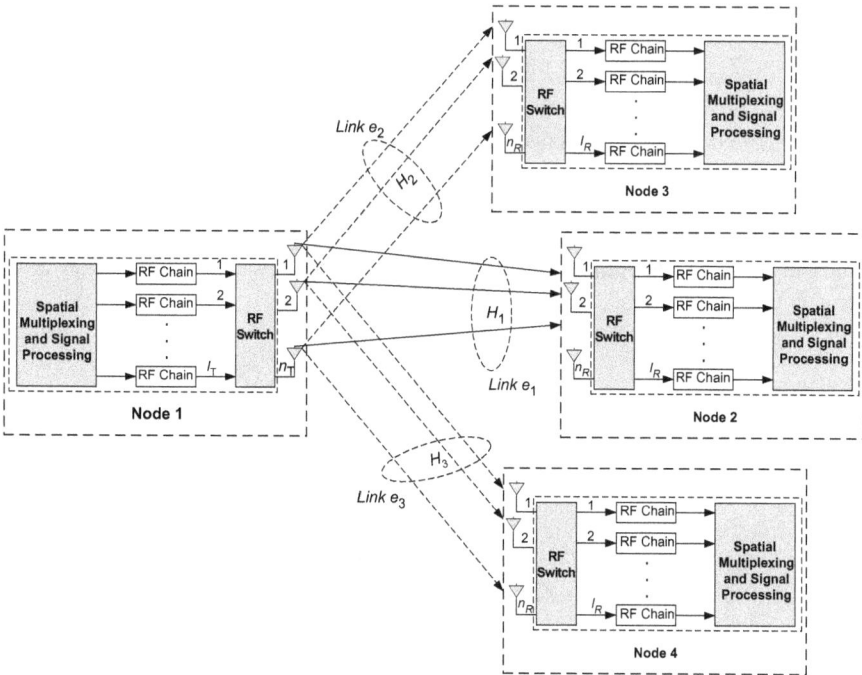

FIGURE 2.3 Multi-user (MU) MIMO transceiver system.

channels. A series of practical strategies, including channel-aware precoding and multiuser reception, have been devised.

Combined with effective scheduling schemes, the achievable rate of MU-MIMO approaches the theoretical limit. Later, MU-MIMO has been extended to the cooperative MIMO, where distributed transceivers cooperate with each other to exploit the spatial efficiency. A creative idea named interference alignment was proposed, which achieves the rate of $K/2 \log (1 + \text{SNR})$ when serving K users. The gap between the sum capacity of SU-MIMO and the rate of cooperative MIMO has greatly narrowed. With the advent of artificial intelligence, new MIMO systems are emerging with native AI/ML/DL support.

NETWORK MIMO, COOPERATIVE MIMO, DISTRIBUTED MIMO, OR VIRTUAL MIMO

ADVANTAGES

Terms like "network MIMO," "cooperative MIMO," "distributed MIMO," and "virtual MIMO" [1–4] are used in the literature for the same technology. In conventional systems, such as point-to-point MIMO or collocated MIMO, require both the transmitter and receiver of a communication link to be equipped with multiple antennas. Many wireless devices cannot support multiple antennas due to size, cost, and/or hardware limitations. More importantly, the separation between antennas on a mobile device and even on fixed radio platforms is often insufficient to allow meaningful performance gains. Furthermore, as the number of antennas is increased, the actual MIMO performance falls farther behind the theoretical gains.

Network/cooperative/distributed/virtual MIMO effectively exploits the spatial domain of mobile fading channels to bring significant performance improvements to wireless communication systems. This technology is a family of cooperative multiantenna transmission/reception techniques for wireless communications. With these kinds of MIMO systems, multiple wireless access points share user data and CSI via high-capacity backhaul links and perform joint precoding to transmit to multiple-user devices in the DL and jointly decode the received signals for different users in the UL.

DISADVANTAGES

The disadvantages of cooperative MIMO come from the increased system complexity and the large signaling overhead required for supporting device cooperation. The advantages of cooperative MIMO, on the other hand, are its capability to improve the capacity, cell edge throughput, coverage, and group mobility of a wireless network in a cost-effective manner. These advantages are achieved by using distributed antennas, which can increase the system capacity by decorrelating the MIMO subchannels and allow the system to exploit the benefits of macro-diversity in addition to microdiversity. In many practical applications, such as cellular mobile and wireless ad hoc networks, the advantages of deploying cooperative MIMO technology outweigh the disadvantages.

AI/ML/DL APPLICATION IN MIMO

AI/ML/DL is being utilized in MIMO for optimization of MIMO capacity, cell edge throughput, coverage, and group mobility. In addition, it has also been incorporated in MIMO for automatic beam formation as explained below.

Deep reinforcement learning (DRL) beam:

- The DRL block will learn the dynamics of the environment given the current channel measurement and historical beam training data.
- The DRL block will switch between different beam training methods for different channel conditions.
- The goal is to achieve a good trade-off between the average beam training overhead and the average spectral efficiency for a mobile mm-wave channel.

MERITS OF MIMO IN COMMUNICATION SYSTEMS

In recent years, cooperative MIMO technologies have been adopted into the mainstream of wireless communication standards. We are now summarizing merits of MIMO communication systems and sensing systems as follows:

- Merits of MIMO in communication systems
  a. **Spatial diversity**: Antenna spacing is used for spatial diversity that combats fading, because it averages the deep fading probability from one path to multiple independent paths. Moreover, related studies have proven that if the antenna spacing is more than ten wavelengths, signals transmitted or received by different antennas experience independent fading. MIMO provides a maximum of $(n_T * n_R)$ independent transmitting and receiving paths. Using this property, if transmit antennas emit the same signal, the receiver thus obtains $(n_T * n_R)$ independent replicas. The probability that these replicas all go through deep fading is the $(n_T * n_R)$ multiplications of the probability of one path. Therefore, the outage probability is largely reduced.
  b. **Spatial multiplexing**: Spatial multiplexing increases spectrum efficiency (SE), while the spatial diversity aims to combat fading and spatial multiplexing and utilizes fading to improve the system rate. If transceiver pairs experience independent fading, the corresponding channel matrix H is more likely to be well conditioned and full rank. This channel can thus be decomposed into multiple parallel paths. Using this property, multiple streams are allowed to be transmitted concurrently using overlapped time and frequency resources.
  c. **Flexible beamforming**: MIMO supports different beamforming strategies for different scenarios with multiple choices exploiting spatial diversity and spatial multiplexing. We can take transmit beamforming as an example. The high directional beam is used to improve the received SNR when serving one user. Multiple spatially separable beams can be

applied to support multistream transmissions for serving multiple users. Concerning co-channel interference, zero-forcing precoding provides interference-free channels for different users. Note that the pencil beam limits the power leakage to adjacent cells in the cellular networks, and, thereby, the background noise is controlled. In essence, different beamforming schemes exploit different degrees of usage of spatial diversity and spatial multiplexing.

MERITS OF MIMO IN SENSING SYSTEMS

a. **Spatial diversity**: Like radar communications systems, MIMO radar has the outstanding merits due to its spatial diversity. As described above, using properly spaced antenna arrays, $(n_T * n_R)$ independent observations could be obtained at the receiver. This reduces the probability of missing targets and makes the detection robust to cluttering effects. However, the collocated MIMO radar, which allows correlations among different transceiver paths, has great parameter identifiability and high resolution. Compared with the phased-array radar, MIMO radars could improve the parameter identifiability at most n_R fold.

b. **Adaptive waveform manipulation**: MIMO radar can use unique features like adaptive waveform designs. The reliability of the target detection is substantially improved using the probing beam that emits high power toward the target while nulling surrounding clutters. The multitarget identifiability is improved using correlations among different sub-beams that can be surpassed by adjusting the transmit covariance matrix **Covariance** [H] matrix. Multiple directional beams can be generated simultaneously to illuminate the entire target area for detecting the high-speed target that is usually missed by the single-beam scanning. For target tracking, the dynamic beam is used to follow this target. Like communication, different waveform designs are achieved by the different degrees of usage of spatial diversity and spatial multiplexing.

SUMMARY

We can clearly see that there are conflicts in integration of two functions in MIMO: communications and sensing. There are similarities between communications and sensing exploiting spatial diversity for reliability; even then both have their own emphasis on the role of MIMO. Communication mainly uses MIMO to adapt to channels, with the emphasis on directional beamforming, to point to different users and control interference. Radar functions that are used for sensing mainly use MIMO to adapt to even with three-dimensional (3D) targets mapping the whole area, with the emphasis on the waveform, to facilitate signal processing and target information extraction. Many research and development (R&D) have proposed several schemes, but 6G JCAS prototype products are still far away to design. Due to different emphases, the fundamental interest of R&D will be how these two functions can be integrated in the balanced JCAS MIMO designs.

REFERENCES

1. 6G Next G Alliance Report: 6G Technologies, June 2022.
2. https://www.3gpp.org/news-events/3gpp-news/partner-pr-6g
3. https://futurenetworks.ieee.org/podcasts/ingr-executive-overview
4. IEEE/CIC International Conference on Communications in China (ICCC Workshops), 2022.

MASSIVE MIMO

mMIMO is the most captivating technology for 6G wireless communications networks. Present MIMO technologies are not sufficient to accommodate the ever-increasing demands. The number of wireless users has increased exponentially in the last few years; these users generate trillions of data that must be handled efficiently and with more reliability. Additionally, there are billions of internet-of-things (IoT) devices, having various applications to smart healthcare, smart homes, and smart energy, which contribute to the data traffic. It is predicted that there will be trillions of connected devices in the future. mMIMO has emerged as the promising technology to address this. Many studies on mMIMO have been analyzing their benefits removing the present problems.

Compared with MIMO, mMIMO uses a much larger array. The quantitative change in antenna numbers brings amazing qualitative changes. mMIMO is the advancement of contemporary MIMO systems used in current wireless networks, which groups together hundreds and even thousands of antennas at the BS and serves tens of users simultaneously. The extra antennas that mMIMO uses helps to focus energy into a smaller region of space to provide better spectral efficiency and throughput. As the number of antennas increases in a mMIMO system, radiated beams become narrower and spatially focused to the user. These spatially focused antenna beams increase the throughput for the desired user and reduce the interference to the neighboring user.

Research shows that if the transmitter uses an infinite number of antennas to serve only a few users, the randomness of fast fading would vanish, and the channels tend to be orthogonal. As a result, the inter-user interference disappears, and the scalability of MIMO is greatly improved. Moreover, the coherent processing of mMIMO could combat the high path loss of millimeter wave channels. By leveraging the sparsity of millimeter wave channels, the techniques of hybrid digital–analog precoding and compressed sensing were ingeniously applied to overcome the high cost of mMIMO. The combination of mMIMO and millimeter wave communication promoted the progress of wireless communication. mMIMO offers an immense advantage over the traditional MIMO system, which are summarized in Table 2.1.

TABLE 2.1
Capabilities Comparison between MIMO and Massive MIMO

Description	MIMO	Massive MIMO
Number of antenna	≤ 8	≥ 16
Pilot contamination	Low	High
Throughput	Low	High
Antenna coupling	Low	High
Bit error rate	High	Low
Noise resistance	Low	High
Diversity/capacity gain	Low	High
Energy efficiency	Low	High
Cost	Low	High
Complexity	Low	High
Scalability	Low	High
Link stability	Low	High
Antenna correlation	Low	High

Present Research also shows some limitations of mMIMO:

- It works only with TDD mode, where you change between UL and DL on the same frequency, and, for that reason, one can measure the channel in the UL and use it also for DL transmission.
- In FDD systems, channel estimation and feedback for many antennas present a challenge. Unless the channel structure is available at the BS, the prohibitive DL channel training and feedback in FDD systems set an upper limit on the number of BS antennas.
- The performance of mMIMO is limited by the finite and correlated scattering given the space constraints. The degrees of freedom of the system, solely determined by the spatial resolution of the antenna array, can reach saturation point.
- With mMIMO, there is a challenge of manufacturing many low cost, low-precision components which also affect how to approach testing and verification of the performance of these antennas since over-the-air test methods must generally be applied.

3 High-Level Capacity Estimation for Massive MIMO with Single Antenna-Based Multi-User

The detail capacity estimation of Massive MIMO is very complex. We will only consider some simplified configurations as depicted in Figure 3.1. We assume that many users with single-input single-output (SISO) antenna are communicating via the Massive MIMO-based base station (BS), wireless local area network WLAN access point (AP), or satellite gateway using uplinks and downlinks. We will offer a simple example for calculating the calculating the capacity of the Massive MIMO for the downlink only (Figure 3.1).

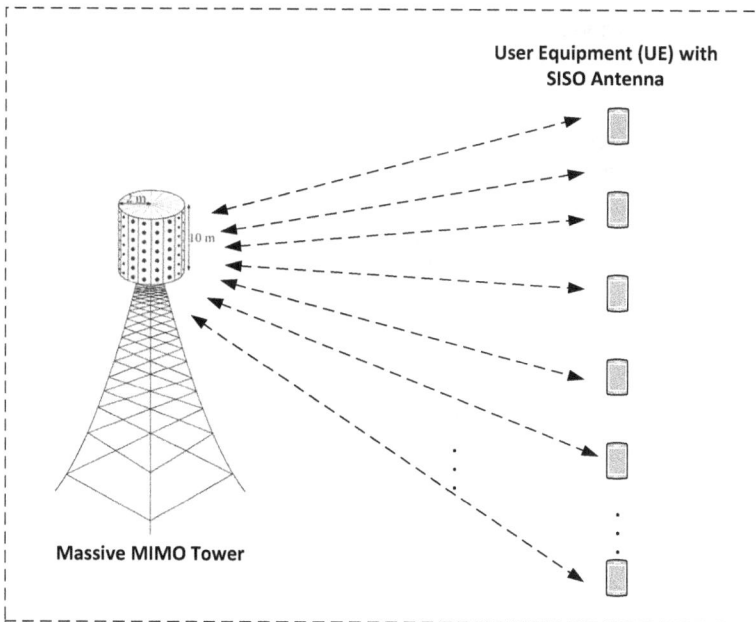

FIGURE 3.1 Massive MIMO network with M antennas communicating with N-single antenna users.

DOI: 10.1201/9781003499480-3

We are assuming that the Massive MIMO system has very large M numbers of antennas with time division duplex (TDD) system, where the channel has the small-scale fading (SSF) owing to slowly fading Rayleigh channel (assumption) due to large numbers of antennas, and the magnitude of M is very large in Massive MIMO system. However, users will have large-scale fading (LSF) factor because of the distance between the users and using of the SISO antenna and the Massive MIMO network. Users are assumed to be randomly and uniformly distributed in the specified circular geographical area. We are considering only the downlink for capacity analysis, and only λ_s of users are selected from N users. Precoding is needed for removing interferences. Linear precoding schemes, such as zeros forcing (ZF), minimum mean square error (MMSE), or maximum ratio transmission (MRT) can be applied for the reduction of inter-user interferences. Under these conditions, the ergodic capacity C_{mMIMO} of the TDD-based downlink Massive MIMO system is provided by:

$$C_{mMIMO} = \left(1 - \frac{\lambda_s}{\tau}\right) \sum_{k=1}^{\lambda_s} \log_2 \left(1 + \frac{|h_k P_k|^2}{\Psi_k + \sum_{i \neq k, k \in \lambda_s} c d_k^{-l} |h_k P_i|^2}\right)$$

where,

- C_{mMIMO} is the Massive MIMO channel capacity.
- τ is the channel coherence time.
- $\lambda_s \ll M$ is the number of users selected to serve simultaneously, and M is the number of antennas of the Massive MIMO system.
- Ψ_k denotes the ratio of total noise to transmitting power given by $\Psi_k = \frac{w_k^2}{\rho_k}$, ρ_k is the Kth active user's transmitting power scaling factor, and w_k^2 noise power in the Kth active user receiver.
- h_k is the channel matrix of the Kth active user.
- P_k is the κth column vector of the precoding matrix for the respective precoding schemes that can be, for example, ZF, MMSE, or MRT for reduction of inter-user interferences. $P_{k,ZF} = \mathbf{H}^H(\mathbf{H}\mathbf{H}^H)^{-1}\Gamma$ where $\Gamma = \text{diag}(\gamma_1, \gamma_2, \ldots, \gamma_k)$ is the diagonal matrix of $P_{k,ZF}$. $P_{k,MMSE} = H^H(\mathbf{H}\mathbf{H}^H + \Psi_k I_k)^{-1}$, where I_k is the identity matrix of the kth column and $P_{k,MRT} = \mathbf{H}^H$.
- c is the LSF factor.
- l is the pathloss exponent.

The capacity of a Massive MIMO cellular network depends on user and antenna selection algorithms, and on the acquisition of perfect channel state information (CSI). The channel state is estimated at regular intervals in the cellular system. This interval is independent of the coherence time τ of the users but defines the shortest coherence time that the users may have. Note that the coherence time is roughly inversely proportional to the speed of the user, so the time interval corresponds to a maximum speed of mobility. The above equation shows the effect of coherence time τ for Massive MIMO capacity C_{mMIMO}. Longer coherent time improves the capacity of the Massive MIMO system. Low computational cost algorithms for user and antenna selection significantly may enhance the system capacity, as they would consume a smaller bandwidth out of the total bandwidth for downlink transmission. However, it is expected that 6G Massive MIMO will have native AI/ML/DL support, and thereby computation algorithms will take advantage of the simplicity of algorithms and higher speeds of computing nodes of the network. We have taken a simple example for Massive MIMO capacity calculations. In practice, users might also have multiple MIMO-based antennas. In this situation, the capacity calculation for Massive MIMO will be more complicated even for the downlink.

As explained earlier, the objective of this chapter is to maximize the system sum-rate capacity with efficient user and antenna scheduling algorithms and linear precoding. We consider a slowly fading Rayleigh channel with perfect acquisition of CSI to explore the system sum-rate capacity of a Massive MIMO network. The users are characterized with both SSF due to slowly fading Rayleigh channel and LSF due to the distance dependent large-scale path loss from the BS to users. Further, we use linear precoding techniques such as ZF, MMSE, and MRT, to reduce interference, thereby improving average system sum-rate capacity. We also assume that the cellular network is a circular geographical area with outer radius $r_{max} = 300$ m and inner radius $r_{min} = 45$ m.

We assume the coherence time $\tau = 256$, wireless path-loss exponent $l = 2$, and LSF factor $c = 10^{-3}$. Users are randomly and uniformly distributed in the cell to schedule $\lambda_s = 8$ number of users and select equal number of BS antennas. We consider the noise power level is $w_k^2 = -174$ dBm, the bandwidth is 3.5 GHz, throughout this chapter, for cellular service operations in 5G, and the downlink transmission power is in the range of 5 – 20 dBm.

The simulation results showing average system sum-rate vs. downlink transmission power are shown in Tables 3.1 and 3.2 using the semi-orthogonal user scheduling and antenna selection (SUS-AS) algorithm. We have explored how the average system sum-rate of a Massive MIMO cellular mobile service depends on the user selection and antenna scheduling algorithms with different precoding schemes. The main objective is to reduce inter-user interference by adopting appropriate user selection and antenna scheduling algorithms to enhance the average system sum-rate. We consider a slowly fading Rayleigh channel with perfect acquisition of CSI to explore the system sum-rate capacity of a Massive MIMO network. The users are characterized with both SSF due to slowly fading Rayleigh channel and distance dependent

TABLE 3.1
Average Sum-Rate for Downlink for Different SNR

SNR (dBm)	Average Sum-Rate for Downlink (Bits/Hertz/Sec)		Remarks
	ZF	MMSE	
5	~ 185	~ 185	SUS-AS algorithm
10	~ 196	~ 194	Non-precoding
15	~ 206	~ 201	Number of BS antennas: $M = 23$
20	~ 215	~ 206	Number of single antenna users: $N = 64$

TABLE 3.2

Average Sum-Rate for Downlink for Different Number of BS Antennas

Number of BS Antennas	Average Sum-Rate for Downlink (Bits/Hertz/Sec)		Remarks
	ZF	MMSE	
40	~ 188	~ 186	SUS-AS algorithm
60	~ 195	~ 198	Non-precoding
80	~ 205	~ 203	SNR = 15 dBm
100	~ 208	~ 205	Number of single antenna
120	~ 210	~ 206	users: $N = 64$
140	~ 212	~ 209	

large-scale path loss. We studied the system sum-rate variation with variation in the downlink transmitting power, number of BS antenna, and the number of users.

We applied linear precoding schemes, such as ZF, MMSE, and user SUS-AS scheduling algorithms. The computational complexity of algorithm-2 and algorithm-3 is $O(1)$, and that of algorithm-1 is $O\left(N_{\lambda_s} M_{\lambda_s}^2\right)$, where λ_s is the number of users scheduled with an equal number of antennas for simultaneous data transmission using TDD mode. The antennas are selected based on maximum SNR to scheduled users. We observe that the SUS-AS user scheduling with maximum SNR-based antenna selection technique can enhance the average system sum-rate to the highest extent when ZF precoding scheme is used. We also observe that for a given value of N, the increase of M increases the system sum-rate for the scheduling algorithm with any precoding schemes. The same is observed when M is fixed and N is increased. This shows that the increase of M or N increases the diversity gain, thereby enhancing the system sum-rate.

HIGH-LEVEL CAPACITY ESTIMATION FOR MASSIVE MIMO WITH MULTIPLE ANTENNAS-BASED MULTI-USER

In this section, we have analyzed the performance of Massive MIMO systems with N-antenna users (Figure 3.2). The benefit is that N streams can be multiplexed per user, at the price of increasing the channel estimation overhead linearly with N. Uplink and downlink spectral efficiency (SE) expressions are derived for any N, and these are achievable using estimated channels and per-user-basis MMSE-SIC detectors [1]. Large-system approximations of the SEs are obtained. This analysis shows that minimum mean-squared error–successive interference cancellation (MMSE-SIC) has similar asymptotic SE as linear MMSE detectors, indicating that the SE increase from having multi-antenna users can be harvested using linear detectors. We generalize the power scaling laws for Massive MIMO to handle arbitrary N and show that one can reduce the multiplication of the pilot power and payload power as $\dfrac{1}{M}$, where M is the number of BS antennas, and still notably increase the SE with M before reaching a non-zero asymptotic limit. Simulations testify our analysis and

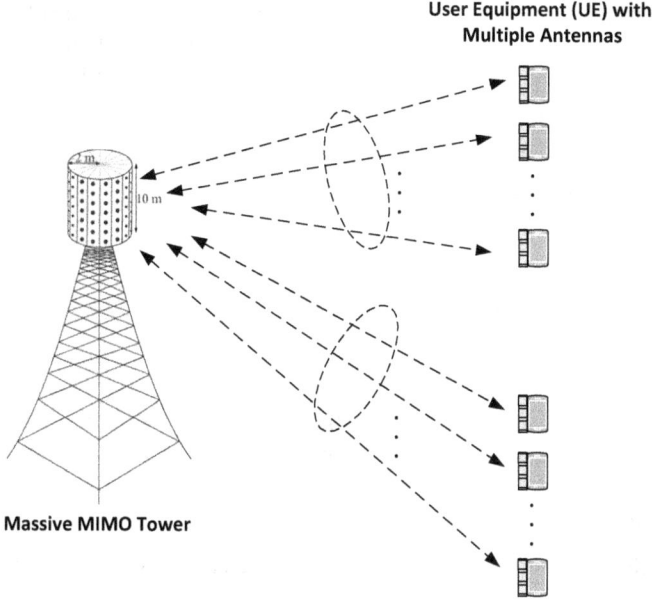

FIGURE 3.2 Massive MIMO base station with multi-antenna user.

show that the SE increases with N. We also note that the same improvement can be achieved by serving N times more single-antenna users instead; thus, the additional user antennas are particularly beneficial for SE enhancement when there are few active users in the system.

We analyze the SE of a Massive MIMO system with estimated CSI and any number of antennas N per user. Lower bounds on the sum capacity are derived for the uplink and downlink, which are achievable by per-user basis minimum mean-squared error successive interference cancellation (MMSE-SIC) detectors and only uplink pilots. Large-system approximations of the lower bounds are further obtained, which are tight as M grows large. Furthermore, we generalize the power scaling laws from [2, 3] to handle arbitrary N. The analysis shows that equipping users with multiple antennas can greatly enhance the SE, particularly in lightly loaded systems where there are too few users to exploit the full multiplexing capability of Massive MIMO with $N = 1$, and the benefits can be harvested by linear processing.

We consider a single-cell system in TDD mode, where the BS has M antennas and serves K users within each time-frequency coherence block. Each user is equipped with N antennas. We assume that each coherence block contains S transmission symbols and the channels of all users remain unchanged within each block. Let $G_k \in \mathbb{C}^{M \times N}$ denote the channel response from k^{th} user to the BS within a coherence block. The fading can be spatially correlated, due to insufficient spacing between antennas and insufficient scattering in the channel. We use the classical Kronecker model to describe the spatial correlation:

$$G_k = R_{r,k}^{\frac{1}{2}} G_{\omega,k} R_{t,k}^{\frac{1}{2}}$$

Where entries of $G_{\omega,k} \in \mathbb{C}^{M \times N}$ follow independent and identically distributed (i.i.d.) zero-mean circularly symmetric complex Gaussian distributions. $R_{t,k} \in \mathbb{C}^{M \times N}$ represents the spatial correlation at user k, and $R_{r,k} \in \mathbb{C}^{M \times N}$ describes the spatial correlation at the BS for the link to user k. The LSF parameter is included in $R_{r,k}$ and can be extracted as $\frac{1}{M} \mathrm{tr}(R_{r,k})$. Let $R_{t,k} = U_k \Lambda_k U_k^H$ be the eigenvalue decomposition of $R_{t,k}$, where $U_k \in \mathbb{C}^{M \times N}$ is an unitary matrix and $\Lambda_k = \mathrm{diag}\{\lambda_{k,1}, \dots, \lambda_{k,N}\}$ contains the eigenvalues.

UPLINK ACHIEVABLE CHANNEL CAPACITY

During the uplink pilot signaling, $B = NK$ orthogonal pilot sequences are needed to estimate all channel dimension at the BS. The pilot matrix of user k is defined as $F_k \in \mathbb{C}^{M \times N}$. The CSI, $R_{t,k} \in \mathbb{C}^{n_T n_R \times n_T n_R}$ is the positive defined on the column stacking of the channel matrix where mean Rician fading channel matrix is represented by $H \in \mathbb{C}^{n_T \times n_R}$ and $H \in \mathcal{CN}(x, R)$ is used to denote circularly symmetric complex Gaussian random vectors with x as the mean and R the covariance matrix. The pilot matrix $R_{t,k}$ minimizes the mean squared error (MSE) of the channel estimation under the pilot energy constraint $\mathrm{tr}(F_k F_k^H \leq BP_k)$, which has the form of $F_k = U_k L_k^{\frac{1}{2}} V_k^T$, where P_k is the maximum transmit power of user k, $L_k = \mathrm{diag}\{l_{k,1}, \dots, l_{k,N}\}$ distributes the power among the N channel dimensions, and $V_k \in \mathbb{C}^{M \times N}$ satisfies $V_k^H V_k = BI_N$ and $V_k^H V_l = 0$ if $k \neq l$. Thus, the received signal at the BS is:

$$Y = \sum_{k=1}^{K} G_k F_k + N = \sum_{k=1}^{K} \left(R_{r,k}^{\frac{1}{2}} G_{\omega,k} R_{t,k}^{\frac{1}{2}} \right) \left(U_k L_k^{\frac{1}{2}} V_k^T \right) + N = \sum_{k=1}^{K} H_k D_k^{\frac{1}{2}} V_k^T + N \in \mathbb{C}^{M \times N}$$

Where $H_k \equiv R_{r,k}^{\frac{1}{2}} G_{\omega,k} U_{t,k}$ and $D_k = \Lambda_k L_k$ with $d_{k,i}$ is its ith diagonal element. $\mathbb{N}$ is the receiver noise that follows $\mathbb{N} \sim \mathcal{CN}(0, \sigma^2 I_{BM})$. If the BS knows the statistical information, then MMSE estimate of h_k:

$$h_k = \left(D_k^{\frac{1}{2}} \otimes F_k \right) \left((D_k \otimes R_{r,k}) + \frac{\sigma^2}{B} \right)^{-1} b_k$$

Where $b_k = \frac{1}{B} Y_k V_k^* = H_k D_k^{\frac{1}{2}} + \frac{1}{\sqrt{B}} I_{M,N}$ and $\otimes$ denotes the Kronecker product. Let $h_{k,i}$ be the ith column of H_k, then:

$$E\{h_{k,i} h_{k,j}^H\} = \begin{cases} \phi_{k,i} & i = j \\ 0, & i \neq j \end{cases}$$

Where $\phi_{k,i} = d_{k,i} R_{r,k} \left(d_{k,i} R_{r,k} + \frac{\sigma^2}{B} I_M \right)^{-1} R_{r,k}$

When the receiving BS knows the perfect CSI of all users while each transmitter has only its own statistical CSI, the precoding directions of each user that maximize the sum capacity coincide with the eigenvectors of their own spatial correlation matrix. The uplink achievable channel capacity or SE with the MMSE detector from the kth user is provided as follows:

$$C_{ul,k}^{MMSE} = \sum_{i=1}^{N} E\left\{\log_2\left(1 + \eta_{k,i}^{ul}\right)\right\}$$

Where the signal-to-interference-plus-noise ratio (SINR) of the ith stream is:

$$\eta_{k,i}^{ul} = \frac{\lambda_{k,i} p_{k,i}\left|f_{k,i}^{H} h_{k,i}\right|^2}{E\left\{f_{k,i}^{H}\left(yy^H - \lambda_{k,i} p_{k,i} h_{k,i} h_{k,i}^H\right)f_{k,i}\middle| H\right\}}$$

where

$$f_{k,i} = \left(\sqrt{\lambda_{k,i} p_{k,i}}\right)\Sigma\left(h_{k,i}\right)$$

$$\Sigma = \left(\Sigma_k^{-1} + H_k Q_k H_k^H\right)^{-1}$$

$$\Sigma_k^{-1} = \text{Covariance Matrix}$$

$$H_k \equiv R_{r,k}^{\frac{1}{2}} G_{\omega,k} U_{t,k} Q_k$$

$$h_{k,i} = \left(D_{k,i}^{\frac{1}{2}} \otimes F_{k,i}\right)\left(\left(D_{k,i} \otimes R_{r,k,i}\right) + \frac{\sigma^2}{B}\right)^{-1} b_{k,i}$$

$p_{k,i}$ is the dignal element of $P_k = \text{diag}\left\{p_{k,i}, \ldots, p_{k,N}\right\}$

$\lambda_{k,i}$ is the eigenvalue containing in $\Lambda_k = \text{diag}\left\{\lambda_{k,1}, \ldots, \lambda_{k,N}\right\}$

$$y = \sum_{k=1}^{K} G_k F_k x_k + n = \sum_{k=1}^{K} H_k \Lambda_k^{\frac{1}{2}} P_k^{\frac{1}{2}} x_k + n$$

$\otimes$ represents Kronecker product

Where $x_k \sim CN(0, I_N)$ is the transmitted data symbol from user k and $n \sim CN(0, \sigma^2 I_M)$ is additive receiver noise.

Since interference from the user's own streams is not suppressed by $f_{k,i}$, it is intuitive that $C_{ul,k}^{SIC} \geq C_{ul,k}^{MMSE}$.

Uplink Achievable Channel Capacity

To limit the estimation overhead, we assume no downlink pilot or CSI feedback from the BS to users. This is common practice in Massive MIMO since only the BS needs CSI to achieve channel hardening. Hence, the users have no instantaneous CSI except to learn the average effective channel, $H_k \equiv \Lambda_k^{\frac{1}{2}} E\{H_k^H W_k\} \Omega_l^{\frac{1}{2}}$, and covariance matrix of the interference term. Let $W_k \in \mathbb{C}^{M \times N}$ be the downlink precoding matrix associated with user k and let $\Omega_k = \text{diag}\{\omega_{k,i}, ..., \omega_{k,N}\}$ allocate the total transmit power P'_k among N streams. Then the total transmit power from the BS is $\sum_{k=1}^{K} P'_k$. The received signal at user k is:

$$y_k = G_k^H \sum_{l=1}^{K} W_l \Omega_l^{\frac{1}{2}} x_l + n_k \in \mathbb{C}^{M \times N}$$

Where $x_l \sim \mathcal{CN}(0, I_M)$ is the downlink signal intended for user l and $n_k \sim \mathcal{CN}(0, \sigma^2 I_N)$ is the additive receiver noise.

Without loss of generality, let user k use U_k^H (the eigenvector matrix of its own correlation matrix) as a first-step detector to adapt to the channel correlation, then the processed received signal is:

$$z_k = U_k^H y_k = \Lambda_k^{\frac{1}{2}} H_k^H \sum_{l=1}^{K} W_l \Omega_l^{\frac{1}{2}} x_l + U_k^H n_k$$

A lower bound on the mutual information $I(z_k; x_l)$ is given as follows:

$$I(z_k; x_l) \geq \log_2 \left| I_N + H_k^H \mathfrak{M}_k H_k \right| \equiv C_{\text{dl},k}^{SIC}$$

Where $\mathfrak{M}_k = \left(\Lambda_k^{\frac{1}{2}} E\left\{ H_k^H \sum_{l \neq k} \left(H_k^H \Omega_l \right) H_k \right\} \Lambda_k^{\frac{1}{2}} + \sigma^2 I_N \right)^{-1}$

The lower bound can be achieved if user k applies MMSE-SIC detection to z_k when regarding H_k as the true channel and the uncorrelated term $(z_k - H_k x_l)$ is treated as worst-case Gaussian noise in the detector. The conventional SE analysis of Massive MIMO from $N = 1$ to arbitrary N. The user can also apply a linear MMSE detector for symbol detection based on z_k. Let us denote $h_{k,i}$ as the ith column of H_k, then with knowledge of H_k the MMSE detector for the ith stream of user k that maximizes the corresponding downlink SE is as follows:

$$r_{k,i} = \mathfrak{M}_k h_{k,i}$$

where

$$\mathfrak{M}_k = \mathfrak{M}_k^{-1} + H_k H_k^H$$

The achievable SE of user k is provided as follows:

$$C_{dl,k}^{MMSE} = \sum_{i=1}^{N} E\left\{\log_2\left(1+\eta_{k,i}^{dl}\right)\right\}$$

Where the SINR $\eta_{k,i}^{dl}$ of its ith stream is:

$$\eta_{k,i}^{dl} = \frac{\left|\mathbf{r}_{k,i}^{H}\mathbf{h}_{k,i}\right|^2}{\mathbf{r}_{k,i}^{r}E\left\{\mathbf{z}_k\mathbf{z}_k^{H}\right\}\mathbf{r}_{k,i} - \left|\mathbf{r}_{k,i}^{H}\mathbf{h}_{k,i}\right|^2}$$

Intuitively, the MMSE-SIC detector will have a higher performance than the MMSE detector in the downlink. To compare their performance in Massive MIMO systems, we derive their asymptotic SEs in the large system limit in the next section.

ASYMPTOTIC ANALYSIS

We consider the large system regime where M and K go to infinity while N remains constant since the users are expected to have a relatively small number of antennas. In what follows, the notation $M \to \infty$ refers to $K, M \to \infty$ such that $\lim \sup_M(K/M) < \infty$ and $\lim \inf_M(K/M) > 0$. The correlation coefficients between adjacent antennas at the BS and at the users are $a_r e^{j\theta_{r,k}}$ and $a_t e^{j\theta_{t,k}}$, respectively, with $\theta_{r,k}$ and $\theta_{t,k}$ are uniformly distributed in $[0, 2\pi)$.

POWER SCALING LAWS

It is known that for single antenna $N = 1$ user, the transmit power can be reduced with retained performance as the number of BS antennas grows. However, we have generalized the fundamental result to handle any fixed N numbers of antennas per user as described earlier. Assume that the pilot power is reduced as $L_k = \frac{1}{M^\alpha}L_k^{(0)}$ and the payload powers are $P_k = \frac{1}{M^{1-\alpha}}P_k^{(0)}$ and $\Omega_k = \frac{1}{M^{1-\alpha}}\Omega_k^{(0)}$, where $0 \le \alpha \le 1$ and the matrices $(\blacksquare)^{(0)}$ are fixed.

SIMULATION RESULTS

We consider a cell with a radius of 500 meters. The user locations are uniformly distributed at distances to the BS of at least 70 meters. Statistical channel inversion power control is applied in the uplink, equal power allocation is used in the downlink, and the power is divided equally between the N streams of each user such that the cell-edge SNR (without shadowing) is −3 dB. The exponential correlation models for $R_{t,k}$ and $R_{r,k}$ are used. The correlation coefficients between adjacent antennas at the BS and at the users are $a_r e^{j\theta_{r,k}}$ and $a_t e^{j\theta_{t,k}}$, respectively, with $a_r = a_t = 0.4$; $\theta_{r,k}$ and $\theta_{t,k}$ are uniformly distributed in $[0, 2\pi)$. The coherence block length of 200 is used to support high user mobility.

TABLE 3.3

Achievable Uplink and Downlink Channel Capacity with Number of BS Antennas having Single- and Multi-Antenna per User

Uplink/Downlink	Number of Base Station Antennas (M)	Achievable sum SE (bits/sec/Hertz)		Total Number of Users (K)
		Number of Antennas Per User		
		$N = 1$	$N = 3$	
Uplink	50	~40	~60	12
	100	~50	~100	
	200	~60	~120	
	300	~65	~135	
	400	~66	~140	
Downlink	50	~22	~30	
	100	~30	~50	
	200	~40	~70	
	300	~45	~85	
	400	~50	~90	

The uplink and downlink sum SE of the MMSE-SIC and MMSE detectors are shown in Table 3.3. It shows that the two detectors achieve almost the same SEs, which verifies the conclusion that a linear detector can achieve most of the SE improvements from equipping users with multiple antennas in Massive MIMO. Moreover, although the pilot overhead increases, 90% and 75% performance gains are achieved for the uplink and the downlink, respectively, by increasing N from 1 to 3 for $M = 200$.

Table 3.4 testifies the power scaling laws. Results for $\alpha = 0.5$ and $\alpha = 1$ are shown. It is observed that, even with a $1/M$ reduction of the multiplication of pilot and payload powers, a notable increase of SE can still be obtained for an extremely wide range of M before reaching the limit, especially for $M \in [50, 1000]$, which is of practical interest.

Recall that the channel estimation overhead NK equals the number of data streams that are transmitted. For a fixed number of data streams NK, the system can schedule NK single-antenna users and send one stream to each user, or schedule fewer multi-antenna users and send several streams to each. It is seen that for any given NK, scheduling NK single-antenna users are always (slightly) beneficial. The optimal NK is found around 100, which requires 100 active users per coherence block if $N = 1$. With multi-antenna users, more realistic user numbers are sufficient to reach the sweet spot of $NK \approx 100$. Therefore, additional user antennas are beneficial to increase the spatial multiplexing in light- and medium-loaded systems.

We analyzed the achievable SE of single-cell Massive MIMO systems with multi-antenna users. With estimated CSI from uplink pilots, lower bounds on the ergodic sum capacity were derived for both the uplink and the downlink, which are achievable by per-user MMSE-SIC detectors. Large system SE approximations were derived and shows that the MMSE-SIC detector has an asymptotic performance

TABLE 3.4

Achievable Uplink and Downlink Channel Capacity with Number of BS Antennas with Power Control with Multi-Antenna User

Uplink/Downlink	Number of Base Station Antennas (M)	Achievable Sum SE (bits/sec/Hertz)		Power Law Parameters
		Power Control Parameter		
		$\alpha = 1$	$\alpha = 0.5$	
Uplink	10^2	~ 10	~ 25	Total number of
	10^3	~ 10	~ 38	users: K = 10,
	10^4	~ 10	~ 46	Number of antennas
	10^5	~ 10	~ 48	per user: N=3,
	10^6	~ 10	~ 49	and correlation
	10^7	~ 10	~ 50	coefficient between adjacent antennas
Downlink	10^2	~ 8	~ 17	at the base station
	10^3	~ 8	~ 23	and at the users are
	10^4	~ 8	~ 26	$a_r = 0$ and $a_t = 0.4$,
	10^5	~ 8	~ 27	respectively.
	10^6	~ 8	~ 28	
	10^7	~ 8	~ 30	

similar to the linear MMSE detector, indicating that linear detectors are sufficient to handle multi-antenna users in Massive MIMO. We generalized the power scaling laws for Massive MIMO from $N = 1$ to arbitrary N. We showed that the SE increases with N, but for a fixed value of NK the highest SE is achieved by having NK single-antenna users. Hence, additional user antennas are mainly beneficial to increase the spatial multiplexing in systems with few users.

MASSIVE MIMO WITH MULTIPLE ANTENNAS-BASED MULTI-USER

Multi-user MIMO, or MU-MIMO, is an enhanced form of MIMO technology that is gaining acceptance. MU-MIMO enables multiple independent radio terminals to access a system enhancing the communication capabilities of each individual terminal [3]. Accordingly, it is often considered as an extension of space division multiple access (SDMA). MU-MIMO exploits the maximum system capacity by scheduling multiple users to be able to simultaneously access the same channel using the spatial degrees of freedom offered by MIMO. To enable MU-MIMO to be used, there are several approaches that can be adopted, and several applications/versions that are available.

MU-MIMO vs. SU-MIMO

Both single-user-MIMO and MU-MIMO systems can be used within wireless and cellular telecommunication systems. Each form of MIMO has its advantages and disadvantages. Table 3.5 provides the comparison of MU-MIMO vs. SU-MIMO.

TABLE 3.5

Comparison of MU-MIMO vs. SU-MIMO

Feature	MU-MIMO	SU-MIMO
Main feature	For MU-MIMO, the base station can separately communicate with multiple users	Base station communicates with a single user
Key aspect	Using MU-MIMO provides capacity gain	Provides increased data rate for the single user
Key advantage	Multiplexing gain	Interference reduction
Data throughput	MU-MIMO provides a higher throughput when the signal-to-noise ratio is high	Provides a higher throughput for a low-signal-to-noise ratio
Channel state	Perfect CSI is required	No CSI needed

MU-MIMO Basics

MU-MIMO provides a methodology whereby spatial sharing of channels can be achieved. This can be achieved at the cost of additional hardware – filters and antennas – but the incorporation does not come at the expense of additional bandwidth as is the case when technologies such as FDMA, TDMA, or CDMA are used. When using spatial multiplexing, MU-MIMO, the interference between the different users on the same channel is accommodated using additional antennas, and additional processing when the spatial separation of the different users is enabled.

There are two scenarios associated with MU-MIMO:

- **Uplink – multiple access channel (MAC):** The development of the MIMO-MAC is based on the known single-user MIMO concepts broadened out to account for multiple users.
- **Downlink – broadcast channel (BC):** The MIMO-BC is the more challenging scenario. The optimum strategy involves pre-interference cancellation techniques known as "Dirty Paper Coding," DPC – see below. This is complemented by implicit user scheduling and a power loading algorithm.

MU-MIMO offers some significant advantages over other techniques:

- MU-MIMO systems enable a level of direct gain to be obtained in a multiple access capacity arising from the multi-user multiplexing schemes. This is proportional to the number of BS antennas employed.
- MU-MIMO appears to be affected less by some propagation issues that affect single-user MIMO systems. These include channel rank loss and antenna correlation – although channel correlation still affects diversity on a per user basis, it is not a major issue for multi-user diversity.
- MU-MIMO allows spatial multiplexing gain to be achieved at the BS without the need for multiple antennas at the UE. This allows to produce cheap remote terminals – the intelligence and cost are included within the BS.

- The advantages of using MU-MIMO come at a cost of additional hardware – antennas and processing – and obtaining the CSI, which requires the use of the available bandwidth.

MIMO-MAC

- This form of MU-MIMO is used for a MAC and is used in uplink scenarios.
- For the MIMO-MAC, the receiver performs much of the processing – here, the receiver needs to know the channel state and uses CSI at the receiver (CSIR). Determining CSIR is generally easier than determining CSIT, but it requires significant levels of uplink capacity to transmit the dedicated pilots from each user. However, MIMO MAC systems outperform point-to-point MIMO particularly if the number of receiver antennas is greater than the number of transmit antennas at each user.

MIMO-BC

- This form of MU-MIMO is used for the MIMO broadcast channels, that is, the downlink. Of the two channels, BC and MAC, it is the broadcast channel that is the more challenging within MU-MIMO.
- Transmit processing is required for this, and it is typically in the form of pre-coding and SDMA, SDMA-based downlink user scheduling. For this, the transmitter has to know the CSI at the transmitter (CSIT). This enables significant throughput improvements over that of ordinary point-to-point MIMO systems, especially when the number of transmit antennas exceeds that of the antennas at each receiver.

Dirty Paper Coding

- DPC is a technique used within telecommunications scenarios, particularly wireless communications, to provide efficient transmission of digital data through a channel that is subject to interference, the nature of which is known to the transmitter.
- The DPC technique consists of precoding the data so the interference data can be read in the presence of the interference. The pre-coding normally uses the CSI.
- To explain DPC, an analogy of writing on dirty paper can be used. Normally black ink would be used, but if the paper is dirty, i.e., black, then the writing cannot be read. However, if the writing was in white, although it could not be read on white paper, it would be perfectly legible on black, or dirty paper. The same technique is used on the data transmission, although the nature of the interference must be known so that the pre-coding can be incorporated to counter the effect of the interference.

Single-user Massive MIMO and MU-Massive MIMO are still in its infancy, and many developments are underway to determine the optimum formats for its use. Coding types as well as levels of channel state indication are being determined as

TABLE 3.6

Comparison between SU and MU Massive MIMO

Capabilities	SU Massive MIMO	MU-Massive MIMO
Beamforming	Possible in both uplink and downlink	Possible in both uplink and downlink
Spatial multiplexing	Possible in both uplink and downlink	Not possible
Spectral efficiency	High spectral efficiency per user	Low spectral efficiency per user
Directivity	High directivity in both uplink and downlink in beamforming mode	High directivity in downlink in beamforming mode
Modulation schemes and power consumption	Consume significant low power for modulation schemes, for example quadrature phase shifting Keying (QSPK)	Consume large amount of power due to use of M-ary constellation

TABLE 3.7

Conflicting Options of Beamforming and Spatial Multiplexing

Capabilities	Beamforming	Spatial Multiplexing
Directivity	High directivity in beamforming	Little directivity for spatial multiplexing
Spectral efficiency	Low spectral efficiency per user since the same signal is transmitted from each antenna element	High spectral efficiency per user since different signals are transmitted from each antenna element
Spectral efficiency improvement	Spectral efficiency can be improved by increasing the constellation size, resulting in high peak-to-average power ratio (PAPR)	QPSK constellations with PAPR 0 dB can be used
Coding	Difficult to turbo/low-density parity check (LDPC) code for large constellations	Easy to turbo/LDPC code QPSK
Bit-error-rate (BER)	Large BER at SNR per bit close to 0 dB	Small BER at SNR per bit close to 0 dB

these use up valuable resources and can detract from the overall data throughput available. However, the significant gains that can be made by using MU-Massive MIMO mean that it will be introduced in the foreseeable future (Table 3.6).

However, there are conflicting options between the beamforming and the spatial multiplexing. It is depicted in Table 3.7.

SUMMARY

We have discussed Massive MIMO with single- and multiple-antenna-based multi-user. In both cases, we have derived uplink- and down-link achievable channel capacity. In addition, power scaling laws are also analyzed. A comparison between single- and multiple-antenna multi-user is provided. The analytical models have been validated using modeling and simulation.

REFERENCES

1. Sheikh, T. A. et al., "Capacity maximizing in massive MIMO with linear precoding for SSF and LSF channel with perfect CSI", *Digital Communications and Networks* 7 (2021) 92–99.
2. Li, X., et al., "Massive MIMO with Multi-Antenna Users: When are Additional User Antennas Beneficial?" March 2016, arXiv:1603.09052v1 [cs.IT] 30 Mar 2016.
3. He, X., et al., "On the Multi-User Multi-Cell Massive Spatial Modulation Uplink: How Many Antennas for Each User?" March 2017, *IEEE Transactions on Wireless Communications* 16(3).

4 Ultra-Massive MIMO
Ultra-Massive MIMO Platforms

Ultra-Massive MIMO (UM-MIMO) technologies are envisioned to operate in the millimeter-wave (30–300 GHz) and Terahertz-band communications (0.3–10 THz) for meeting the traffic demand of 6G technologies having Terabit-per-second (Tbps) links. However, this very large available bandwidth in this ultra-broadband frequency range comes at the cost of a very high propagation loss, which combined with the low power of millimeter-wave and THz-band transceivers limits the communication distance and data rates. UM-MIMO platforms have been proposed to increase the communication distance and data rates at millimeter-wave and THz-band frequencies. Fortunately, it has been researched that capabilities of novel intelligent plasmonic antenna arrays can operate in transmission, reception, reflection, and waveguiding, as well as the peculiarities of the millimeter-wave and THz-band multipath channel. Recently, plasmonic-based UM-MIMO antennas have been developed and tested. Based on the developed model, extensive quantitative results for different scenarios are provided to illustrate the performance improvements in terms of both achievable distance and data-rate in UM-MIMO environment.

The UM-MIMO platforms consist of reconfigurable plasmonic antenna arrays both at the transmitting and receiving nodes, operating as plasmonic transmit-receive arrays, and in the transmission environment, in the form of plasmonic reflect-arrays and able to operate in different modes, including transmission, reception, reflection, and waveguiding. Plasmonic antennas leverage the physics of surface plasmon polariton (SPP) waves to efficiently radiate at the target resonant frequency while being much smaller than the corresponding wavelength. This particular property allows them to be integrated in very dense arrays, beyond traditional antenna arrays, and enables the precise radiation and propagation control of electromagnetic (EM) waves with sub-wavelength resolution.

The MU-MIMO model estimates the performance of intelligent communication environments considering the transmitter, the receiver deploying through the communication channel in a synergistic operation manner to overcome the main challenge at millimeter wave and THz frequencies UM-MIMO platforms, enables the creation of intelligent communication environments in both indoor and outdoor scenarios. With the derivations readily extendable to the outdoor case, here we focus our analysis on the indoor case.

The intelligent environments consist of two major parts: the plasmonic transmit-receive arrays at the nodes and the plasmonic reflect array systems in the propagation environment. Plasmonic reflect arrays can be embedded or applied to surfaces of indoor objects (e.g., walls and ceilings) like adhesive foil papers with low energy cost

 DOI: 10.1201/9781003499480-4

and allow signal transmissions through reflections on the plasmonic layer (i.e., the top layer) or waveguiding on the waveguiding layer (i.e., the bottom layer). The control layer (i.e., implemented in the middle layer) estimates the channel, coordinates with the transmitting and receiving arrays at the nodes, and assigns the operation modes to individual or groups of plasmonic units. The UM-MIMO platforms are powered by batteries in mobile transceivers and AC power supply for wallcovering reflect arrays.

PLASMONIC TRANSMIT–RECEIVE ARRAYS

The physics of plasmonic material allows to build antenna arrays with much denser elements and go beyond the conventional $\lambda/2$ sampling of space toward more precise space and frequency beamforming. On that basis, we have demonstrated that, at THz frequencies, graphene can be used to build nano-transceivers and nano-antennas with maximum dimension $\lambda/20$, allowing them to be densely integrated in very small footprints (1,024 elements in less than 1 mm²). Recently, UM-MIMO capacity has been derived using a conceptual design depicted in the figure shown below.

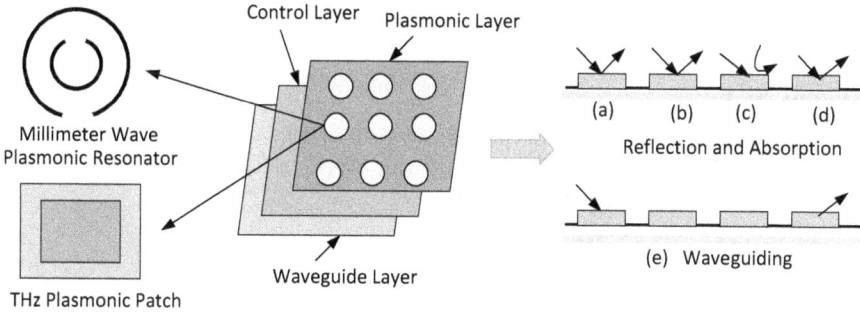

FIGURE 4.1   Conceptual Design of Ultra-Massive (UM) MIMO using Plasmonic Transmit-Receive Arrays.

In a downlink channel, the received signal vector y can be expressed as a function of transmitted signal vector x, downlink channel matrix $\mathbf{H}$, optimal transmitter power P_t, and complex Gaussian noise with zero-mean and unit-variance CN as $y = \sqrt{P_t}\mathbf{H}_x + CN$. Based on the assumption that the given resource optimization, the UM-MIMO platforms have perfect knowledge of the channel, the achievable system throughput capacity is provided as follows [1]:

$$C = B\log_2 \det\left(\mathbf{I} + P_t\mathbf{H}^H\mathbf{H}\right) = \log_2\left(\mathbf{I} + P_t\mathbf{H}^H\mathbf{H}\right)$$

which is upper bounded by the size of the antenna arrays in the plasmonic transmit–receive arrays and the plasmonic reflect arrays, when the number of users is fewer in the same environment. Note that we have not shown the complicated derivation of the throughput capacity of UM-MIMO for sake of brevity. Interested readers will find the details in paper [1]. Therefore, interference is asymptotically zero in this case. Many numerical results have been published in the paper [Nie]. It provides the

TABLE 4.1
UM-MIMO Capacity

M_s	Transmit Power (dBm)	System Throughput (bits/sec) $\times 10^{11}$
50	1–10	~ 7.8 to 9.3
100		~ 8.4 to 10
200		~ 8.9 to 10.5
400		~ 9.4 to 11
800		~ 9.8 to 11.5
1200		~ 10.2 to 11.6

distance enhancement and achievable data rates with different sizes in UM-MIMO platforms at a center frequency f_c = 300 GHz with a bandwidth of 50 GHz. By utilizing a size of N_a = 1024 × 1024 (i.e., transmission ×reception antennas), plasmonic transmit–receive arrays, and with various numbers of units M_s ranging from 50 to 1,200 of plasmonic reflect arrays in the environment, the sum data-rate achievable in that environment with UM-MIMO platforms can reach the Tbps level, as shown in Table 4.1.

REFERENCE

1. Nie, S. et al, "Intelligent Environments-based on Ultra-Massive MIMO Platforms for Wireless Communication in Millimetre Wave and Terahertz Bands," 2019. arXiv:1904.07958v1 [eess.SP] 16 Apr 2019

5 Vehicle-to-Everything Ultra-Massive MIMO Communications

6G vehicle-to-everything (V2X) communication will be combined with vehicle automatic driving technology and play an important role in automatic driving. However, in 6G V2X systems, vehicle users have the characteristics of high-speed movement. Therefore, how to provide stable and reliable wireless link quality and improve channel gain has become a problem that must be solved. To solve this problem, a new multi-user scheduling algorithm based on block diagonalization (BD) precoding for 6G ultra-massive multiple-input multiple-output (UM-MIMO) systems is proposed in this chapter [1]. The algorithm takes advantage of the sensitive nature of BD precoding to channel correlation, uses the Pearson coefficient after matrix vectorization to measure the channel correlation between users, defines the scheduling factor to measure the channel quality according to the user noise enhancement factor, and jointly considers the influence of the correlation between user channels and channel quality, ensuring the selection of high-quality channels while minimizing channel correlation. In this section, we consider a typical 6G V2X communication scenario and the UM-MIMO downlink system in the 6G V2X autonomous scenario shown in Figure 5.1 [2].

The system consists of a base station (BS), a roadside unit (RSU), and vehicles connected to the BS. BS and RSU are connected by optical fiber. The RSU (transmitting end) is equipped with N_t antennas, and the set of vehicles is $N = \{1, 2, ..., k\}$, where each vehicle user is equipped with N_r receiving antennas and multiple sensors. The system can schedule $R = \{1, 2, ..., R\}$, $R \subset N$ vehicles each time. In the downlink of the system with the assumption that the channel state information (CSI) is available, the maximum number of transmission data streams that each user can support is N_s and the total number of data streams sent by the BS is RN_s.

It can be seen from Figure 5.1 that the original signal is transmitted through the precoder at the antenna end and received by the receiving antenna of the receiving end through the channel. The received signal passes through the combiner to eliminate the interference between the user sub-channels, and finally undergoes signal processing to obtain the transmitted signal vector and user sub-channels:

$$s = \sum_{k=1}^{R} F_k s_k$$

where $s \in \mathbb{C}^{(R \times N_s) \times 1}$ denotes the original transmission symbol vector and is such that $E\{ss^H\} = I_{RN_s}$, $s_k \in \mathbb{C}^{N_s \times 1}$ is the original vector sent by the RSU to use k with $E\{s_k s_k^H\}$,

DOI: 10.1201/9781003499480-5

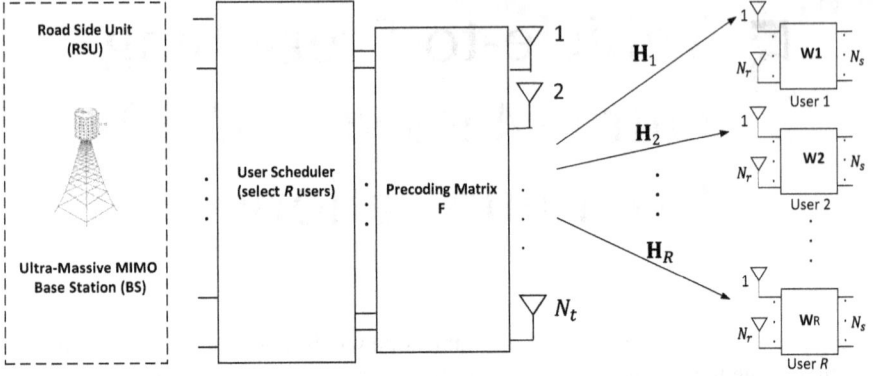

FIGURE 5.1 Multi-user scheduling for ultra-massive MIMO base station with multi-antenna mobile users.

and $F_k \in \mathbb{C}^{N_t \times N_s}$ is the precoding matrix of user k with $\|F_k\|^2 = 1$, which is the unitary matrix.

If the vehicle has perfect time and frequency synchronization for each time slot t, the received signal of user k after passing through the channel is expressed as:

$$y_k(t) = H_k(t) F_k s_k + \sum_{i=1, i \neq k}^{R} H_k(t) F_i s_i + n_k(t), \ k = 1, 2, \ldots, R$$

Where $H_k(t) F_k s_k$ is the kth user useful data, $H_k(t) \in \mathbb{C}^{N_r \times N_t}$ is the channel matrix between the RSU and user k in the tth time slot, and each element of the matrix obeys a Gaussian distribution with zero mean and unit variance. $F_i s_i$ denotes interference from other user data to the given user data. $n_k(t) \in \mathbb{C}^{N_r \times 1}$ is the additive white Gaussian noise of $\mathcal{CN}(0, \sigma_k^2 I)$. The received signal obtained after the combiner is given by:

$$\tilde{y}_k(t) = W_k^H H_k(t) F_k s_k + \sum_{i=1, i \neq k}^{R} W_k^H H_k(t) F_i s_i + W_k^H n_k(t), \ k = 1, 2, \ldots, R$$

Where W_k is the combiner of the kth user. If proper precoding is used, the data interference between users can be eliminated at the receiving end; the above equation is expressed as follows:

$$\tilde{y}_k(t) = W_k^H H_k(t) F_k s_k + W_k^H H_k(t) F_i s_i + W_k^H n_k(t), \ k = 1, 2, \ldots, R$$

Assuming that BD precoding algorithm is used at the transmitting end for M numbers of users and the power P is equally distributed, the channel capacity of user k having bandwidth B is provided by:

$$C_k = B \log_2 \det \left(I_{kN} + \frac{P}{M\sigma^2} H_i^H(t) H_j^H(t) \right)$$

The use set channel capacity C can be expressed as:

$$C\left(\left[H_k(t)H_k^H(t)\right]^H\right) = B\log_2 \det\left(I_N + \frac{P}{M\sigma^2}\left[H_k(t)H_k^H(t)\right]^H\left[H_k(t)H_k^H(t)\right]\right)$$

SYSTEM PERFORMANCE COMPARISON

To verify the feasibility and effectiveness of the algorithm proposed in this chapter, and to study the impact of vehicles on the algorithm performance in different mobile environments, this chapter uses MATLAB simulation software to build a multi-user UM-MIMO system to simulate the proposed algorithm and the comparison algorithm, and finally to compare the simulation results. In the simulation, it is assumed that the receiving end and the transmitting end can obtain perfect CSI. There are M users, the number of each user RF chains is $N_t^{RF} = N_t^{RF} \times M$. Other main simulation parameters are summarized in Table 5.1.

The simulation result mainly compares the bit error rate (BER) performance and spectrum efficiency of the proposed algorithm and the algorithm based on subspace correlation, the geometric angle algorithm, and the condition number algorithm. The vehicle speed is 80 km/h, and the vehicle density is determined by the speed of the vehicle.

Tables 5.2 and 5.3, respectively, show the BER and spectral efficiency comparison diagrams of the proposed algorithm and the comparison algorithm when 256 transmitting antennas are configured at the BS and two vehicle users are scheduled. We can see from Tables 5.2 and 5.3 that the BER and channel capacity of the four algorithms are relatively close at low signal-to-noise ratio (SNR). With the increase in

TABLE 5.1
Simulation Parameters

Simulation Parameters	Values
Carrier frequency	28 GHz
Number of transmit antennas, N_t	8
Number of single-user receiving antennas, N_s	2
Number of user, M	2/4
Channel model	Time-varying geometric model
Number of clusters, N_c	3
Number of paths in each cluster, N_{ray}	7
Azimuth mean distribution	$[0, 2\pi]$
Mean elevation angle distribution	$[-2/\pi, 2/\pi]$
Antenna array structure	Uniform linear array (ULA)
Antenna spacing, d	0.5λ
Vehicle antenna height	1.5 m
Vehicle density	0.0025/0.005/0.0083 vehicle/m
Vehicle moving direction	Move in a straight line along the road
Channel estimation	idle

TABLE 5.2

Signal-to-Noise Power (SNR) of Power and Bit-Error-Rate (BER) for Different Algorithms

	BER				
SNR/dB	Multi-User Scheduling (MUS) Algorithm Based on Diagonalization (BD) Precoding for UM-MIMO System	Subspace Correlation (SC) Algorithm	Geometric Angle Algorithm (GA) Algorithm	Conditional Number (CN) Algorithm	Remarks
-40	$\sim\!4 \times 10^{-1}$	$\sim\!4 \times 10^{-1}$	$\sim\!4 \times 10^{-1}$	$\sim\!4 \times 10^{-1}$	Number of user: $M = 2$ Number of transmit antenna: $N_t = 256$
-35	$\sim\!3.5 \times 10^{-1}$	$\sim\!4 \times 10^{-1}$	$\sim\!4 \times 10^{-1}$	$\sim\!4 \times 10^{-1}$	
-30	$\sim\!3.0 \times 10^{-1}$	$\sim\!3.2 \times 10^{-1}$	$\sim\!3.3 \times 10^{-1}$	$\sim\!3.4 \times 10^{-1}$	
-25	$\sim\!2.0 \times 10^{-1}$	$\sim\!3.0 \times 10^{-1}$	$\sim\!4.0 \times 10^{-1}$	$\sim\!4.0 \times 10^{-1}$	
-20	$\sim\!9.0 \times 10^{-2}$	$\sim\!1.3 \times 10^{-1}$	$\sim\!1.6 \times 10^{-1}$	$\sim\!1.8 \times 10^{-1}$	
-15	$\sim\!2.0 \times 10^{-2}$	$\sim\!3.5 \times 10^{-2}$	$\sim\!5.0 \times 10^{-2}$	$\sim\!8.0 \times 10^{-2}$	
-10	$\sim\!2.8 \times 10^{-3}$	$\sim\!5 \times 10^{-3}$	$\sim\!1.2 \times 10^{-2}$	$\sim\!1.9 \times 10^{-3}$	

TABLE 5.3

Signal-to-Noise Power (SNR) of Power and Spectral efficiency for Different Algorithms

	Spectral Efficiency (Bits Per Second/Hz)				
SNR/dB	Multi-User Scheduling (MUS) Algorithm Based on Diagonalization (BD) Precoding for UM-MIMO System	Subspace Correlation (SC) Algorithm	Geometric Angle (GA) Algorithm	Conditional Number (CN) Algorithm	Remarks
-40	$\sim\!0.5$	$\sim\!0.5$	$\sim\!0.5$	$\sim\!0.5$	Number of user: $M = 2$ Number of transmit antenna: $N_t = 256$
-35	$\sim\!1.0$	$\sim\!0.9$	$\sim\!0.85$	$\sim\!0.84$	
-30	$\sim\!2.0$	$\sim\!1.9$	$\sim\!1.8$	$\sim\!1.7$	
-25	$\sim\!4.5$	$\sim\!4.3$	$\sim\!4.2$	$\sim\!4.1$	
-20	$\sim\!7.8$	$\sim\!7.6$	$\sim\!7.5$	$\sim\!7.1$	
-15	$\sim\!13.5$	$\sim\!12.0$	$\sim\!11.5$	$\sim\!11.0$	
-10	$\sim\!19.5$	$\sim\!17.8$	$\sim\!17.0$	$\sim\!16.5$	

SNR, the difference of system BER and total system capacity realized by different scheduling algorithms gradually increases. In general, the proposed algorithm is better than the comparison algorithm in terms of BER and spectral efficiency. This is mainly because the conditions of the channel itself have a great impact on the information transmission effect. In general, the proposed multi-user scheduling algorithm with 256 transmitting antennas is better than the comparison algorithm in terms of

BER and spectral efficiency. This is mainly because the conditions of the channel itself have a great impact on the information transmission effect.

There will be a better transmission effect when the conditions of the channel itself are good. The comparison algorithm only considers the correlation of the channel matrix and ignores the influence of the conditions of the channel itself. The algorithm in this chapter considers not only the channel correlation but also the conditions of the channel itself. Adding a scheduling factor to measure the conditions of the channel can effectively eliminate the system interference and improve the system performance. The simulation results (no table included) also show that the spectrum efficiency achieved by several algorithms in the scenario where the number of vehicle users is four is greater than that of the same algorithm in scenario where the number of vehicle users is two. This is because with the increase in number of scheduling users, the system capacity increases, and the transmitter can have higher diversity gain, so the spectral efficiency performance increases.

Table 5.4 shows the comparison of BER and spectrum efficiency performance of the above algorithms when the number of scheduled users is two when 128 antennas are configured at the BS. Compared with Tables 5.5 and 5.6, the frequency efficiency of the system is also improved when the number of transmitting antennas is increased with the number of scheduling users fixed and the system resources limited. This is because increasing the number of antennas at the transmitting end can improve the antenna array gain and effectively improve the signal-to-interference-plus-noise ratio (SINR) at the receiving end of the system, to improve the spectrum efficiency of the system. However, with the increase in the number of transmitting antennas, the BER performance of the system decreases. The main reason is that when the number of transmitting antennas increases, the interference between channels also increases, which reduces the effective transmission signal of users. It can also be seen from Tables 5.4 and 5.5 that the performance of the algorithm proposed in this chapter is

TABLE 5.4
Speed of Vehicles and Bit-Error-Rate (BER) for Different Algorithms

	Bit Error Rate (BER)				
Speed (km/h)	Multi-User Scheduling (MUS) Algorithm Based on Diagonalization (BD) Precoding for UM-MIMO System	Subspace Correlation (SC) Algorithm	Geometric Angle (GA) Algorithm	Conditional Number (CN) Algorithm	Remarks
90	$\sim1.5 \times 10^{-5}$	$\sim7 \times 10^{-5}$	$\sim6 \times 10^{-4}$	$\sim1.3 \times 10^{-3}$	Number
100	$\sim4 \times 10^{-5}$	$\sim3.5 \times 10^{-4}$	$\sim2.3 \times 10^{-3}$	$\sim5.5 \times 10^{-3}$	user: 2
110	$\sim9.0 \times 10^{-5}$	$\sim9 \times 10^{-4}$	$\sim6 \times 10^{-3}$	$\sim1.8 \times 10^{-2}$	Number of
120	$\sim0.8 \times 10^{-4}$	$\sim1.8 \times 10^{-3}$	$\sim0.4 \times 10^{-3}$	$\sim4.0 \times 10^{-2}$	transmitting
130	$\sim1.2 \times 10^{-4}$	$\sim2 \times 10^{-3}$	$\sim2.2 \times 10^{-2}$	$\sim7 \times 10^{-2}$	antenna: 128
140	$\sim2.0 \times 10^{-4}$	$\sim4 \times 10^{-3}$	$\sim4 \times 10^{-2}$	$\sim1.5 \times 10^{-1}$	
150	$\sim3 \times 10^{-4}$	$\sim4 \times 10^{-3}$	$\sim5 \times 10^{-2}$	$\sim2 \times 10^{-1}$	

TABLE 5.5
Speed of Vehicles and Spectral Efficiency for Different Algorithms

	Spectral Efficiency (Bits Per Second/Hz)				
Speed (km/h)	Multi-User Scheduling (MUS) Algorithm Based on Block Diagonalization (BD) Precoding for UM-MIMO System	Subspace Correlation (SC) Algorithm	Geometric Angle (GA) Algorithm	Conditional Number (CN) Algorithm	Remarks
90	~14.5	~13	~11.7	~11.5	Number
100	~14.4	~12.8	~11.6	~11.4	user: 2
110	~14.2	~12.8	~11.4	~11.2	Number of
120	~14	~12.5	~11.2	~11	transmitting antenna: 128
130	~13.6	~12.3	~10.9	~10.7	
140	~13.2	~11.7	~10.5	~10.2	
150	~12.5	~11.2	~10.2	~9.8	

the best regardless of BER or spectral efficiency. This is because the proposed algorithm considers not only the correlation between channels but also the conditions of the channel itself. This can effectively eliminate the interference between selected users, which also proves that the proposed algorithm is useful in the UM-MIMO system in the 6G V2X scenario.

Tables 5.4 and 5.5, respectively, show the impact of vehicle speed on the proposed algorithm and the performance of the comparison algorithm (including BER and spectral efficiency) under different vehicle speed environments. The vehicle speed is set from 90 to 150 km/h, where $SNR = -10$ dB, $M = 2$, and $N_t = 128$. It can also be seen that the BER of the proposed MUS algorithm and the comparison algorithm gradually increases, and the spectral efficiency performance of the system gradually decreases with the continuous increase in vehicle speed. This is because the Doppler effect becomes more serious with an increase in vehicle speed, resulting in the decline of system performance. At the same time, the simulation results show that the proposed MUS algorithm has better robustness than the comparison algorithm. This is because the proposed MUS algorithm makes full use of the time dimension information of the time-varying channel. Therefore, compared with the comparison algorithm, the performance of the proposed algorithm does not decline significantly and has high stability when vehicle speed is increasing.

APPLICATION OF ARTIFICIAL INTELLIGENCE AND MACHINE LEARNING INTO ULTRA-MASSIVE MIMO WIRELESS SYSTEMS

Applying machine learning in Massive MIMO certainly has some advantages. The ever-emerging machine learning-related research is ongoing, and much of the well-developed tools can boost its application here in Massive MIMO. We briefly list some design considerations specifically for Massive MIMO. In this topic, we introduce learning mechanisms in which we investigate the application of machine learning to realize Massive MIMO with low implementation complexity and little prior information. The significant needs include: (1) resource allocation based on machine learning; (2) channel estimation based on machine learning; (3) signal detection based on machine learning; (4) interference management based on machine learning; (5) physical layer design based on machine learning. The challenges are presented as follows:

- **Resource allocation-based machine/deep learning:** The application of machine learning into resource allocation has the potential to achieve low complexity implementation and decrease operational costs. The development of resource allocation based on machine learning would improve spectral efficiency and energy efficiency, increase the number of users, and decrease energy consumption as well as the time delay. Advanced machine learning methods would also be developed for resource allocation. How does one develop resource allocation based on machine learning to improve spectral efficiency, the energy efficiency, the number of users, and decrease the energy consumption and time delay? That kind of resource allocation requires accurate modeling and enough test data. Besides, it is also important to evaluate its performance.
- **Channel estimation based on machine/deep learning:** To achieve an efficient estimation of the channel, the challenges are how to appropriately establish, efficiently train, and adjust the deep neural networks and how to develop unsupervised learning methods.
- **Signal detection based on machine/deep learning:** Different from the conventional linear and nonlinear detection methods, the challenges of designing signal detection based on machine learning are overcoming the issues of overfitting and underfitting when training the deep neuron networks due to complicated channel distortion and interference. Enabling massive connectivity with massive MIMO.
- **Interference management based on machine/deep learning:** To efficiently manage the inter-cell or inner-cell interference, the challenges are how to determine the number of interference sources, the interference levels, and how to overcome interference by using machine learning.
- **Physical layer design based on machine/deep learning:** To achieve low complexity design of the physical layer, the challenges are how to design modulation and demodulation, precoding scheme based on machine learning.
- **Detailed design considerations:** How to appropriately establish, efficiently train, and adjust the deep neural networks and how to develop unsupervised learning methods. This detailed consideration here poses some challenges in selecting the best-fit tools, the programming language, the supervised and unsupervised selection.
- **Overcome overfitting and underfitting:** This problem not only is happening in Massive MIMO-related applications but also is the design factor for all other applications. Thus, the feeding data should be carefully designed, and the number of deep learning networks should be chosen appropriately.

- **System modeling:** How to determine the number of interference sources, the interference levels, and how to overcome interference by using machine learning? The aim should be to model every setting but should select those that are the most relevant to performance improvements.
- **Modeling of modulation and demodulation:** How to design modulation and demodulation, precoding scheme based on machine learning. Massive MIMO has a lot of data for users and operators to process. Selecting the appropriate design factor, which includes modulation demodulation, might be a hard one.

SUMMARY

Simulation results show that compared with the proposed multi-user scheduling algorithm based on subspace correlation, condition number, and geometric angle, the proposed MUS algorithm can obtain higher user channel gain, effectively reduce the system BER, and can be applied to 6G V2X communication. The proposed 6G V2X multi-user scheduling algorithm is based on BD precoding for UM-MIMO systems. Aiming at the high-speed movement and millimeter wave characteristics of the vehicle in the 6G V2X scheme, a time-varying geometric channel model is established. On this basis, the BD precoding is sensitive to channel correlation, and the Pearson coefficient after matrix vectorization is used to measure the channel correlation between users, comprehensively consider the two factors of channel correlation and channel conditions, and ensure that the channel correlation is minimized while ensuring the selection of high-quality channels to achieve multi-user scheduling. BD precoding is then used to eliminate data stream interference. The system simulation results show that the proposed algorithm can effectively reduce the BER and improve the system spectrum efficiency and is suitable for 6G V2X communication. We have also pointed out how ML/DL can be used for UM-MIMO design and operations.

REFERENCES

1. Nie, S. et al., "Intelligent Environments based on Ultra-Massive MIMO Platforms for Wireless Communication in millimeter wave and Terahertz Bands," April 2019, arXiv: 1904.07958v1 [eess.SP] 16 Apr 2019.
2. He, S. et al., "Multi-User Scheduling for 6G V2X Ultra-Massive MIMO System," *Sensors*, 2021, 21, 6742. https://doi.org/10.3390/s21206742.

6 Reconfigurable Intelligent Surface

A reconfigurable intelligent surface (RIS) is a passive planar structure, made of several sub-lambda spaced reflective elements, whose electromagnetic characteristics and in particular the phase offsets are imposed when reflecting impinging waves can be controlled and adaptively changed [1]. Figure 6.1 depicts a high-level view of the RIS communications architecture.

RISs can be thus employed to extend coverage, and/or to improve the signal-to-interference-plus-noise ratio (SINR) at the intended receiver location. Several studies have recently appeared showing the benefits that RISs can bring in different applications such as mobile edge computing networks [4], physical layer security systems, cognitive radio networks, and radar systems. Recently, RISs have emerged as one of the most striking innovations for the evolution of 5G systems into 6G systems. RIS could have full-duplex (FDX) communications, and the RIS FDX technology has the potential to radically evolve wireless systems, facilitating the integration of both communications and radar functionalities into a single device, thus enabling joint communication and sensing (JCAS). The simulation results demonstrate that the RIS FDX scheme can achieve significant performance improvement for both communications and sensing. It is also possible that jointly designing the MIMO beamformers and for the RIS phase configuration to be system interference (SI)-aware can significantly loosen the requirement for additional SI cancellation.

DISTRIBUTED RIS COMMUNICATIONS

The distributed RIS-assisted millimeter-wave multi-user multi-input multi-output (MIMO) beam space system with beam space systems offer a low-scattering communication to users through beams. To enrich the propagation environment, multiple reconfigurable intelligent surfaces are installed on top of the surrounding buildings to provide an additional communication link, between the base station and users, by introducing phase shifts in the incoming signal arriving at it. Such systems allow the selection of two distinct beams for each user. The first beam is directly steered toward the user, and the second beam is steered to the user via RIS. The selection of desirable beams is required to reduce the number of radio frequency (RF) chains at the transmitter. In this chapter, we propose a beam selection algorithm that can assign two distinct and unshared beams to each user after optimally adjusting the phases of each RIS. Additionally, the a max–min fairness power allocation scheme obtains a minimum achievable rate for each user.

DOI: 10.1201/9781003499480-6

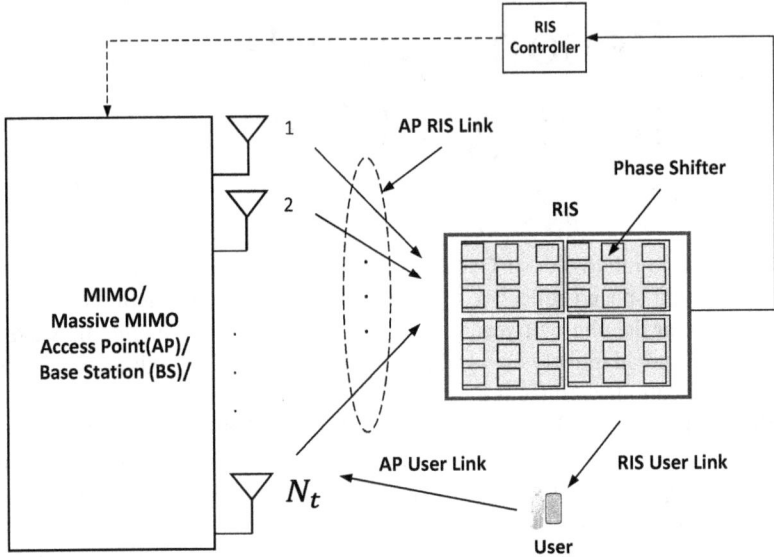

FIGURE 6.1   RIS communications architecture.

COOPERATIVE BEAMFORMING IN RIS COMMUNICATIONS

RIS has emerged as a cost-effective and promising solution to extend the wireless signal coverage and improve the performance via passive signal reflection. The cooperative beamforming in RIS accounts for the cooperation between RISs and provides full space coverage, which are not available in noncooperative RIS systems. The marriage of cooperative double-RIS with simultaneously transmitting and reflecting RIS (STAR-RIS) in the Massive MIMO setup provides superior channel capacity under correlated Rayleigh fading conditions.

JCAS FOR RIS AND MASSIVE MIMO

Empowering cellular networks with augmented sensing capabilities is a key research area in sixth-generation (6G) communication systems. Both sensing and communication capabilities of millimeter-wave orthogonal frequency-division multiplexing (OFDM) JCAS systems in the presence of RISs improve when the RIS is adequately designed.

RIS AND MASSIVE MIMO-ENABLED APPLICATIONS

The RIS is deployed to help conventional Massive MIMO networks serve the users in the dead zone. These surfaces consist of many passive elements of metamaterials whose impedance can be controllable to change the phase, amplitude, or other characteristics of wireless signals impinging on them. An antenna structure where a (nonlarge) array of radiating elements is placed at short distance in front of an RIS can emulate the performance of traditional Massive MIMO arrays with much lower hardware complexity, i.e., with a significantly smaller number of active antennas and RF chains. The analysis of channel estimation scheme that exploits spatial correlation characteristics at both the Massive MIMO base station and the planar RISs along with other statistical characteristics of multi-specular fading in a mobile environment are required for using RIS in conjunction with Massive MIMO-based applications. This scheme improves the spectral efficiency (SE) of the cell-edge mobile users substantially in comparison to a conventional single-cell Massive MIMO system.

SUMMARY

We considered an RIS-aided Massive MIMO system in which an RIS is placed at short distance and in front of an active antenna array, with a nonlarge number of elements. The user equipment (UE) to be served are placed in the backside of the active antenna array. We provided a signal and channel model suitable for the analysis of such a system and proposed a channel estimation procedure which, using different configurations of the RIS phase shifts, is capable to estimate the channel from the large number of RIS elements and all the UEs exploiting only the active antennas. Please see the closed-form expression for an achievable downlink SE per user in paper [1], by using the popular hardening lower-bound, and formulated a generic optimization problem which can be used for two purposes. In the numerical results, we have shown the effectiveness of the considered architecture assuming both omni-directional and directional active antennas and of the phase shifts optimal configuration algorithm in paper [1].

REFERENCE

1. Buzzi, S. et al., "RIS-aided Massive MIMO: Achieving Large Multiplexing Gains with non-Large Arrays," October 2021, arXiv:2110.12800v1 [cs.IT] 25 Oct 2021.

7 Cell-Free Communications

In sixth-generation (6G) networks, it is expected that the massive devices and mobile applications will be connected across the wireless communication networks. The dramatic rate variations and inter-cell interference inherent in classical cell-based structures impose a heavy burden on the existing wireless network infrastructure [1]. As a result, cell-free (CF) ultra-dense (UD) massive multiple-input multiple-output (CF UD-Massive MIMO (mMIMO)) has been identified as a candidate wireless networking technology to cater to the continuous data traffic surge, remedy the cell-edge performance issues, and guarantee unprecedented wireless network link reliability. However, the transitioning toward tether-less connectivity for a fully mobile-networked society including body area network, achieving multiple orders of energy efficiency (EE) gains and quality throughput, poses an increasingly important design criterion.

The traditional MIMO systems cannot provide very high spectral efficiency (SE), EE, and link reliability, which are critical to guaranteeing the desired quality of experience (QoE) in 5G and beyond 5G wireless networks. CF mMIMO has been envisioned to provide uniform SE and ubiquitous connectivity for 6G wireless network. The key idea of CF mMIMO systems is that through spatial multiplexing on the same time-frequency resource, many geographically distributed access points (APs) connected to a central processing unit (CPU) serve the user equipment (UE) coherently. Therefore, the huge macro-diversity gain of CF mMIMO systems brought by the joint transmission and reception has made academic interest in this area grow exponentially. Compared to traditional cellular systems, the CF mMIMO system-based joint signal processing can effectively alleviate the influence of nonideal factors, such as channel aging, hardware impairments, and other interferences. In addition, UD CF-mMIMO systems are exploited to boost cell-edge performance and provide ultra-low latency in emerging wireless communication systems. Figure 7.1 shows the 6G CF communications architecture.

The real-time requirements of connected devices and capacity demands of the current wireless network infrastructure are increasing rapidly. The landscape of upcoming wireless networks is envisaged to support pervasive interconnectivity, high transmission rates, ultra reliability, and low latency, spurring novel information and communication technologies (ICTs) and revolutionary UD wireless infrastructure. The envisioned next-generation wireless communication systems will integrate virtually everything into the internet while accommodating novel technologies like virtual reality, machine-to-machine (M2M) communications, vehicle-to-everything (V2X), and device-to-device (D2D) communications.

Essentially, the next-generation wireless networks will rely on the availability of higher-frequency bands, which brings more opportunities for multi-gigabit throughput and extreme capacity. UD networks (where the density of the base stations (BSs) are

 DOI: 10.1201/9781003499480-7

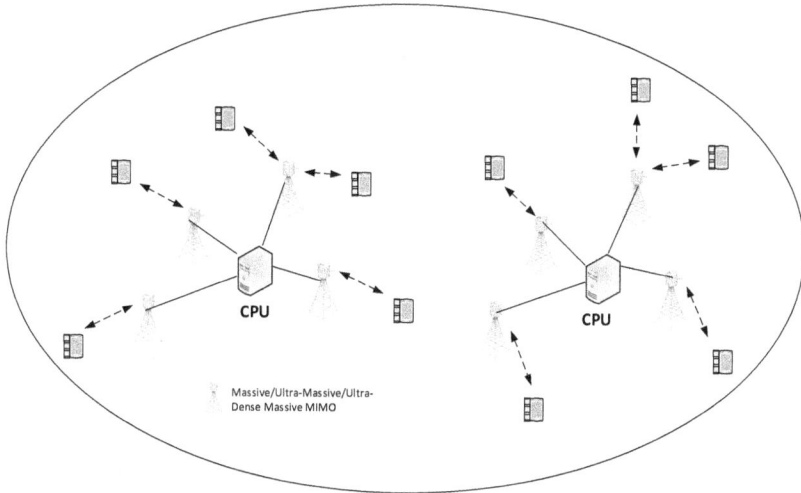

FIGURE 7.1   Cell-free communications architecture.

much higher than that of the network users) and/or mMIMO (where large antenna arrays enhance the BSs) have been identified as enabling technology to satisfy the vast requirements of next-generation wireless networks. However, the service-quality variations and cell-edge issues inherent in the traditional cellular network architecture pose a significant bottleneck for the mobile network operators.

CF mMIMO system, a practical incarnation of distributed mMIMO, has become an intensive research topic in industry and academia. CF UD-mMIMO can potentially mitigate significant path loss variations and cell-edge performance issues inherent in conventional cellular networks. Despite the enormous merits of CF UD-mMIMO, the outcry for environment-friendly designs and greener networking solutions is alarming. Adopting a distributed antenna array in CF UD-mMIMO systems could dramatically increase the power consumption and the overall energy emissions of wireless communication systems.

Consequently, the geometric outburst of digital signal processing (DSP) power consumption, prohibitive radiofrequency (RF) circuit power costs, and energy costs of data processing units have become a critical concern for mobile network operators. As such, EE, expressed in bits/joule, has emerged as a dominant performance index for benchmarking wireless communication systems. Depicted in Figure 7.1 is a typical illustration of the next-generation communication system. In this configuration, a dearth of UE is jointly served by a large chunk of arbitrarily allocated APs, all connected to a CPU that handles complex signal processing.

ULTRA-DENSE CELL-FREE MASSIVE MIMO SYSTEMS

In the last decade, the geometric outburst of various innovative communication technologies has been overwhelming. Primarily, mMIMO network architecture has evolved over the years to support the explosive increase in wireless data traffic.

In the regime of large-scale antenna arrays at the BSs and by exploiting spatial multiplexing, spatial diversity, and advanced beamforming, many UEs can be served with the same time/frequency resources. However, large service-quality variations and cell-edge performance issues constitute a significant setback for the effective deployment of cellular network infrastructure. Thus, UD CF-mMIMO has appeared as a promising network paradigm to cater to the continuous data traffic surge and alleviate the mediocre cell-edge problems inherent in cellular networks. In these networks, an excess number of geographically allocated APs coordinated by a CPU coherently serve a smaller number of UEs distributed across a wide serving area.

UD CF-mMIMO is essentially a practical embodiment of mMIMO and network MIMO, where cooperation between low-complexity APs deployed in a distributed manner helps to potentially minimize the effects of intercell interference and offer considerable gains in coverage probability. As in cellular networks, UD CF-mMIMO reaps the benefits of channel hardening and favorable propagation through simple signal processing. Moreover, compared with small cells (SCs) and co-located systems, UD CF-mMIMO promises multifold improvement in SE, EE, and 95% likely per-user throughput. The fundamental structure of a typical UD CF-mMIMO network comprises the CPU and the associated APs linked via a fronthaul network to the CPU without the intervening cells.

SYSTEM MODEL

A UD CF-mMIMO system operated in time division duplex (TDD) mode is investigated. Let M specify the number of randomly deployed single-antenna APs simultaneously serving k single-antenna UEs at the same time/frequency block. All APs are linked via a fronthaul network to a CPU, wherein network information is exchanged. Let h_{mk} reflect the channel between the mth AP and the kth UE. The channel model is defined as shown below:

$$h_{mk} = \sqrt{\beta_{mk}} d_{mk}$$

where β_{mk} accounts for the large-scale fading and d_{mk} accounts for the small-scale fading. It is assumed that d_{mk} are independent identically distributed (i.i.d) random variables, $d_{mk} \sim \mathcal{CN}(0, 1))$ for $m = 1, \ldots, M, k = 1, \ldots, K$. What is more, concerning frequency, β_{mk} coefficients are assumed to be constant and known as a priori at any moment. In this context, two communication protocols – UL training and DL payload data transmission – are analyzed.

UPLINK TRAINING

For each coherence interval, let $\tau_{u,p}$ reflect the length of UL training duration and let $\sqrt{\tau_{u,p}} \psi_k \in \mathbb{C}^{\tau_{u,p} \times 1}$ reflect the pilot sequence forwarded by the kth user, $k = 1, \ldots, K$. Further, we assume that these assigned pilot sequences are mutually orthonormal.

$\psi_k^D \psi_{k'} = 0$ for $k' \neq k$, and $\psi_k^2 = 1$, which necessitates that $\tau_{u,p} > K$. The mth AP receives a pilot signal defined by:

$$x_{up,m} = \sqrt{\tau_{u,p} \rho_{u,p}} \sum_{k=1}^{K} h_{mk} \psi_k + z_{up,m}$$

where $\rho_{u,p}$ accounts for the normalized transmit signal-to-noise ratio (SNR) and $z_{up,m} \sim \mathbb{C}N(0, NI\tau_{u,p})$ accounts for the additive noise. The received pilot at the mth AP is processed as:

$$\check{x}_{up,mk} = \psi_k^D x_{up,m} = \sqrt{\tau_{u,p} \rho_{u,p}} h_{mk} + \psi_k^D z_{up,m}$$

By adopting the minimum mean squared error (MMSE) technique, the channel h_{mk} is estimated. The channel estimation of h_{mk} is expressed as:

$$\hat{h}_{mk} = \frac{E\{\check{x}_{up,mk}^*\}}{E\{|\check{x}_{up,mk}|^2\}} \check{x}_{up,mk} = c_{mk} \check{x}_{up,mk}$$

where c_{mk} is reflected by:

$$c_{mk} \equiv \frac{\sqrt{\tau_{u,p} \rho_{u,p}} \beta_{mk}}{\tau_{u,p} \rho_{u,p} \beta_{mk} + 1}$$

The associated channel estimation error is reflected by:

$$\tilde{h}_{mk} = h_{mk} - \hat{h}_{mk}$$

Downlink Payload Data Transmission

In this phase, conjugate beamforming (CB) is employed. The transmitted signal from the mth AP to the K users is expressed as:

$$\omega_m = \sqrt{\rho_d} \sum_{k=1}^{K} \sqrt{\eta_{mk}} h_{mk}^* v_k$$

where m^{th} specifies the normalized transmit SNR corresponding to the data symbol and v_k specifies the symbol assigned to the kth UE, satisfying $E\{|v_k|^2\} = 1$. Moreover, by exploiting $m = 1, ..., M, k = 1, ..., K$, and h_{mk} power control coefficients, the average power constraint $E\{|\omega_m|^2\} \leq \rho_d$ is satisfied. Thus, the power constraint can be remodeled:

$$\sum_{k=1}^{K} \eta_{mk} \phi_{mk} \leq 1, \text{for all } m$$

where ϕ_{mk} reflects the variance of the channel estimate and is given by:

$$\phi_{mk} \triangleq E\left\{\left|\hat{h}_{mk}\right|^2\right\} = \sqrt{\tau_{u,p}\rho_{u,p}}\,\beta_{mk}c_{mk}$$

The received data signal at the kth UE is defined as:

$$y_{d,k} = \sum_{m=1}^{M} h_{mk}\omega_m + z_{d,k} = \sqrt{\rho_d}\sum_{k=1}^{K} a_{kk'}v_{k'} + z_{d,k}$$

where $z_{d,k} \sim CN(0, 1)$ and $a_{kk'} \triangleq \sum_{m=1}^{M}\sqrt{\eta_{mk'}}h_{mk}\hat{h}_{mk}^*, k' = 1,\ldots,K$.

POWER CONSUMPTION MODEL

According to recent reports [1], the total power consumption is defined:

$$P_{\text{total}} = \sum_{m=1}^{M}P_{ac,m} + \sum_{m=1}^{M}P_{b,m}$$

where $P_{ac,m}$ accounts for the amplifier and circuit power consumption at the mth AP and $P_{b,m}$ accounts for the backhaul link power consumption associated with the mth AP. The power consumption $P_{ac,m}$ is obtained as:

$$P_{ac,m} = (\sigma_m)^{-1}\rho_d N_0\left(N\sum_{k=1}^{K}\eta_{mk}\phi_{mk}\right) + NP_{tc,m}$$

where $0 < \sigma_m \le 1$ depicts the efficiency of the power amplifier, N_0 depicts the noise power, and $P_{tc,m}$ depicts the power needed to run the circuit units of each antenna at the mth AP. As a further advance, the backhaul power consumption $P_{b,m}$ can be modeled as:

$$P_{b,m} = P_{0,m} + BS_e P_{bt,m}(\eta_{mk})$$

where $P_{0,m}$ reflects the fixed power consumption value of each backhaul link, B reflects the transmission bandwidth, S_e is the sum of the SE, and $P_{b,m}$ specifies the traffic-dependent power. By substituting the values of $P_{ac,m}$ and $P_{0,m}$ into above equations, the total power consumption P_{total} can be remodeled as:

$$P_{\text{total}} = \rho_d N_0\sum_{m=1}^{M}(\sigma_m)^{-1}\rho_d N_0\left(N\sum_{k=1}^{K}\eta_{mk}\phi_{mk}\right)$$
$$+ \sum_{m=1}^{M}(NP_{tc,m} + P_{0,m}) + BS_e\left(\sum_{m=1}^{M}P_{bt,m}(\eta_{mk})\right)$$

Power Control

The transmit power constitutes one of the primary radio resources for a wireless network and should be managed appropriately. Power control encompasses the procedure of intelligently controlling the power of a transmitter to improve the quality of service (QoS) of all the users. During the UL transmission phase, the K users are required to select suitable transmit powers $0 \leq p_k \leq p_{max}$, $k = 1, ..., K$, while the M APs are required to forward their transmit powers $0 \leq p_k \leq p_{max}$, $m = 1, ..., M$ during DL transmission phase. Power control is of prime importance in CF-mMIMO to optimize the power of the desired received signal, reduce pilot contamination, manage the generated interference, and offer considerable improvement in the QoS to the UEs.

Spectral Efficiency

The total achievable EE, measured in bit/joule, is defined as the ratio of the sum throughput (bit/s) and the total consumed power (watt) in the system and is defined as:

$$E_e\left(\eta_{mk}\right) = \frac{BS_e\left(\eta_{mk}\right)}{P_{total}}\left(\frac{bit}{joule}\right)$$

where S_e is the sum of SE and E_e is the EE. Going forward, it is instructive to provide insights into the concept of SE which is a critical performance metric to be considered in keeping pace with an ever-increasing number of wireless data services and high-rate expectations. UD CF-mMIMO is envisioned to provide tremendous gains in SE through efficient utilization of massive and dense antenna arrays, insightful power optimization algorithms, and novel approaches to receiver filter coefficient design. In its ideal form, it will be able to maintain a uniform level of the QoS among all users in the network, especially under mMIMO systems. Another benefit is combating the unfavorable effects of signal propagation, such as signal fading and shadowing.

We briefly analyze power consumption parts in the network as follows:

- **Circuit power**: The deployment of large-scale antenna arrays, as in CF-mMIMO, significantly increases the power consumption of wireless networks owing to an abundance of antenna circuit elements. The circuit power, resulting from the energy dissipated by the circuit elements such as analog devices and residually lossy factors in BSs, grows in tandem with the number of transmit antennas. Consequently, as the size of the hardware of the system grows large, the total circuit power dissipated increases.
- **Signal processing power**: Signal processing power dissipation is increasingly becoming a prime challenge to researchers, due to the greater complexity of signal processing algorithms and architectures per requirement. Advanced wireless communication systems, such as CF-mMIMO, require sophisticated signal processing units to code, precode, and decode symbols at both transmission ends. Although manifold benefits are realized, the total power consumption resulting from signal processing remains a significant issue to be addressed.

- **Signal transmission power**: The optimized data speed and simultaneous transmission of multiple signals in modern wireless communication systems necessitate even higher transmission power to maximize SNR, operating capacity, and coverage area. Besides, the power allocated to the transmitter during networking signals corresponds to the number of distributed users and antenna elements. Thus, signal transmission power has become a big deal in designing energy-efficient network architectures. The power consumption resulting from the signal transmission can be derived mathematically as $P_T = \dfrac{P_a}{\eta}$, where P_a specifies the average transmit power of the BS, and η specifies the efficiency of the average transmit power.
- **The system fixed power**: In addition to achieving tether less connectivity and quality performance, minimizing the static energy (stable power consumption) caused by the hardware components of UD infrastructures, especially when performing different types of communication processes is critical.
- **Power Control**: Power control can help to limit the total power consumption in a CF mMIMO network. Thus, selecting appropriate power control algorithms is critical to maximizing the overall efficiency of the wireless network. It is worth noting that power control algorithms can be dynamically deployed to solve specific system-wide utility functions. Some of the most common utility functions – max EE, max-min fairness, and max sum SE – are discussed extensively in the next section. A concise summary of recent advances in power control algorithms (geometric programming (GP), successive convex approximation (SCA), fractional power control, second-order cone program (SOCP), and machine learning (ML)-based power optimization schemes are described in paper [1]. Essentially, these schemes are dimensioned to maximize the common utility optimization problems.

This analysis shows why CF UD-mMIMO) has been envisioned to provide uniform SE and ubiquitous connectivity for 6G wireless network. CF UD-mMIMO systems provide uniform SE through spatial multiplexing on the same time-frequency resource, where many geographically distributed APs connected to a CPU serve the UEs coherently [4]. Therefore, the huge macro-diversity gain of CF UD-mMIMO systems is brought by the joint transmission and reception [5]. For instance, in terms of 95%-likely per user SE, CF UD-mMIMO systems outperform SC systems due to the joint interference cancellation capability [6]. In addition, compared to traditional cellular systems, the CF UD-mMIMO systems–based joint signal processing has effectively alleviated the influence of nonideal factors, such as channel aging, hardware impairments, and other impairments. We have summarized pros and cons of the CF UD-mMIMO system below:

ADVANTAGES OF CELL-FREE CD-MASSIVE MIMO

- CF system offers macro diversity gain due to aggressive deployment densification. APs are available closer to users, which leads to reduced path loss and shadowing effects, and the reliability of link is improved as each user is served by many Aps, and there is less probability of blockage.

- Inter-cell interference is reduced substantially in CF CD-mMIMO. Hence, it outperforms SC networks.
- CF system offers higher uniform QOS among users due to its user-centric implementation.
- CF provides beamforming and spatial multiplexing gain.
- CF UD-mMIMO combines three technologies, viz. ultra-densification, millimeter/multi-GHz-wave band and UD-mMIMO. This results into higher throughput.

DISADVANTAGES OF CELL-FREE CD-MASSIVE MIMO

- CF UD-mMIMO architecture requires widespread and expensive hardware.
- CF UD-mMIMO architecture requires accurate synchronization and coordination among the APs.
- CF UD-mMIMO requires simple but effective resource allocation schemes to minimize signaling overhead.
- CF UD-mMIMO is not scalable system in its nonstandardized form as opposed to the cell-based system.
- Another difficult issue is that whether CF UD-mMIMO architecture will be implemented in a distributed manner or in a centralized manner.

SUMMARY

We have described the concepts and techniques proposed for energy efficient power control in UD CF-mMIMO systems. An elaborate introduction to the technical foundations and mathematical system model of CF-mMIMO is presented. Next, a comprehensive evaluation of the power consumption model, EE, and power consumption parts are provided. Further, standard EE maximization techniques for the state of the art, including energy-efficient resource allocation schemes, energy harvesting and exchange techniques, switching and sleep mode techniques, and virtualization, are discussed comprehensively. Additionally, a review of recent advances in energy-efficient power control in UD CFmMIMO systems was highlighted. Different power allocation schemes, such as GP, SCA, SOCP, fractional power control, and ML-based power control, target max EE, max sum SE, and max-min fairness in wireless networks are discussed elaborately in paper [1] and we have not included all items here for the sake of brevity. Finally, critical insights on the open issues and critical challenges in guaranteeing sustainable wireless communication are delineated in paper [1]. Key findings from the survey thread evidence that an ever-increasing number of users and high-rate demands are accompanied by a substantial increase in energy consumption. While significant gains in EE have been realized through efficient utilization of energy maximization techniques and dense deployment of antenna arrays, the concept of energy sustainability and green design expectations remains. Future work would examine the design and characterization of optimal energy-efficient power control schemes in UD CF-mMIMO networks for application in the envisioned 6G wireless communication systems.

REFERENCE

1. Imoize, L. A. et al., "A Review of Energy Efficiency and Power Control Schemes in Ultra-Dense Cell-Free Massive MIMO Systems for Sustainable 6G Wireless Communication," *Sustainability* 2022, 14, 11100. https://doi.org/10.3390/su141711100

8 Non-Terrestrial Communications

Non-terrestrial network (NTN) refers to a network, or segments of a network, where spaceborne or airborne vehicles act as a base station (BS) or relay node, with the intent of providing anytime-anywhere connectivity by offering wide-area coverage and ensuring service availability, continuity, and scalability (see Figure 7.1). The key challenges to the realization of a satellite-based cellular network are large propagation delays and round-trip time, large Doppler effects, and mobility handling due to satellite movements [1]. These have profound implications in different aspects of the system, requiring changes in both radio access network (RAN) and core network (CN), to overcome their impact in the link establishment and maintenance. Differential delay in relation to cellular links needs to be handled effectively. Integration of global navigation satellite system (GNSS) capability within the NTN satellite network is an opportunity and a challenge for independence of the network from global positioning system (GPS), GNSS, and Galileo for clock and location.

AERIAL COMMUNICATIONS

The previous sections have identified several technologies that are expected to be key in evolving the system and network architecture. The traditional terrestrial concepts will be expanded with satellites and other aerial platforms to provide alternative coverage. Topologically, mesh network and embedded subnetworks that break and then extend the cellular concept will be introduced by adding new connectivity options. The boundaries where computing happens will dissolve, with computing and data storage occurring where it makes sense [1]. Artificial intelligence (AI) is considered a key technology to enable the network to be both manageable and dynamic enough. Thus, the network must provide support for AI at all the various places it will be used, from applications to network operations. And finally, the rigid allocation of functions must be replaced by more flexible architectures and protocols that accommodate the diverse environments and use cases that 6G will demand.

SATELLITE COMMUNICATIONS

The key challenges to the realization of a satellite-based cellular network are large propagation delays and round-trip time, large Doppler effects, and mobility handling due to satellite movements. These have profound implications in different aspects of the system, requiring changes in both RAN and CN, to overcome their impact in the link establishment and maintenance. NTN capability poses challenges for frequency pre-compensation and timing synchronization in relation to techniques in use for TNs. Differential delay in relation to cellular links needs to be handled effectively.

DOI: 10.1201/9781003499480-8

Integration of GNSS capability within the NTN satellite network is an opportunity and a challenge for independence of the network from GPS, GNSS, and Galileo for clock and location [1].

Accurate beam-steering from a satellite to a ground-based user and the design of antenna array systems with a small enough form factor with the constraints of the satellite size and other constraints are some further challenges. NTN networks should be designed to meet realistic power ranges for the satellite transmitters and requirements for indoor penetration, coverage, and capacity. Moreover, TN-NTN co-existence should be considered. Low-power enhancements and a common device interface for reduced cost and form factors need to be addressed.

The North American region has been and continues to be a cradle of innovation in space technologies and has pioneered the deployment of satellite-based communication networks. There is a tremendous opportunity to capitalize on this momentum and maintain leadership in the NTN space.

Necessary research areas include new waveforms, modulation schemes, and error-correction techniques, which may be required to efficiently overcome the vagaries of the satellite-to-ground airlink. Cognitive or AI/ML-based techniques for performing dynamic spectral allocation to reduce interference and the usage of visible light communication are also fertile areas of investigation.

SUMMARY

We have described briefly the nonterrestrial communications that consist of aerial and satellite communications. However, aerial communications have been described in full details in Chapters 22 and 23. We have not addressed the satellite communications for the sake of brevity.

REFERENCE

1. 6G Next G Alliance Report: 6G Technologies, June 2022.

9 Massive MIMO Radar

A multiple-input multiple-output (MIMO) radar approach employing widely dispersed transmit and receive antennas is studied for the detection of moving targets. The MIMO radar transmits orthogonal waveforms from the different transmit antennas so these waveforms can be separated at each receive antenna. For a moving target in colored Gaussian noise-plus-clutter, we quantify the gains from having widely dispersed antennas that allow the overall system to "view" the target simultaneously from several different directions. The MIMO radar performance is contrasted with that of a traditional phased-array approach, which employs closely spaced antennas for this purpose. The MIMO radar approach is well suited to handle targets that have small radial velocities for scenarios in which co-located sensors cannot separate the target from the background clutter. Both a centralized processing and a simple distributed processing form of the MIMO radar approach are developed and studied, and the gains from the centralized version, which come at the price of additional complexity, are clearly demonstrated, and explained intuitively. The constant false alarm rate (CFAR) property of an adaptive version of the MIMO moving target detector is also demonstrated for homogeneous clutter [1].

MASSIVE MIMO RADAR SYSTEM

Originally motivated by the recent advances in MIMO wireless communications, there have been many recent publications on MIMO radar. In communications, MIMO systems combat the fading effects of the wireless (multipath) channel with spatial diversity. Further, the scattering environment can be used by such systems to achieve spatial multiplexing.

In radar, complex targets consisting of several scatterers take the place of the multipath channel. A target's radar cross section (RCS), which determines the amount of returned power, varies greatly with the considered aspect angle. Those variations significantly impair the detection and estimation performance of conventional radar. Distributed MIMO radar systems observe a target simultaneously from different aspect angles resulting in spatial diversity. This diversity countervails the fluctuations in received power. In MIMO radar, one has M multiple transmit and N multiple receive antennas, and all receive antennas pick up the signals from all transmit antennas, which sets up MN paths. This allows for gains in performance that scale with MN. We also have explored the improvement MIMO radar can achieve in target detection and angle-of-arrival (AOA) estimation under certain conditions relative to the performance of conventional (phased-array) radar systems. The reader is referred to these publications for an extensive treatment of the target's spatial RCS fluctuations and its effects in the MIMO radar context. In this chapter, we analyze a MIMO radar approach using orthogonal waveforms for moving target detection.

DOI: 10.1201/9781003499480-9

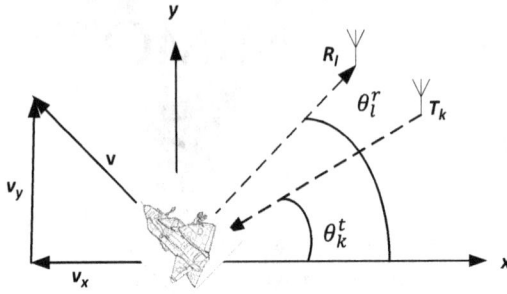

FIGURE 9.1   Radar and moving target.

$$v_x = |v|\cos\Omega$$

$$v_y = |v|\sin\Omega$$

In radar systems, it is common to exploit frequency shifts of the returned signals to distinguish between the returns of moving objects of interest and the ones due to the omnipresent clutter. Clutter leads to only small frequency shifts, whereas the target's radial movement results in larger frequency shifts. These differences have been exploited in classical radar systems to separate targets from clutter. If a target has a given absolute speed $|v|$ and an angle Ω, in an arbitrary coordinate system centered at the target, the target's velocity components in the x and y direction are given by (Figure 9.1).

If a target is moving in a direction orthogonal to the direction of illumination from a conventional single antenna radar whose single antenna is employed for both transmit and receive, this leads to a lack of Doppler shift in the received signal, impairing the ability of the radar to detect the target. In MIMO radar with widely dispersed antennas, the target moves from several directions simultaneously, which overcomes this limitation of single antenna radar systems. We have discussed these ideas and demonstrated these gains using analysis and simulations. A multistatic airborne radar consisting of several phased-array platforms is considered for moving target detection, while we employ the MIMO radar architecture defined as that which exploits all the possible MN paths from the M available transmit antennas to the N available receive antennas. As each platform only receives its own signal in the radar architecture, all the possible MN paths are not exploited. Due to this, the radar architecture is therefore unable to fully exploit the diversity gains and the "geometry" gains available with our MIMO radar architecture. Here we have shown these gains, diversity, and "geometry."

We have provided analysis of a MIMO radar system with widely spaced multiple transmit and receive antennas using orthogonal waveforms for moving target detection in colored noise-plus-clutter. We illustrate the advantages of systems that use multiple transmit antennas over those using a single transmit antenna. We provide a comparison between conventional phased-array radar and MIMO radar. We use receiver operating characteristic (ROC) curves of the test statistics to show the significant gains of the MIMO radar systems due to the distributed antennas. We describe both centralized and distributed processing implementations of our MIMO radar architecture. We show that the centralized MIMO system performs better than the distributed one by comparing the cumulative distribution function (cdf) curves. We give analysis of the improvements due to spreading out the transmit antennas for MIMO radars. We also investigate CFAR detectors for the MIMO radar systems we consider. Note that RCS is the amount of power reflected to the source.

MIMO SIGNAL MODEL

In this section, we develop two separate models, one for MIMO radar and one for conventional phased-array radar. We assume the target does not leave the given cell-under-test during K consecutive pulse transmissions. Furthermore, it is assumed that the M transmit elements employ pulse waveforms that maintain orthogonality under a variety of mutual delays and frequency shifts. This allows each of the N receive elements to separate the pulses transmitted by different transmit elements. Under the H_1 hypothesis (target present in the cell-under-test) and the H_0 hypothesis (target absent), respectively, the K consecutive signal samples at the lth, $l = 1, ..., N$ receiver due to the pulses from the kth, $k = 1, ..., M$ transmitter are given by the vector:

$$r_{k,l} = \begin{cases} c_{k,l} + n_{k,j} & H_0 \\ \alpha_{k,l} d_{k,l}\left(v_x, v_y\right) + c_{k,l} + n_{k,l} & H_1 \end{cases}$$

The scalar $\alpha_{k,l}$ represents the unknown complex target reflectivity. We assume $\alpha_{k,l}$ does not vary during the K consecutive scans for any given k and l. We define this as follows:

$$d_{k,l}\left(f_{k,l}\right) = \begin{bmatrix} 1 \\ e^{j2\pi f_{k,l}T_{\text{PRF}}} \\ . \\ . \\ . \\ e^{j2\pi f_{k,l}(K-1)T_{\text{PRF}}} \end{bmatrix}$$

where T_{PRF} is the pulse repetition interval and $f_{k,l}$ denotes the observed Doppler shift at the lth receiver due to the signal transmitted by the kth transmitter. Note that:

$$f_{k,l} = \left(\cos\theta_k^t + \cos\theta_l^r\right)\frac{f_c v_x}{c} + \left(\sin\theta_k^t + \sin\theta_l^r\right)\frac{f_c v_y}{c}$$

where f_c is the carrier frequency and c the speed of light. The transmit and receive elements are located at angles θ_k^t and θ_l^r, respectively, when viewed from an origin located at the target. Figure 9.1 illustrates the geometry of this scenario. Now we define:

$$d_{k,l}\left(v_x, v_y\right) = d_{k,l}\left(f_{k,l}\right)\Big|_{f_{k,l} = \left(\cos\theta_k^t + \cos\theta_l^r\right)\frac{f_c v_x}{c} + \left(\sin\theta_k^t + \sin\theta_l^r\right)\frac{f_c v_y}{c}}$$

It shows that $c_{k,l}$ represents the clutter echo and $n_{k,l}$ the thermal noise at the receiver. We assume the clutter to be zero-mean complex Gaussian distributed with the following properties:

$$E\left\{c_{k,l}c_{k,l}^{H}\right\} = C'$$

and

$$E\left\{c_{k,l}c_{k',l'}^{H}\right\} = 0_{K\times K}, \quad \forall k \neq k' \text{ or } l = l'$$

The thermal noise at the receive element is assumed to be zero-mean complex white Gaussian noise with the correlation matrix:

$$E\left\{n_{k,l}n_{k,l}^{H}\right\} = \sigma_n^2 I_{K\times K}$$

The noise terms corresponding to different transmit–receive paths are assumed to be independent so that:

$$E\left\{n_{k,l}n_{k',l'}^{H}\right\} = 0_{K\times K}, \quad \forall k \neq k' \text{ or } l = l'$$

In the signal model just described, we ignore any different path losses between different transmitter–receiver pairs. Thus, we assume that all transmitters and receivers have roughly the same distance to the cell-under-test but view this cell from different angles θ_k^t and θ_l^r.

PHASED-ARRAY SIGNAL MODEL

For comparison with conventional radars, we consider a radar system consisting of M transmit and N receive elements, co-located, jointly forming a linear array with small inter-element spacing. The transmit elements do not use different orthogonal pulse waveforms but instead transmit the same waveform. The transmit elements steer a beam toward the direction ϕ_{tsteer} by imposing phase shifts on the transmit pulse. The phase shift at the kth transmit element corresponds to $b_k(\phi_{tsteer}) = e^{j2\pi k \sin \phi_{tsteer}\Delta_t/\lambda}$, where Δ_t is the inter-element spacing on the transmit side and λ_s is the considered wavelength. The received K consecutive signal samples at the lth, $l = 1, \ldots, N$ array element are given by the vector:

$$r_l = \begin{cases} \sum_{k=1}^{M} c_{k,l}b_k\left(\varnothing_{tsteer}\right) + n_j, & H_0 \\ \sum_{k=1}^{M} \left(a_l\left(\phi_r\right)\alpha d\left(v_x,v_y\right)b_k^*\left(\varnothing_t\right) + c_{k,l}\right)b_k\left(\varnothing_{tsteer}\right) + n_j, & H_1 \end{cases}$$

Here, the fading coefficient α is the same for all elements as they always view the same RCS aspect of the target. Further, the vector $d(v_x, v_y))$ is the same across all elements as they have the same perspective with respect to the target's motion. The angle ϕ_t between the normal vector to the transmit array and the vector from the array to the target results in phase shifts of the impinging pulses given by $b_k(\phi_t) = e^{j2\pi k \sin \phi_t \Delta_t / \lambda}$. Furthermore, the angle ϕ_r between the normal vector to the receive array and the vector from the array to the target results in phase shifts across the receive array elements, as given by $a_l(\phi_r) = e^{j2\pi l \sin \phi_r \Delta_r / \lambda}$, where ϕ_r is the inter-element spacing on the receive side. The noise components have the same properties as described for MIMO radar. The clutter properties are discussed next.

CLUTTER MODEL

We employ a simple clutter model. Clutter exists everywhere, unlike a target which has a limited extent. The clutter background is assumed spatially homogeneous, i.e., scatterers are governed by identical statistical distributions. Since the clutter echo received at each antenna is the result of a large sum of contributions from different clutter scatterers, it is asymptotically Gaussian. Such a model may apply to the returns from the forest, grassland, or other homogeneous surfaces. Motivated by this, we adopt the usual assumptions for clutter. The clutter is also subject to internal motion such as wind. Thus, the received clutter echoes are complex Gaussian, and the temporal clutter fluctuations are slow compared with the observation interval of K pulses.

A complete model of the clutter requires the specification of temporal and spatial correlation properties. The temporal correlation is described through the matrix C'. Since our processing is valid for any given C', we wait until the numerical results section to give the explicit form assumed there. In subsequent sections, we denote, for conciseness, the thermal noise and the clutter return as $x_{k,l} = c_{k,l} + n_{k,l}$. The temporal correlation matrix of $x_{k,l}$ is then:

$$C = E\left\{x_{k,l}x_{k,l}^H\right\} = C' + \sigma_n^2 I_{K \times K}$$

It is assumed, as before, that the clutter returns corresponding to different transmit–receive paths are uncorrelated. This assumption is justified for widely spread antennas. For simplicity, we also employ this assumption for closely spaced antennas in the case of the conventional phased-array radar. This is likely overly optimistic, thus overstating the true performance of the conventional phased-array radar. Thus, we should expect even larger gains for MIMO radar in practice over conventional phased-array radar compared to those shown here. We assume that for a given transmitter–receiver pair, the clutter temporal correlation is known or estimated a priori.

MOVING TARGET DETECTOR

Using the signal models that defined earlier equations, we derive the detectors for the different schemes based upon the generalized likelihood ratio test (GLRT). We assume that in the cell-under-test, there is either a single target with a velocity vector $v = (v_x, v_y)^T$ or no target at all.

CENTRALIZED MIMO MOVING TARGET DETECTOR

In standard MIMO radar, we assume centralized processing such that all received signals corresponding to the MN transmit–receive paths are transmitted to a central station for joint processing. Stacking the MN received vectors $r_{k,l}$ in one vector $r = \left[r_{1,1}^T, \ldots, r_{M,N}^T \right]^T$ and the MN target coefficients $\alpha_{k,j}$ in the vector $\alpha = [\alpha_{1,1}, \ldots, \alpha_{M,N}]^T$, we can write the joint probability density function (pdf) of the received vectors conditioned on the hypotheses and parameters as:

$$f(r \mid \alpha, v_x, v_y, H_1)$$
$$= \prod_{k=1}^{M} \prod_{l=1}^{N} \frac{1}{\pi^K \sqrt{\det[C]}} \cdot e\left[-\left(r_{k,l} - \alpha_{k,l} d_{k,l}(v_x, v_y) \right)^H C^{-1} \cdot \left(r_{k,l} - d_{k,l}(v_x, v_y) \right) \right]$$

and

$$f(r \mid H_0) = \prod_{k=1}^{M} \prod_{l=1}^{N} \frac{1}{\pi^K \sqrt{\det[C]}} \cdot e^{-r_{k,j}^H C^{-1} r_{k,j}}$$

where we used $C = C' + \sigma_n^2 I_{K \times K}$. The GLRT is then defined as:

$$\hat{\xi} = \ln \left(\frac{\max\limits_{\alpha, v_x, v_y} f(r \mid \alpha, v_x, v_y, H_1)}{f(r \mid H_0)} \right) \begin{array}{c} H_1 \\ > \\ < \\ H_0 \end{array} \hat{\gamma}$$

We note that maximizing $f(r|\alpha, v_x, v_y, H_1)$ with respect to α, v_x, and v_y is equivalent to minimizing:

$$\sum_{k=1}^{M} \sum_{l=1}^{N} \left(r_{k,l} - \alpha_{k,l} d_{k,l}(v_x, v_y) \right)^H C^{-1} \left(r_{k,l} - d_{k,l}(v_x, v_y) \right)$$

which in turn is equivalent to maximizing:

$$\sum_{k=1}^{M} \sum_{l=1}^{N} 2\mathcal{R}\left\{ \alpha_k^* d_{k,l}^H(v_x, v_y) \right\} - |\alpha_{k,l}|^2 d_{k,l}^H(v_x, v_y) C^{-1} d_{k,l}(v_x, v_y)$$

Moreover, the expression is maximized for any given pair v_x and v_y by:

$$\alpha_{k,l} = \frac{d_{k,l}^H(v_x, v_y) C^{-1} r_{k,l}}{d_{k,l}^H(v_x, v_y) C^{-1} d_{k,l}(v_x, v_y)}$$

Using (8) and (10) results in the following decision rule:

$$\xi = \max_{v_x,v_y} \sum_{k=1}^{M}\sum_{l=1}^{N} \frac{\left| d_{k,l}^{H}\left(v_x,v_y\right)C^{-1}r_{k,l}\right|^2}{d_{k,l}^{H}\left(v_x,v_y\right)C^{-1}d_{k,l}\left(v_x,v_y\right)} \overset{H_1}{\underset{H_0}{\gtrless}} \gamma$$

We emphasize that the centralized MIMO moving target detector employs a joint estimate of the true velocity vector $v = (v_x, v_y)$ formed using all elements of r. In the next section, we discuss the implications of this joint estimate when compared to local velocity estimates determined at each receive antenna.

DISTRIBUTED MIMO MOVING TARGET DETECTOR

When distributed processing is adopted in MIMO radar, the system complexity is dramatically reduced. Here we investigate a simple type of distributed processing and compare its complexity and performance with those of centralized processing. In this case, each receive station (receive antenna) processes its received signal autonomously and transmits a local decision to a fusion center. We implement a distributed MIMO detector utilizing soft decisions as described by the following test and global test statistic:

$$\xi = \sum_{k=1}^{M}\sum_{l=1}^{N} \max_{f_{k,l}} \frac{\left| d_{k,l}^{H}\left(v_x,v_y\right)C^{-1}r_{k,l}\right|^2}{d_{k,l}^{H}\left(v_x,v_y\right)C^{-1}d_{k,l}\left(v_x,v_y\right)} \overset{H_1}{\underset{H_0}{\gtrless}} \gamma$$

This expression implies that the lth receive station estimates the Doppler shift $f_{k,l}$ based on the signal transmitted by the kth transmit station only. The soft decisions are transferred to the fusion center that sums all the soft decisions to compute the final test statistic. Thus, instead of transmitting K samples per test for each transmitter–receiver pair to the central station, as in the centralized MIMO case, only one sample is sent to the fusion center. It can be argued that binary or multibit local decisions are a form of quantization of those samples. Following the common observation that the fusion of soft decisions outperforms the fusion of hard decisions, we can infer that a detector performs at least as well as a detector based on binary or multibit decisions.

For distributed MIMO, one needs to perform a one-dimensional search to find the Doppler shift $f_{k,l}$ that corresponds to each of the MN paths. Let us denote the complexity of this search as η, which stands for the average number of computations needed to perform the one-dimensional search. At each receive antenna, one must perform the search M times for each transmitted signal, so the complexity at each receiver is $M\eta$. The total complexity at all N receive antenna is $MN\eta$ for the distributed MIMO system. For the centralized MIMO system, one needs to perform a two-dimensional search as per (14) to find the two-dimensional vector (v_x, v_y), and the complexity would be on the order of η^2. If M and N are reasonably small, the complexity of the distributed MIMO system will be lower. Of course, since the Doppler shift estimate, on which the GLRT test statistic is based, is produced for each path individually and independently while physically all the Doppler shifts correspond to a common target, the performance of the distributed system will be inferior.

PHASED-ARRAY MOVING TARGET DETECTOR

A conventional phased-array radar will steer a transmit beam and a receive beam toward the cell-under-test. The beamformer on the transmit side is already incorporated in the signal model. The beamformer on the receive side sums the N received signal vectors into a single received vector r:

$$r = \sum_{l=1}^{N} \alpha_l^* \left(\Phi_{rsteer} \right) r_l$$

where ϕ_{rsteer} is defined like ϕ_r.

We can show our signal-to-interference-plus-noise ratio (SINR) scales with N. The moving target detection procedure is based on the single $K \times 1$ vector r. Assume that the steering angle ϕ_{rsteer} matches the AOA ϕ_r. Following the same derivation, the test statistic is derived as:

$$\max_{f_d} \frac{\left| d^H \left(f_d \right) C^{-1} r \right|^2}{d^H \left(f_d \right) C^{-1} d \left(f_d \right)} \overset{H_1}{\underset{H_0}{\gtrless}} \gamma$$

We note again that the conventional phased-array radar observes only one Doppler shift f_d resulting from the target's radial velocity. Accordingly, the GLRT test statistic is based on an estimate of this frequency shift.

ADAPTIVE MIMO MOVING TARGET DETECTOR

All the decision statistics previously described have been derived under the assumption of known noise-plus-clutter (called interference in the sequel) temporal correlation matrix C. However, this matrix might be unknown. Moreover, different transmitter–receiver pairs in the centralized or decentralized MIMO system might be subject to different interference levels or characteristics and thus have different interference matrices. Accordingly, we use $C_{k,l}$ to denote the interference matrix for the path between of the kth transmitter and the lth receiver.

It is possible to convert all the previously described detectors to adaptive detectors that maintain a CFAR. This is accomplished by replacing the true clutter covariance matrix by its estimate:

$$\hat{C}_{k,l} = \frac{1}{L} \sum_{i=1}^{L} r_{k,l} \left(i \right) r_{k,l}^H \left(i \right)$$

based on $L > K$ "secondary" data vectors $r_{k,l}(i)$. We note that the $r_{k,l}(i)$ are measured in adjacent range cells, for which target absence is assumed. The adaptive centralized MIMO test statistic is then given as:

$$\xi = \max_{v_x,v_y} \sum_{k=1}^{M} \sum_{l=1}^{N} \frac{\left| d_{k,l}^H \left(v_x, v_y\right) \hat{C}^{-1} r_{k,l} \right|^2}{d_{k,l}^H \left(v_x, v_y\right) \hat{C}^{-1} d_{k,l} \left(v_x, v_y\right)} \begin{matrix} H_1 \\ > \\ < \\ H_0 \end{matrix} \gamma$$

It is shown that this test statistic is indeed subject to CFAR behavior. The same result can easily be extrapolated to the adaptive distributed MIMO and phased-array cases. In this context, it is noted that the clutter is assumed to have the same statistical characteristics for all transmitter–receiver pairs in the phased-array system. The CFAR property of the test statistic is demonstrated with numerical results, and the performances of the MIMO and phased-array system in adaptive mode are compared using simulation experiments.

SIMULATION RESULTS AND INFERENCES

In this section, we discuss simulation results and compare the performance of the different systems and setups. First, we describe the choice of the clutter covariance matrix C'. Assuming internal motion of the clutter scatterers due to, for example, wind affecting a forest or grassland, the temporal correlation of such clutter can be described by its power spectral density (PSD) [1]:

$$S_{cc}\left(f\right) = \frac{P_{cc}\lambda}{\sqrt{2\pi}\,2\sigma_v} e^{-f^2\lambda^2/8\sigma_v^2}$$

where P_{cc} is the clutter power, λ the wavelength, and σ_v the rms of clutter velocity. This relates to a continuous autocorrelation function (ACF) of:

$$\varphi_{cc}\left(\tau\right) = \int_{-\infty}^{\infty} \frac{P_{cc}\lambda}{\sqrt{2\pi}\,2\sigma_v} e^{-f^2\lambda^2/8\sigma_v^2} e^{j2\pi f\tau} df = P_{cc}e^{-\pi^2 8\sigma_v^2/\lambda^2}$$

Sampling this $\tau = kT_{PRF}$, $k = 0, \ldots, K - 1$, we find the correlation coefficients $\rho_{cc}(k)$ of the K consecutive samples, i.e., $\rho_{cc}(k) = \varphi_{cc}(kT_{PRF})$. The clutter temporal correlation matrix is then given as:

$$C' = \begin{bmatrix} \rho_{cc}\left(0\right) & \rho_{cc}\left(1\right) & \cdots & \rho_{cc}\left(K-1\right) \\ \rho_{cc}\left(1\right) & \rho_{cc}\left(0\right) & \cdots & \cdots \\ \cdots & \cdots & \cdots & \rho_{cc}\left(1\right) \\ \rho_{cc}\left(k\right) & \cdots & \rho_{cc}\left(1\right) & \rho_{cc}\left(0\right) \end{bmatrix}$$

In the simulations, we assume the carrier frequency $f_c = 1$ GHz, the number of time samples $K = 10$, and the rms of clutter velocity $\sigma_v = 1.25$ m/s. The search procedures to obtain the GLRT statistics for different systems are performed over a certain velocity range corresponding to the unambiguous velocities determined by the chosen pulse repetition frequency (PRF).

RECEIVER OPERATING CHARACTERISTICS

Here, we compare the different radar systems according to their ROC. Assume the PRF is 500 Hz, the clutter-to-noise ratio is 40 dB, and each radar system has two transmitters and two receivers, where the transmitters are located at:

$$\left\{\theta_k^t\right\}_k^2 = \left\{0°, 65°\right\}$$

and the receivers are located at:

$$\left\{\theta_l^r\right\}_l^2 = \left\{-30°, 40°\right\}$$

Consider a target with $\alpha_{k,l} = \alpha$=constant (nonrandom) in (2), where $k = 1, 2$ and $l = 1, 2$. Further assume the target moves with random direction and velocity magnitude of 68 km/h. Choosing $\alpha_{k,l} = \alpha$=constant allows us to isolate the "geometry gains" from the diversity gains by essentially turning off the diversity gains by eliminating target fluctuations.

The ROCs of the centralized MIMO (CMIMO), distributed MIMO (DMIMO), and conventional phased-array radar systems for different target-to-clutter (t_c) ratios have been computed. It is observed that the ROC curves for the centralized MIMO are on top of each other, and the corresponding detection probabilities are close to 1, which shows that the performance of the centralized MIMO systems is quite good in all these cases.

Due to the "geometry gains," the MIMO radars outperform the conventional phased-array radars, which observe the target only from one aspect. The conventional phased-array radar is limited to observing the target from only one viewing angle per array which can lead to poor performance, in comparison with MIMO radar, with random target direction.

Note that the effect of random target direction is like that of target RCS fluctuation. The reason is that the conventional radars perform well only for certain target directions that are consistent with their viewing angles. For these target directions, the effect is like a large RCS return. For other target directions, the effect is like a small RCS return.

We have also accounted for the target's RCS fluctuations. In this case, we select $\alpha_{k,l}$, $k = 1, \ldots, M$ and $k = 1, \ldots, N$ as a set of complex Gaussian random variables. MIMO radars again outperform the conventional phased-array radars. However, this time the improvement is due to both "geometry" and spatial diversity gains. The spatial diversity gains reduce the effects of the RCS fluctuations over different transmitter–receiver paths.

Taken together, we have demonstrated two separate gains for MIMO radar when compared with the conventional phased-array radar. MIMO radar can view the moving target from different directions, while the phased-array views only from one direction per array. Thus, if the target moves perpendicular to the phased-array viewing direction, performance is degraded. This is called "geometry gain" here. The other gain is the diversity gain. The first gain leads to the improvements, while the two gains together lead to the much larger gains.

For completeness, we remind the reader that, MIMO radar is not always better than the conventional phased-array approach. MIMO radar is better for cases with sufficiently high signal-to-noise ratio (SNR). At very low SNR, the conventional phased-array approach is better. The interested reader is referred to [1] to see the detail for performance curves, analysis, and discussion, which is not repeated here in the interest of brevity. Thus, if we were to significantly decrease the signal strength in the cases studied here, we would eventually encounter this phenomenon. However, since we consider moving targets, we find that MIMO radar is better for even lower signal strengths than predicted in other papers [1]. The reason is the geometry-based MIMO radar gains over the conventional phased-array approach, which come in addition to the diversity gains studied when one employs widely separated antennas. Extensive simulation results, which are very similar in nature to those given in [1], have verified this.

COMPARING CENTRALIZED AND DISTRIBUTED MIMO RADARS

In this section, we give more insight into the centralized and distributed MIMO detectors by comparing the cumulative density functions (cdfs) of their test statistics with a focus on the "geometry gains" only. Assume the PRF is 2 kHz, the clutter-to-noise ratio is 30 dB, a single transmitter is located at $\theta_1^t = 0°$, and the locations of the receivers are as follows:

- when $N = 1$, $\theta_1^r = 0°$;
- when $N = 4$, $\left\{\theta_l^r\right\}_{l=1}^{4} = \left\{-39°, -13°, 13°, 39°\right\}$;
- when $N = 8$, $\left\{\theta_l^r\right\}_{l=1}^{8} = \left\{-39°, -26°, -13°, 0°, 13°, 26°, 39°, 51°\right\}$;

Assume the target-to-clutter ratio is 0 dB. The cdf of the test statistics is computed for centralized and distributed MIMO radars under hypothesis H_1, where a target with $\alpha_{k,l} = \alpha$=constant (nonrandom) and absolute velocity 300 km/h is present. Note that for this choice of target-to-clutter ratio, the test statistics for both systems follow a similar distribution. The cdf of the test statistics for centralized and distributed MIMO radars under hypothesis H_0 has been computed, where the target is absent, and the received signal contains only noise-plus-clutter. These plots show that for a given cdf value $F(\xi|H_0)$, the test statistic ξ_{H0} for centralized MIMO radar is smaller than the one for distributed MIMO radar.

Recall that for a threshold γ, the detection probability is $P_d = 1 - F(\gamma|H_1)$ and the false alarm probability is $P_{fa} = 1 - F(\gamma|H_0)$. Comparing the cdf curves, we find that to get the same detection probability P_d, the centralized MIMO detector and the distributed MIMO detector need a threshold of similar size, while given a threshold of similar size, the former has lower P_{fa}. This explains the gains we saw earlier already for the centralized MIMO detector in terms of the statistics of the tests involved. Unlike the distributed detector which performs a one-dimensional search for each transmit–receive path, the centralized detector searches over a two-dimensional space jointly based on all available samples. The joint maximization approach in centralized MIMO radar imposes more restrictions on the test statistic and thus effectively reduces the chances for false alarms. This is why the centralized MIMO radar outperforms the distributed MIMO radar. Keep in mind that the two-dimensional search for centralized MIMO radar demands higher computational complexity.

DISTRIBUTED TRANSMITTER ELEMENTS

The target Doppler shift $f_{k,l}$ caused by the k, lth transmit–receive path can be represented as [1]:

$$f_{k,l} = \frac{2 f_c |v|}{c} \cos\left(\emptyset_{k,l}^{bs}\right) \cos\left(\frac{\theta_k^t - \theta_l^r}{2}\right)$$

where $|v|$ denotes the target absolute velocity. The transmitter and receiver have aspect angles θ_k^t and θ_l^r referenced to the horizontal axis. The angle $\emptyset_{k,l}^{bs}$ is measured between the target velocity vector and the bisector of the transmitter-target-receiver angle. From earlier equation, for a given absolute target velocity, the magnitude of the Doppler shift $f_{k,l}$ is determined by $\emptyset_{k,l}^{bs}$ and $\left(\theta_k^t - \theta_l^r\right)$. On the one hand, widely dispersed antennas in MIMO radar ensure that some antenna will yield $\cos\left(\emptyset_{k,l}^{bs}\right) \approx 1$. On the other hand, to get large Doppler shifts, the difference between the transmitter and the receiver aspects should be kept small. The validity of this argument is demonstrated by comparing a 1×8 and a 2×4 centralized MIMO radar. Assume the PRF is 2 kHz, and the clutter-to-noise ratio is 30 dB. The element locations of the 1×8 system is:

- $\theta_1^t = 45°$ and $\left\{\theta_l^r\right\}_{l=1}^8 = \left\{0°, 13°, 26°, 38°, 50°, 62°, 75°, 90°\right\}$.

The 2×4 system consists of transmitter and receiver elements located at:

- $\left\{\theta_k^t\right\}_{k=1}^2$ and $\left\{\theta_l^r\right\}_{l=1}^4 = \left\{0°, 30°, 60°, 90°\right\}$

The total transmit energy of the 2×4 system is twice that of the 1×8 system. According to the analysis presented in [2], these two systems should perform identically in a nonmoving target situation since they have the same number of "diversity branches" (i.e., $N \times M = 8$) and emit the same power per transmitter.

However, this shows that the system with two transmitters outperforms the single transmitter system under the moving target model considered in this paper [1]. This is an interesting observation, as it contrasts possible inferences from the analysis found for stationary targets. Further, noting that the 1×8 system has a total of nine transmit or receive stations, whereas the 2×4 system has only six stations, the preference toward systems with multiple transmitter elements may be strengthened by an economical argument.

CFAR PROPERTIES OF THE TEST STATISTICS

In this section, we discuss 1×4 and 1×8 centralized MIMO and phased-array systems. Assume the PRF is 2 kHz, and the clutter-to-noise ratio is 30 dB. We have seen the empirical cdfs of the adaptive centralized MIMO test statistic of earlier equation under H_0 for a 1×4 system and different numbers of secondary vectors L. The transmit and receive elements have the same positions.

The solid curve is the cdf for the known clutter covariance matrix. The dashed curves reflect the empirical cdf of the test statistic ξ under 1×4 for $L = 160$, and the dotted ones denote L = 40. For each L, the following two scenarios are considered.

- Scenario 1 (sc. 1): The clutter covariance matrix is the same for all receive elements and has the previously described form (see (23), assuming the PRF is 2 kHz, and the clutter-to-noise ratio is 30 dB).
- Scenario 2 (sc. 2): Two of the four receive elements observe clutter returns with the same covariance matrix as before. For the other two, the clutter-to-noise ratio is reduced by 20 dB and the complete interference power raised by 6 dB.

In either case, secondary data vectors are used to estimate the covariance matrix. The almost identical empirical cdfs for the two scenarios illustrate that the adaptive test statistic has a distribution that is independent of the underlying clutter characteristic and therefore can be used to achieve at least approximate CFAR properties. However, the number of secondary vectors used in the estimate of the covariance matrix has an impact on the test statistic distribution. For small values of L, the variance of the test statistic under H_0 increases. This leads to a reduced P_d and is a well-observed fact for CFAR detection.

Consider the ROCs for 1×4 MIMO and phased-array radar systems for CFAR detection. Here $L = 200$ secondary data vectors are used. A target with a speed of $|v|= 300$ km/h is considered for H_1. In this case, though the performance of both centralized MIMO radar and phased-array radar is slightly impaired from not having exact knowledge of C, the centralized MIMO system still outperforms the phased-array system.

REFERENCES

1. He, Q. et al., "MIMO Radar Moving Target Detection in Homogeneous Clutter," *IEEE Transaction on Aerospace and Electronic Systems*, Vol. 46, No. 3, July 2010.
2. Fortunati, S. et al., "Massive MIMO Radar for Target Detection," January 2020, arXiv:1906.06191v4 [eess.SP] 15 Jan 2020.

10 Massive MIMO Radar for Target Detection

Since the seminal paper by Marzetta from 2010, the Massive MIMO paradigm in communication systems has changed from being a theoretical scaled-up version of MIMO, with an infinite number of antennas, to a practical technology. Its key concepts have been adopted in the 5G new radio standard and base stations, where 64 fully digital transceivers have been commercially deployed. Motivated by these recent developments, this chapter considers a co-located MIMO radar with M_T transmitting and M_R receiving antennas and explores the potential benefits of having a large number of virtual spatial antenna channels $N = M_T M_R$. Particularly, we focus on the target detection problem and develop a robust Wald-type test that guarantees certain detection performance, regardless of the unknown statistical characterization of the disturbance. Closed-form expressions for the probabilities of false alarm and detection are derived for the asymptotic regime $N \to \infty$. Numerical results are used to validate the asymptotic analysis in the finite system regime with different disturbance models. Our results imply that there always exists a sufficient number of antennas for which the performance requirements are satisfied, without any a priori knowledge of the disturbance statistics. This is referred to as the Massive MIMO regime of the radar system [1].

BACKGROUND

Consider a multiple antenna radar system characterized by N spatial channels collecting K temporal snapshots $\{x_k\}_{k=1}^K \in \mathbb{C}^N$ from a specific resolution cell, defined in an absolute reference frame. The primary goal of any radar system is to discriminate between two alternative hypotheses: the presence (H_1) and absence (H_0) of the target, in the resolution cell-under-test. Among others, a common model for the signal of interest is $\alpha_k v_k$, where $v_k \in \mathbb{C}^N$ is known at each time instant $k \in \{1, \ldots, K\}$ and $\alpha_k \in \mathbb{C}$ is a deterministic, but unknown, scalar that may vary over k. Any measurement process involves a certain amount of disturbance. In radar signal processing, the disturbance is produced by two components, the clutter (unwanted echoes) and white Gaussian measurement noise, and it is modeled as an additive random vector, say c_k, whose statistics may vary over k. Formally, the detection problem can be recast as a composite binary hypothesis test (HT):

$$H_0: x_k = c_k \quad k = 1, \ldots, K$$
$$H_0: x_k = \alpha_k v_k + c_k \quad k = 1, \ldots, K$$

DOI: 10.1201/9781003499480-10

To solve above the equation, a decision statistic $\Lambda(X)$ of the dataset $X \triangleq [x_1, \ldots, x_K]$ is needed and its value must be compared with a threshold λ:

$$\Lambda(X) \underset{< H_0}{\overset{> H_1}{\gtrless}} \lambda$$

to discriminate between the null hypothesis H_0 and the alternative H_1. A common requirement in radar applications is that the probability of false alarm has to be maintained below a preassigned value, say P_{FA}. Consequently, the threshold λ should be chosen to satisfy the following integral equation:

$$\Pr\{\Lambda(X) > \lambda | H_0\} = \int_{\lambda}^{\infty} p_{\Lambda|H_0}(a | H_0) da \triangleq P_{FA}$$

where $p_{\Lambda|H0}$ is the probability density function (pdf) of decision statistics $\Lambda(X)$ of dataset X under the null hypothesis H_0.

TECHNICAL CONSIDERATIONS

Finding a solution to P_{FA} (3) is in general a challenge. The common way out relies upon some "ad-hoc" assumptions on the statistical model of the dataset X. To clarify this point, let us have a closer look at the steps required to solve (3). Firstly, a closed-form expression for $p_{\Lambda|H_0}$ is needed. By definition, $p_{\Lambda|H_0}$ is a function of the chosen decision statistic $\Lambda(X)$ and of the joint pdf $p_X(X)$ of X. If all the $\alpha_k \, \forall \, k$ are modeled as deterministic unknown scalars, $p_X(X)$ is fully determined by the joint pdf $p_C(C)$ of the disturbance $C = [c_1, \ldots, c_K]$. A first simplification comes from the assumption that the disturbance vectors $\{c_k\}$ are independent and identically distributed (i.i.d.) random vectors such that $p_C(C) = \prod_{k=1}^{K} p_C(c_k)$.

This assumption is, however, not always valid in practice. A second simplification that is commonly adopted in the radar literature is to assume that the functional form of $p_C(c_k) \equiv p_C(c_k; \gamma)$ is perfectly known, up to a possible (finite-dimensional) deterministic nuisance vector parameter, for example the (vectorized) covariance matrix. To obtain a consistent estimate $\hat{\gamma}$ of γ, a secondary dataset must be exploited. In radar terminology, a secondary dataset is a set of "signal-free" snapshots H_0 collected from resolution cells adjacent to the one under test and sharing the same statistical characterization. Note that the required $p_{\Lambda|H_0}$ is a function of $\hat{\gamma}$ as well. A third simplifying assumption is that the signal parameters α_k remain constant over k, i.e., $\alpha_k = \sigma, \, \forall \, k$. Under these three assumptions, a possible choice for the decision statistic is the generalized likelihood ratio (GLR) $\Lambda_{GRL}(X)$.

However, a closed-form solution can be found only for a very limited class of disturbance models for which the Gaussianity assumption needs also to be imposed. An asymptotic approximation for the solution can be obtained by exploiting a well-known asymptotic property of the GLR. Under the hypothesis H_0 and for $K \to \infty$, the pdf of $\Lambda_{\mathrm{GRL}}(X)$ converges to the one of a central χ-squared random variable with 2 degrees of freedom, denoted as $\chi_2^2(0)\_22(0)$.

Hence, by using the properties of the χ-squared distribution, it is immediate to verify that is asymptotically satisfied by $\lambda = -2 \ln P_{\mathrm{FA}}$. This is a particularly simple result that has received a lot of attention in the literature. However, it relies on the four simplifying assumptions previously introduced and summarized as follows:

- A1 The disturbance vectors $\{c_k\}_{k=1}^{K}$ are independent and identically distributed (i.i.d.) over the observation interval.
- A2 The pdf $p_C(c_k; \gamma)$ of the disturbance is perfectly known, up to an unknown nuisance parameter vector γ.
- A3 The target complex amplitude α_k is maintained constant over the observation interval, i.e., $\alpha_k \equiv \alpha, \forall k$.
- A4 The number of temporal snapshots K is assumed to be much larger that the spatial channels N.

Even if these assumptions make the (asymptotic) analysis of $\Lambda_{\mathrm{GRL}}(X)$ analytically tractable, they are seldom satisfied in practical applications. Spatial frequency describes the periodic distributions of light and dark in an image. High spatial frequencies correspond to features such as sharp edges and fine details, whereas low spatial frequencies correspond to features such as global shape.

System Model

Consider a co-located MIMO radar system equipped with M_T transmitting antennas and M_R receiving antennas. The transmitting array is characterized by the array manifold, also called steering vector $a_T(\phi)$, where ϕ is the position vector defined in an absolute reference frame. Similarly, the receiving array can be characterized by the steering vector $a_R(\phi)$ since the positions of the antennas in the absolute reference frame are known; see Figure 10.1.

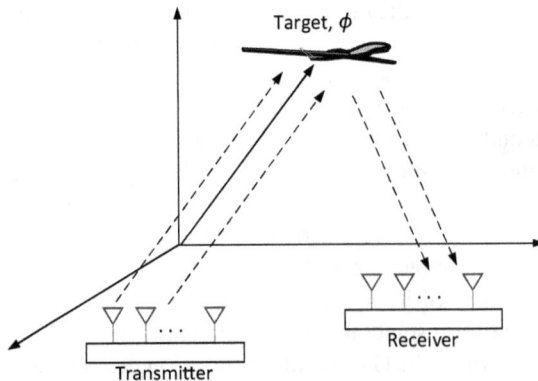

FIGURE 10.1   Co-located MIMO radar.

Given a target with position vector, the signal collected at the receiving array can be modeled:

$$x(t) = \alpha a_R(\phi) a_T^T s(t - \tau) e^{j\omega t} + n(t), \quad t \in [0, T]$$

where $x(t) \in \mathbb{C}^{M_R}$ is the array output vector at time t, $s(t) \in \mathbb{C}^{M_T}$ is the vector of transmitted signals, and $\alpha \in \mathbb{C}$ accounts for the radar cross section (RCS) of the target and the two-way path loss, which is the same for each transmitter and receiver pair. This is generally verified in co-located MIMO radars. The parameters τ and ω represent the actual time delay and Doppler shift, due to the target position and velocity. The complex, vector-valued, random process $n(t) \in \mathbb{C}^{M_R}$ accounts for the disturbance. We assume that $s(t)$ is obtained as a linear transformation of a set of nearly orthonormal signals $s_0(t)$, $\in \mathbb{C}^{M_T}$, where $W = [w_1, \ldots, w_{M_T}]^T \in \mathbb{C}^{M_T \times M_T}$ and $w_m \in \mathbb{C}^M$ is the weighting vector of the transmit antenna element m with power $\|w_m\|^2$.

Let $l = \{1, \ldots, L\}$ and $k = \{1, \ldots, K\}$ be the indices characterizing each time $l\Delta t$ and frequency $k\Delta t$ samples, respectively. The output $X(l, k) \in \mathbb{C}^{M_R \times M_T}$ of the linear filter matched to $s_0(t)$ can be expressed as:

$$X(l,k) = \alpha a_T^T(\phi) W S(l,k) + C(l,k)$$

where

$$S(l,k) \triangleq \int_0^T s_0(t - \tau) s_0^H(t - l\Delta t) e^{-j(k\Delta\omega - \omega)t} dt$$

considers potential "straddling losses," which are losses due to a not-precise centering of the target in a range-Doppler gate or to a not-exact orthogonality between waveforms, and

$$C(l,k) = \int_0^T n(t) s_0^H(t - l\Delta t) e^{-jk\Delta\omega t} dt$$

After omitting the indexes (l, k) for ease of notation, we may rewrite (5) in vectorial form as:

$$x = X = \alpha v(\phi) + c, \quad x \in \mathbb{C}^N$$

where $N \triangleq M_R \times M_T$ and

$$v(\phi) = S^T \otimes I_{M_R} \left[W^T a_T(\phi) \otimes a_R(\phi) \right]$$

and $c \triangleq C$.

While $a_T(\phi)$ and $a_R(\phi)$ depend on the geometry of the transmitting and receiving arrays, respectively, the matrix W can be designed to shape arbitrarily the transmitting beam (see [29, Ch. 4] and references therein). For example, with $W = I_{MT}$, the transmitted power is uniformly distributed over all possible directions. On the other hand, with $W = a_T^H(\phi)a_T^H(\phi) = \Sigma(t-\tau)$, it is fully directed toward the direction ϕ. Intermediate cases can be obtained.

We assume that $n(t)$ is zero-mean and wide-sense stationary; that is, $E\{n(t)\} = 0$, $\forall\, t$ and $E\{n(t)n(\tau)^H\}$. Hence, from (7) it easily follows that $E\{c\} = 0$ and

$$\Gamma \triangleq E\{cc^H\} = \int_0^T\int_0^T \left[s_0^H\left(t-l\Delta t\right)\otimes I_{M_R}\right]\sum\left(t-\tau\right)\times\left[s_0^H\left(t-l\Delta t\right)\otimes I_{M_R}\right]^H$$

$$e^{-jk\Delta\omega(t-\tau)}dtd\tau = \int_0^T\int_0^T [s_0^H\left(t-l\Delta t\right)s_0^H\left(t-l\Delta t\right)\otimes \sum\left(t-\tau\right)]\times e^{-jk\Delta\omega(t-\tau)}dtd\tau$$

As seen, Γ is a function of $\Sigma(t-\tau)$, that is, the covariance matrix of $n(t)$ and $s_0(t)$. In the literature, a simple model for $n(t)$ is to assume that its samples are uncorrelated in both spatial (along the receiving array) and temporal (along T) domains. This implies that:

$$\sum\left(t-\tau\right) = \sigma^2\delta I_{M_R}\left(t-\tau\right)$$

If $s_0(t)$ is a vector of orthonormal waveforms, it is thus immediate to verify that Γ reduces to $\Gamma = \sigma^2 I_N$. Under the assumption of uncorrelated samples in the time domain only and perfect orthonormality of the transmitted waveforms, we have:

$$\Gamma = I_{M_T} \otimes \sum_R$$

where $\sum_R$ denotes the receive spatial covariance matrix. However, the above two conditions may not be satisfied in practice. This is why in this chapter we do not make any a priori assumption on the structure of Γ. We only assume that its $(i-j)^{th}$ entry goes to zero at least polynomially fast as $|i-j|$ increases.

The autoregression moving average (ARMA) is a model of forecasting in which the methods of autoregression (AR) analysis and moving average (MA) are both applied to time series data that is well behaved. In ARMA, it is assumed that the time series is stationary, and when it fluctuates, it does so uniformly around a particular time.

The ARMA(p, q) model is a linear combination of two linear models and thus is itself still linear. Let us consider a time series model $\{x\}$ is an ARMA model of order p, q, ARMA(p, q), if:

$$x_t = \alpha_1 x_{t-1} + \alpha_2 x_{t-2} + \ldots + n_t + \beta_1 n_{t-1} + \beta_2 n_{t-2} + \ldots + \beta_q n_{t-q}$$

where $\{n_t\}$ is white noise with $E\{n_t\} = 0$ and variance σ^2.

We can straightforwardly see that by setting $p \neq 0$ and $q = 0$, we recover the AR(p) model. Similarly, if we set $p = 0$ and $q \neq 0$, we recover the MA(q) model. One of the key features of the ARMA model is that it is *parsimonious* and *redundant* in its parameters. That is, an ARMA model will often require fewer parameters than an AR(p) or MA(q) model alone.

Disturbance Model

As previously discussed, many simplified models have been proposed in the literature to statistically characterize the disturbance vector c at the output of the matched filter bank of a (co-located) MIMO radar system. We refer to [2] for a comparison among various statistical models and for a discussion about the physical simplifying assumptions underlying them. Here, we simply note that the following two main hypotheses are usually made about the statistical characterization of the disturbance vector:

- c is temporally and spatially white, and
- c is Gaussian-distributed. As we will show, advances in robust and misspecified statistics allow us to drop these two strong assumptions in favor of much weaker conditions.

To formally characterize the class of random processes to which the results of this chapter apply, we need to introduce the concepts of uniform and strong mixing random sequences. Roughly speaking, the uniform and strong mixing properties characterize the dependence between two random variables extracted from a discrete process separated by m lags. Without any claim of completeness or measure-theoretic rigor, the results of this chapter apply to any random process that satisfies a restriction on the speed of decay of its autocorrelation function. Specifically, we limit ourselves to the following class of random processes:

- The true representation of a stationary discrete and circular complex-valued unknown disturbance process can be expressed as $\{c_n : \forall n\}$.
- Then, it implies that the autocorrelation function of this kind of disturbance is provided by $r_C\left[m\right] \triangleq E\left\{c_n c_{n-m}^H\right\} = O\left(|m|^{-\gamma}\right)$, $m \in \mathbb{Z}$, $\gamma > \dfrac{\wp}{\left(\wp - 1\right)}$, $\wp > 1$.

It implies that the volume of the MIMO radar must increase with the number of transmitting (M_T) and receiving (M_R) antennas. This means that the results for the radar provided here do not apply to "space-constrained" array topologies. That said, this property of radar disturbance model is very general and allows to account for most practical disturbance models. Indeed, any (not necessary Gaussian) stable second order of statistics (SOS) ARMA(p, q), and consequently any stable SOS AR(p) [2], also satisfies this condition since the auto-correlation function of any stable SOS ARMA decays exponentially. The generality of the ARMA model is an important property because it can approximate, for p and q sufficiently large, the second-order statistics of any complex discrete random processes having a continuous power spectral density (PSD) [40, Ch. 3]. Moreover, a non-necessarily Gaussian ARMA(p, q) can model the "spikiness" of heavily tailed data as well. Another disturbance model of practical interest satisfying this condition is the Compound-Gaussian (CG) model [2]. Indeed, recall that any CG-distributed random vector c admits a representation:

$$c =_d \sqrt{\tau} m$$

for some real-valued positive random variable τ, called texture, independent of the zero-mean, N-dimensional, circular, complex Gaussian random vector, called speckle, $m \sim \mathbb{CN}(0, \Gamma)$, where Γ is its scatter matrix and $=_d$ stands for "has the same distribution as." Under a condition on the structure of Γ, it easily follows that the N entries of m can be interpreted as N random variables extracted from a circular, Gaussian, SOS ARMA(p, q) process $\{m_n : \forall n\}$, with $p, q < N$. We conclude by noticing that this condition representing disturbance is more general than the one adopted in the previous work. In fact, the asymptotic results derived in [1–2] are obtained by assuming an AR disturbance model, driven by innovations with possibly unknown pdf. Unlike [1–2], this work does not require any a priori information on the specific disturbance model; as stated above, only the polynomial decay of its autocorrelation function is needed.

Numerical Results

Numerical results are now used to validate the theoretical findings on the asymptotic properties of Wald-type Λ_{RW}. We consider a uniform linear array at the transmitter and receiver, and a single target located in the far-field. We assume that $W = I_{MT}$ and the transmitted waveforms are orthogonal, i.e., $S = I_{MT}$. We choose the radar geometry that maximizes the parameter identifiability. This is achieved by using a receiving array characterized by M_R antenna elements with an inter-element spacing of d and a transmitting array whose M_T elements are spaced by $M_R d$. This implies that:

$$a_R(\phi) = \left[1, e^{j2\pi v}, \ldots, e^{j2\pi(M_R-1)v} \right]^H$$

$$a_T(\phi) = \left[1, e^{j2\pi M_R v}, \ldots, e^{j2\pi(M_T-1)M_R v} \right]^H$$

where the spatial frequency v is provided by:

$$v \triangleq \frac{f_0 d}{c} \sin(g(\phi))$$

where f_0 is the carrier frequency of the transmitted signal, c is the speed of light and $g(.)$ is a known function of the position vector ϕ. By substituting, the vector $v(\phi) \in \mathbb{C}^N$ can be expressed as, for $i = 1, \ldots, M_R M_T$:

$$\left[v(\phi) \right]_i = e^{j2\pi(i-1)v}$$

which represents the steering vector of an equivalent phased array with N elements. Numerical results are obtained by averaging over 10^6 Monte Carlo simulations. Moreover, the truncation lag is chosen as $l = \left\lfloor N^{\frac{1}{4}} \right\rfloor$.

Scenario 1

Two different models are considered for the disturbance. The disturbance vector c is generated according to an underlying circular, SOS autoregression AR(p):

$$c_n = \sum_{i=1}^{p} \rho_i c_{n-i} + \omega_n, \quad n \in (\infty, -\infty)$$

with $p = 3$, driven by i.i.d., t-distributed innovations ω_n whose pdf ρ_ω is [1, 2]:

$$p_\omega(\omega_n) = \frac{\lambda}{\sigma_\omega^2 \pi} \left(\frac{\lambda}{\eta}\right)^\lambda \left(\frac{\lambda}{\eta} + \frac{|\omega_n|^2}{\sigma_\omega^2}\right)^{-(\lambda+1)}$$

where $\lambda \in (1, \infty)$ and $\eta = \lambda / (\sigma_\omega^2 (\lambda - 1))$ are the shape and scale parameters. Specifically, λ controls the tails of p_ω. If λ is close to 1, then p_ω is heavy-tailed and highly non-Gaussian. On the other hand, if $\lambda \to \infty$, then p_ω collapses to the Gaussian distribution. In our simulations, we set $\lambda = 2$ and $\sigma_\omega^2 = 1$. The AR(3) coefficient vector is:

$$\rho = \left[0.5e^{-j2\pi0}, 0.3e^{-j2\pi0.1}, 0.4e^{-j2\pi0.01}\right]^H$$

The normalized PSD can be expressed as:

$$S(\nu) \triangleq \sigma_\omega^2 \left|1 - \sum_{n=1}^{p} \rho_n e^{-j2\pi n\nu}\right|^{-2}, \quad p = 3$$

This is shown in Table 10.1. As seen, Scenario 1 is characterized by a disturbance whose power is mostly concentrated around the spatial frequency $\nu = 0$.

Scenario 2

To prove the robustness of the proposed function Wald-type statistics Λ_{RW} w.r.t. more general disturbance models, in Scenario 2 we increase the order of the AR process generating the disturbance vector c. Particularly, we consider a circular, SOS auto-correlation (AR) of $AR(6)$ driven (as before) by i.i.d., t-distributed innovations ω_n and characterized by the following coefficient vector:

$$\rho = \left[0.5e^{-j2\pi0.4}, 0.6e^{2\pi0.2}, 0.7e^{-j2\pi0}, 0.4e^{-j2\pi0.1}, 0.5e^{2\pi0.3}, 0.5e^{2\pi0.35}\right]^H$$

The normalized PSD is reported in Table 10.2 and shows that, differently from Scenario 1, the disturbance power is spread over the whole range of ν. Moreover, it presents more than a single peak.

TABLE 10.1
PSD of AR(3) in Scenario 1

	$10\log_{10}\left((\text{PSD}(v))/\max_v \text{PSD}(v)\right)$			
v	$v_1 = -0.2$	$v_2 = 0$	$v_3 = 0.2$	PSD [dB]
−0.5	−	−	−	~ − 21
−0.4	−	−	−	~ − 18
−0.2	−	−	−	~ − 10
0.0	−	−	−	~0.0
0.2	−	−	−	~ − 11
0.4	−	−	−	~ − 20
0.5	−	−	−	~ − 21

TABLE 10.2
PSD of the AR(6) in Scenario 2

	$10\log_{10}\left((\text{PSD}(v))/\max_v \text{PSD}(v)\right)$			
v	$v_1 = -0.2$	$v_2 = 0$	$v_3 = 0.2$	PSD [dB]
−0.5	−	−	−	~ − 9
−0.4	−	−	−	~ − 9.5
−0.2	−	−	−	~ − 7
0.0	−	−	−	~0.0
0.2	−	−	−	~ − 4
0.4	−	−	−	~ − 5
0.5	−	−	−	~ − 9

Both Scenarios

In both scenarios, the disturbance process is normalized to have $\sigma^2 = r_C[0] = 1$. Under hypothesis H_1, the SNR is simply defined as SNR $10\log_{10}\left(\frac{|\alpha|^2}{\rho^2}\right)$.

Performance Analysis

Since the disturbance PSD in Tables 10.3 and 10.4 is not constant w.r.t the spatial frequency v, the performance of decision statistics $\Lambda_{RW}(x)$ at different values of x will be evaluated for three different values of spatial frequency v corresponding to different disturbance power density levels (i.e., three different target DOAs): $v_1 = -0.2, v_1 = -0$ and $v_1 = 0.2$. Tables 10.3 and 10.4 provide the PFA of the proposed robust Wald test for the three considered values of v.

As we can see, in both scenarios Λ_{RW} achieves the nominal value of 10^{-4} for $N \geq 10^4$. Moreover, in this Massive MIMO regime, i.e., for $N \geq 10^4$, the progress of

TABLE 10.3

Estimated PFA as a Function of the Virtual Spatial Antenna Channels N in Scenario 1

$N = M_R M_T$	$P_{FA} - AR(3)$			
	$v_1 = -0.2$	$v_2 = 0$	$v_3 = 0.2$	Nominal P_{FA}
10^2	~6×10^{-2}	~2×10^{-3}	~6×10^{-2}	10^{-4}
10^3	~1×10^{-3}	~2.2×10^{-4}	~6×10^{-3}	10^{-4}
10^4	~1.3×10^{-4}	~1.2×10^{-4}	~1.2×10^{-2}	10^{-4}
10^5	~1.05×10^{-4}	~1.05×10^{-4}	~1.05×10^{-4}	10^{-4}

TABLE 10.4

Estimated PFA as a Function of the Virtual Spatial Antenna Channels N in Scenario 2

$N = M_R M_T$	$P_{FA} - AR(6)$			
	$v_1 = -0.2$	$v_2 = 0$	$v_3 = 0.2$	Nominal P_{FA}
10^2	~9×10^{-2}	~3.5×10^{-3}	~1.5×10^{-3}	10^{-4}
10^3	~3×10^{-3}	~6×10^{-4}	~4×10^{-4}	10^{-4}
10^4	~1.2×10^{-4}	~1.5×10^{-4}	~1.2×10^{-2}	10^{-4}
10^5	~1.05×10^{-4}	~1.05×10^{-4}	~1.05×10^{-4}	10^{-4}

the PFA curves for different scenarios and for different values of spatial frequency are almost identical. This provides a numerical validation of the theoretical result provided.

Table 10.5 considers Scenario 1 and illustrates the estimated and the closed-form expression of P_D given in Corollary 1 for three distinct SNR values. Particularly, we assume SNR = -20, -10, and -5 dB. As seen, the P_D tends to 1 as the number of virtual spatial antenna channels N increases.

TABLE 10.5

Estimated and Nominal P_D as a Function of the Virtual Spatial Antenna Channels N in Scenario 1 for Different SNR Values, Spatial Frequency $v = 0$, and Nominal $P_{FA} = 10^{-4}$

$N = M_R M_T$	$P_D - AR(3)$			
	SNR = -20 dB	SNR = -10 dB	SNR = -5 dB	Nominal P_D
10^2	~0	~0.5	~0.45	~0.05
10^3	~0.05	~0.9	~1	~1
10^4	~0.95	~1	~1	~1
10^5	~1	~1	~1	~1

Specifically, for SNR ≥ 20 dB the P_D approaches 1 for $N \geq 10^4$. From the above, it is also immediate to verify that the P_D estimated through Monte-Carlo runs is in perfect agreement with the theoretical one provided in Corollary 1. We conclude by noticing that similar numerical results have been obtained for the CG disturbance model discussed in Subsection II-B. Since they are perfectly in line with the numerical analysis reported above, we decided not to include them here due to lack of space.

SUMMARY

This chapter considers a co-located MIMO radar with M_T transmitting and M_R receiving antennas and aims at deriving a detector that satisfies preassigned performance requirements without relying on the four assumptions above. Inspired by the recent developments in Massive MIMO communications, we aim at exploring the potential benefits of having a very large number of antennas. Particularly, we assume that a single time snapshot, $K = 1$, is collected, and operate in the asymptotic regime where the number of virtual spatial antenna channels $N = M_T M_R$ grows unboundedly, i.e., $N \to \infty$. This makes the three assumptions A1, A3, and A4 no longer needed. Advances in robust and mis-specified statistics are used to dispose of the cumbersome and unrealistic assumption A2. By adopting a very general disturbance model taking into account the spatial correlation structure of the observed samples, we propose a robust Wald-type detector that is asymptotically distributed, when $N \to \infty$, as a χ−squared random variable (under both H_0 and H_1) irrespective of the actual and unknown disturbance pdf $p_C(c)$. This asymptotic result is achieved without the need of any secondary dataset. Although the theoretical findings of this chapter are valid for a very general disturbance model, numerical results are provided for two non-Gaussian, stable auto-regressive disturbance models of order $p = 3$ and 6. It turns out that a preassigned value of probability of false alarm (PFA), $P_{FA} = 10^{-4}$ is achieved for $N = M_T M_R \geq 10^4$ with both models. This number of virtual spatial antenna channels defines what we call the Massive MIMO regime of the radar system. Compared to our previous paper [1], the main difference lies in the absence of any a priori knowledge of the disturbance model. In fact, in [1] the analysis was developed for an auto-regressive model of order 1, but with no a priori knowledge of its statistics.

The MIMO paradigm has been the subject of intensive research over the past 15 years in radar signal processing. Initially introduced in wireless communications as a new enabling technology, the MIMO framework has been recognized to have a great potential in boosting the capabilities of classical antenna array systems. Based on the array configurations used, MIMO radars can be classified into two main types. The first type uses widely separated antennas (so-called distributed MIMO) to capture the spatial diversity of the target's RCS. The second type employs arrays of closely spaced antennas (so-called co-located MIMO) to coherently combine the probing signals in certain points of the search area. Hybrid configurations are also possible.

While the advantages in terms of spatial resolution, parameter identifiability, direction-of-arrival estimation, and interference mitigation have been largely investigated in the MIMO radar literature, the potential benefits that many virtual spatial antenna channels can bring into the target detection problem in terms of robustness with respect to the generally unknown disturbance model have not been explored yet.

Surprisingly, not only the highly desirable robustness property has been somehow disregarded but even the availability of reliable, nontrivial, disturbance models is scarce. Remarkable exceptions to the mainstream Gaussianity assumption have been recently discussed in other papers [1]. Particularly, the performance of the adaptive normalized matched filter (ANMF), exploiting robust estimators for the disturbance covariance matrix, has been investigated with non-Gaussian disturbance.

Specifically, random matrix tools have been used to obtain asymptotic approximations of the probabilities of false alarm and misdetection of the ANMF for the regime in which both N and K go to infinity with a nontrivial ratio N/K. Similar random matrix tools have been adopted in [2] to derive some asymptotic (in random matrix regime) results about the direction-of-departure and direction-of-arrival estimation in a non-Gaussian disturbance setting. Again, the random matrix machinery has been exploited in [2] to investigate the asymptotic performance of a GLRT detector in cognitive radio applications. Specifically, the HT problem is tackled, but the steering vector v_k that represents the set of phase-delays for an incoming wave at each sensor element, and α_k are assumed unknown. In the same spirit of [2] has recently investigated the possibility to derive eigenvalues-based detectors for HT problem for spectrum sensing and sharing in cognitive radio. However, the disturbance was assumed to be a simple white Gaussian process with distribution a priori known, up to its statistical power. The asymptotic analysis requires that both N and K grow unboundedly. This is different from this paper [1] where the temporal dimension K is kept fixed; specifically, we assume to collect a single snapshot vector.

The detection problem in co-located MIMO radar systems was analyzed in this chapter. A robust Wald-type detector was proposed, and its asymptotic performance investigated when the virtual spatial antenna channels $N = M_T M_R$ goes to infinity. Specifically, the CFAR property of the proposed detector for the asymptotic regime $N \to \infty$ and the wide family of disturbance processes satisfying Assumption 1 was mathematically proved and validated through numerical simulations. The purpose of analyzing the asymptotic performance when $N \to \infty$ is not that we advocate the deployment of radars with a nearly infinite number of virtual antennas. The importance of asymptotic is instead what it tells us about practical systems with a finite number of antennas. Indeed, our main results imply that we can always satisfy performance requirements by deploying sufficiently many virtual antennas N, without any a priori knowledge of the disturbance statistics.

Our numerical results showed that a preassigned value of probability of false alarm (PFA) (PFA = 10^{-4}) can be achieved with $N = M_T M_R \geq 10^4$ with non-Gaussian, stable auto-regressive disturbance models of order $p = 3$ and 6. This defines the so-called Massive MIMO regime of the radar. In Massive MIMO communications, linear combining and precoding schemes can eliminate the interference as the number of antennas grows unboundedly even with imperfect knowledge of propagation channels. We showed that a large-scale MIMO radar yields a target detector, which is robust to the unknown disturbance statistics. We foresee that further breakthroughs can be obtained by extending the Massive MIMO concept to other radar problems [2]. Clearly, by using very large arrays for waveform design, one can radically improve the spatial diversity gain and spatial resolution for target detection, parameter estimation, and interference rejection. However, the lesson learned from the last decade of research in communications is that Massive MIMO is not merely a system with many antennas but rather a paradigm shifts with regards to the modeling, operation, theory, and implementation of MIMO systems. Our vision is that this paradigm can be applied also to radars, and that it opens new research directions.

REFERENCES

1. Fortunati, S. et al., "Massive MIMO Radar for Target Detection," January 2020, arXiv:1906.06191v4 [eess.SP] 15 Jan 2020.
2. He, Q. et al., "MIMO Radar Moving Target Detection in Homogeneous Clutter," *IEEE Transaction on Aerospace and Electronic Systems*, Vol. 46, No. 3, 1290–1301, July 2010.

11 Massive MIMO Radar Scaling

OVERVIEW

Massive MIMO radars that utilize many transmit and receive antennas can provide range extension and fine-grained angular resolution in sensing. Current signal processing strategies for MIMO radar impose a strict trade-off between range and angular field of view, preventing efficient scaling to Massive MIMO platforms [1]. However, we present a compressive signal processing framework to sidestep this trade-off. Relying on the sparsity of the scene in angular domain, we take advantage of compressive beam scanning to provide high-resolution direction estimates with a small number of beacons that scales logarithmically with array size. This approach enables scaling to Massive MIMO frontends while maintaining a small, almost constant frame interval, thereby facilitating high-resolution direction estimation and range extension without sacrificing the field of view or imaging speed [1].

Utilizing many transmit and/or receive antennas enables a radar system to provide high-resolution direction information in addition to range and Doppler, which is immensely beneficial to many applications such as environment awareness for autonomous vehicles, target classification, and close-range gesture recognition. Using many transmit elements can also provide power combining gains and expand the sensing range and SNR. At mm-Wave and THz frequencies, Massive MIMO frontends with hundreds or thousands of elements can fit on small platforms, and several prototypes have been built for communication applications. By repurposing these frontends for sensing, the many benefits of Massive MIMO radar can be realized.

Unfortunately, current signal processing strategies for MIMO radar impose a strict trade-off between range and field of view that prevents efficient scaling to Massive MIMO platforms. Conventional MIMO radar systems typically utilize time-interleaved sensing (for digital transmit arrays) or directional beam scanning (for digital or analog arrays) to obtain direction estimates. In the former approach, transmitters take turns broadcasting chirp sequences, and the received measurements are aggregated to emulate the full MIMO response. Since only one transmitter is active at any time, this method does not realize the full transmit power combining gain and range extension of MIMO. This is also true for the case of orthogonalizing transmitter signals in domains other than time. Furthermore, since the number of chirp bursts in each frame scales proportionally to array size, this strategy is not scalable to very large transmit arrays due to the limited coherence time of the scene. The beam scanning approach, on the other hand, utilizes the power combining gain of the array, but it too scales poorly to large arrays with narrow beams, as covering a large (fixed) field of view with narrow beams requires several chirp sequence transmissions that scales linearly with array size. Thus, to allow sufficiently fast scanning, one must limit

DOI: 10.1201/9781003499480-11

either the transmit array size (hence, range extension capacity) or the field of view that is scanned in each frame.

To circumvent these limitations, we propose an alternative approach that, relying on the angular sparsity of the scene in each range-Doppler bin, takes advantage of compressive signal processing to provide high-resolution direction estimates with a small number of subframes (chirp bursts) that scales logarithmically with array size. We borrow ideas from the extensive literature on compressive channel estimation for MIMO communication applications and overlay compressive estimation in the spatial domain with the conventional range and Doppler estimation of continuous wave frequency modulated (CWFM) radar. Our numerical results show that with as few as 10 compressive beacons, target directions can be estimated with the full resolution of a 128-element transmit aperture, providing an order of magnitude reduction in the frame size compared to beamformed scanning.

The wireless channel, especially at higher frequencies where Massive MIMO frontends are feasible, consists of only a small number of paths and is therefore sparse in the spatial frequency domain. Thus, compressive channel estimation from a small number of random projections has been employed in communication applications for low-overhead channel tracking on large, phased arrays with RF beamforming. Since the spatial frequency of each path lies on a continuum, conventional techniques for sparse vector estimation are not suitable for channel estimation. An algorithm is proposed for off grid compressive estimation using orthogonal matching pursuit with Newton refinement of parameters between null space projections. As Massive MIMO frontends are deployed for radar imaging, adapting the compressive signal processing techniques developed in communication literature to sensing platforms is increasingly of interest.

In sensing literature, "compressive" or "sparse" MIMO radar typically refers to one of two cases. In the first case, target sparsity in the angle-Doppler-range space is exploited to compress the received signal via a linear transformation. These methods require transmit elements to transmit independent signals and are therefore incompatible with the large RF-beamformed arrays that we propose to repurpose for sensing. The second case refers to random under-sampling of the spatial domain, i.e., sparsity in terms of hardware realization wherein transmit and receive array elements are placed at random so that a large aperture is sampled compressively with fewer antenna elements.

In contrast, our focus here is on compressive sampling to take advantage of the sparsity of the scene in the spatial frequency domain, which allows us to observe the entire field of view with fewer beams than would be required by beamformed scanning. We note that the two concepts are complementary: compressive sampling in spatial frequency domain using large transmit arrays, as proposed here, can be combined with randomly spaced receive antennas to efficiently synthesize large apertures.

SYSTEM MODEL

We consider a CWFM MIMO radar system which consists of an N_{tx}-element phased-array transmitter and an N_{rx}-element, digital receiver, with the frame structure depicted in Figure 11.1. For each chirp transmission, the return signal at the receiver is mixed

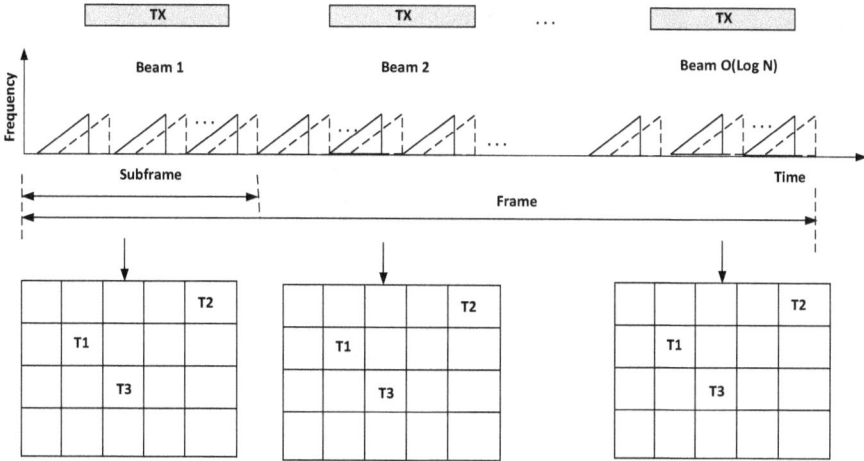

FIGURE 11.1 Frame structure for compressive radar.

with the transmitted chirp and sampled at a rate of B Hz, equivalent to a sampling period of $T_s = 1/B$. The frame is divided into K subframes, each containing a sequence of n_0 chirps. On each subframe, a different randomly generated beamforming weight vector is applied to form a compressive beacon. We use a standard complex Gaussian distribution to draw the beamforming weights, resulting in radiation patterns that spray power randomly in the angular domain, as shown in Figure 11.1.

Thus, in each subframe, the return signal from target t is multiplied by the complex subframe beacon response to spatial frequency:

$$\omega_t^{tx} = \frac{2\pi d_{tx}}{\lambda} \sin \theta_t$$

where θ_t is the direction of target t, d_{tx} is the transmit array's inter-element spacing, and λ is the carrier wavelength. The target's spatial frequency on the fully digital receiver is derived by taking a spatial Fast Fourier Transform (FFT) and satisfies:

$$\omega_t^{rx} = \frac{2\pi d_{rx}}{\lambda} \sin \theta_t$$

We assume here that all targets are sufficiently far away to have angle of departure (AOD) = angle of arrival (AOA) = θ_t.

UNAMBIGUOUS BOUNDS AND RESOLUTION OF PARAMETERS

In the system described above, Targets are separated into range, Doppler (radial speed), and spatial frequency bins with a nominal resolution equal to the corresponding FFT grid length of each domain. In evaluating our algorithm, we regard any estimate of range, Doppler, or spatial frequency that lies within one nominal FFT grid length of the true value as a "successful" detection, and any estimate outside this span is considered incorrect. The nominal resolution of each parameter is described as follows.

Range: The minimum and maximum unambiguously measurable range are defined by the receiver sampling rate, B (Hz), and chirp frequency slope, α (Hz/s), as:

$$R_{\min} = 0, \qquad R_{\max} = \frac{Bc}{2\alpha}$$

where c is the speed of light. The nominal range resolution is determined by the length of the chirp. Assuming the system produces L samples per chirp, the range-FFT grid size will be $g_{\text{range}} = \dfrac{2\pi}{L}$ equivalent to a nominal range resolution of:

$$\delta_R = \frac{R_{\max}}{L} = \frac{Bc}{2\alpha L} = \frac{c}{2T_{\text{chirp}}}$$

where $T_{\text{chirp}} = LT_s$ is the chirp duration in seconds.

Speed: Similarly, the minimum and maximum unambiguous Doppler measurement is determined by the delay between consecutive chirps. Assuming a duration of T_{gap} between the end of one chirp and the start of the next, the largest Doppler frequency is:

$$\frac{\pi}{T_{\text{chirp}} + T_{\text{gap}}}$$

which translates to the radial speed measurement limits,

$$V_{\max} = -V_{\min} = \frac{\lambda}{n_0 \left(T_{\text{chirp}} + T_{\text{gap}}\right)}$$

The nominal Doppler FFT grid size is thus equal to,

$$\frac{2\pi}{n_0 \left(T_{\text{chirp}} + T_{\text{gap}}\right)}$$

corresponding to nominal radial speed resolution,

$$\delta_V = \frac{\lambda}{2n_0 \left(T_{\text{chirp}} + T_{\text{gap}}\right)} = \frac{\lambda}{2T_{\text{subframe}}}$$

where T_{subframe} is the total subframe duration.

Angle: Unambiguous spatial frequency measurements can be made in the range $(-\pi, \pi)$ of with a grid size of,

$$\delta_{\omega,tx} = \frac{2\pi}{N_{tx}}$$

on the transmitter (used for AOD estimation), and

$$\delta_{\omega,rx} = \frac{2\pi}{N_{rx}}$$

on the receiver (AOA estimation). These are translated to angular resolution using above equations, respectively, as

$$\delta_{\theta,tx/rx} = \frac{\lambda}{2\pi d_{tx/rx} N_{rx} \cos\theta} \delta_{\omega,tx/rx}$$

APERTURE EXTENSION WITH SPATIAL SUBSAMPLING

With an appropriate choice of element spacing in the transmitter and receiver, the effective sensing aperture can be increased up to $N_{tx} N_{rx} \lambda/2$. Let us assume that the transmitter has half-wavelength spacing while the receiver is spatially under sampled by a factor of N_{tx} (i.e., the inter-element spacing at the receiver is $N_{tx}\lambda/2$). The translation of transmit spatial frequency to angle is thus the one-to-one transform,

$$AOD = \sin^{-1}\frac{\omega_{tx}}{\pi} \tag{1}$$

whereas the under sampled receiver will see "grating lobes" equivalent to the one-to-N_{tx} mapping,

$$AOA = \sin^{-1}\frac{\omega_{tx} + 2\pi n}{N_{tx}\pi}, \quad \exists n \in \left\{0,\dots,\left(N_{tx}-1\right)\right\} \tag{2}$$

Since AOA = AOD = θ, the angle recovered in (1) can be used to resolve n in (2). Note that the resolution provided by (2) is higher than that of (1) by a factor of (approximately) N_{rx}, meaning the overall direction resolution provided in this way is equivalent to that of an $N_{tx} N_{rx}$-element array with half wavelength spacing.

ALGORITHM

Figure 11.2 shows a sketch of our proposed imaging process for one frame. This process comprises the following steps.

• **Step 1: Range-Doppler measurement**

In this step, we first take a 2D FFT of each subframe's measurement matrix to arrive at a set of K range-Doppler heatmaps, as shown in Figure 11.1. By adding up the power in each range-Doppler bin across subframes, we arrive at the "aggregate power heatmap," an example of which is shown in Figure 11.2. We find the strongest peak in this matrix, obtain an off grid estimate of its range and Doppler frequency, using a

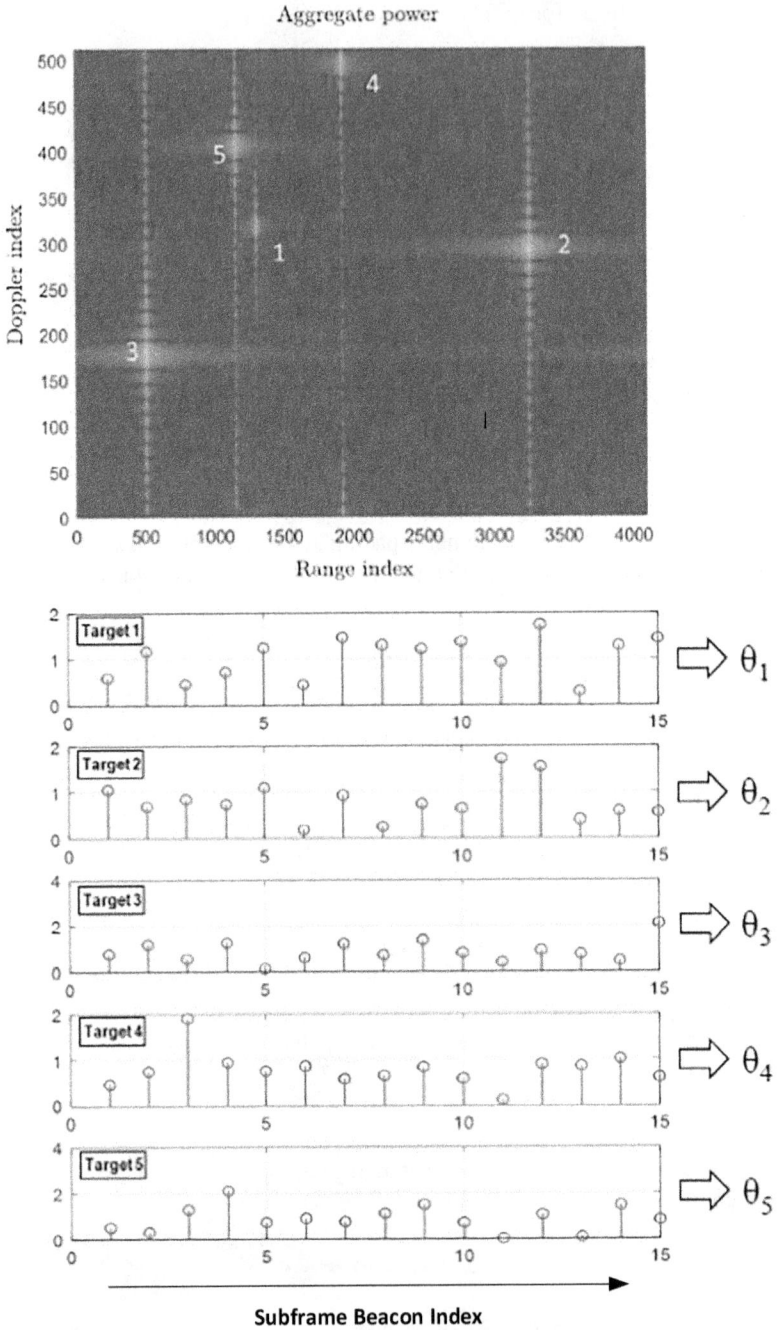

FIGURE 11.2 Aggregated range-Doppler bin powers and compressive direction estimation for a sample scenario.

super-resolution technique such as gradient descent, Newton refinements, and local oversampling of the Fourier transform to maximize the aggregate power cost function. We then subtract its response from the measurement matrices of all subframes and proceed to find the next strongest peak in the aggregate power heatmap of the residual response. This process is repeated until no strong peaks relative to the noise floor are observed in the residual heatmap. We denote by N_{bi} the number of extracted bins, and by $\hat{r}_i$ and $\hat{d}_i$ the super-resolved range and Doppler frequency estimate of bin i, respectively. In principle, each extracted range-Doppler bin can contain one or more targets, which may be separable in the angular domain. However, when simulating point targets randomly dispersed in the range-Doppler-angle space, this is highly unlikely.

- **Step 2: Doppler correction and spatial frequency estimation**

For each significant bin, we observe a set of K complex amplitudes over the compressive subframes. We denote the observation for bin i on subframe k by y_i^k. In a static setting, this complex amplitude is the sum of the response of compressive beacon k to all the targets inside bin i, and therefore the spatial frequencies of all targets in bin i can be found by using conventional an off-grid compressive estimation algorithm, such as NOPM [1], on the observation vector $y_i = \left[y_i^1, \dots, y_i^K \right]^T$. For moving targets, however, the phase of each beacon response is modulated by Doppler drift across subframes, which must be corrected for each subframe measurement to obtain the true compressive spatial projections. For bin i with Doppler frequency d_i, the Doppler phase drift encountered by subframe k is equal to $(k-1)n_0 d_i$. We use our estimate $\hat{d}$ to undo this phase offset and obtain the corrected compressive measurements as

$$\tilde{y}_i^K = y_i^K e^{-j(k-1)n_0 \hat{d}_i}, \quad k = 1, \dots, K$$

The corrected measurement vector $\tilde{y}_i = \left[\tilde{y}_i^1, \dots, \tilde{y}_i^K \right]^T$ is used for compressive estimation of the spatial frequencies of targets inside bin i. Note that, for successful phase correction, the Doppler frequency estimate must be accurate enough to extrapolate well over K subframes, which is why off-grid Doppler estimation in the previous step is crucial for the success of the algorithm.

- **Step 3: Doppler refinement (optional)**

Once a target's direction (spatial frequency) is identified, its compressive beacon response on the K subframes can be calculated. By undoing this spatial beacon response on all subframes, we arrive at an effectively Kn_0-chirp-long subframe that can be used to produce a more accurate off-grid Doppler estimate for that target. This time, instead of using the (noncoherently combined) aggregate power cost function, we can combine likelihoods coherently across subframes and achieve higher estimation accuracy relative to our initial estimate.

NUMERICAL RESULTS

In this section, we present simulation results for a MIMO radar system with an $N_{tx} = 128$ element phased-array transmitter with half-wavelength element spacing, and a single element receiver ($N_{rx} = 1$). Each frame contains K subframes where K is varied between 5 and 15. Spatial beamforming weights for each subframe are drawn independently from a standard complex Gaussian distribution. Each subframe consists of $n_0 = 32$ chirps, and each chirp of $L = 256$ samples. The distance, radial speed, and direction of targets are drawn randomly such that their corresponding range, Doppler, and spatial frequencies are uniformly distributed over the unambiguous range, save for a single FFT-grid-length gap at each end to prevent wraparound ambiguity. In each realization, we model 10 targets with 10 dB dynamic range in target signal strengths. Variations in the strength of target responses arise from differences in radar cross section and distance, both of which are modeled in our simulations.

Simulation result shows the estimation success rate for each parameter as a function of the per-symbol beamformed SNR (which is a factor of N_{tx} higher than the effective SNR for a single-element transmitter) for different values of K. Successful estimation of a parameter implies an absolute estimation error smaller than the nominal FFT grid size. These results are averaged on a per-target basis over 10 realizations, or 100 targets in total. We see that success rate approaches 1 at around −15 dB beamformed *SNR*, and even with 10 subframes, or 320 chirps per frame, target directions are accurately recovered with the full resolution of a 128-element aperture.

By generating point targets randomly, we are all but guaranteeing that each range-Doppler bin contains at most one target, and multi-target bins are very unlikely to occur. This is likely to be true in real-word systems in all but the zero-Doppler bins where a lot of static clutter is present (assuming, of course, that the radar platform is also static). As a higher number of targets are present in one bin, the number of beacons required to accurately estimate their spatial frequencies will increase proportionally. This is perhaps best managed by generating a new set of compressive beacons for each frame (instead of repeating the same predefined set) so that, by combining several frames, the number of subframe measurements can be adapted to the scene coherence time, and even to the radial speed and number of targets on a per-target or per-bin basis.

SUMMARY

We have presented a compressive signal processing framework to sidestep the trade-off between range and field of view that plagues conventional MIMO radar systems. Relying on the angular sparsity of each range-Doppler coordinate of the scene, we take advantage of compressive beam scanning to provide high-resolution direction estimates with a small number of beacons that scale logarithmically with array size. The proposed approach enables scaling to Massive MIMO frontends while maintaining a small, almost-constant frame size, thereby facilitating high-resolution direction estimation and range extension without sacrificing the field of view or agility of the imaging system. Numerical results have been presented for simulated point targets on a CWFM radar system with 128 RF-beamformed transmit elements and a single receiver, showing accurate direction estimates are obtained with as few as ten compressive beacons. In future work, we will investigate the potential for the application of this approach to extended targets with complicated micro-Doppler signatures, and implications for object classification and gesture recognition.

REFERENCE

1. Rasekh, M. E. and Madhow, U., "Scaling Massive MIMO Radar via Compressive Signal Processing," presented at the 2021 55th Asilomar Conference on Signals, Systems, 2021.

12 Ultra-Massive MIMO Radar Sensors and Multiuser Communications

OVERVIEW

Wireless communications and sensing at terahertz (THz) band are increasingly investigated as promising short-range technologies because of the availability of high operational bandwidth at THz. To address the extremely high attenuation at THz, ultra-massive multiple-input multiple-output (MIMO) antenna systems have been proposed for THz communications to compensate propagation losses [1]. However, the cost and power associated with fully digital beamformers of these huge antenna arrays are prohibitive. In this chapter, we develop wideband hybrid beamformers based on both model-based and model-free techniques for a new group-of-subarrays (GoSA) ultra-Massive MIMO structure in low-THz band. Further, driven by the recent developments to save the spectrum, we propose beamformers for a joint ultra-Massive MIMO radar communications system, wherein the base station serves multiantenna user equipment (RX) and tracks radar targets by generating multiple beams toward both RX and the targets. We formulate the GoSA beamformer design as an optimization problem to provide a trade-off between the unconstrained communications beamformers and the desired radar beamformers. To mitigate the beam split effect at THz band arising from frequency-independent analog beamformers, we propose a phase correction technique to align the beams of multiple subcarriers toward a single physical direction. Additionally, our design also exploits second-order channel statistics so that an infrequent channel feedback from the RX is achieved with less channel overhead. To further decrease the ultra-Massive MIMO computational complexity and enhance robustness, we also implement deep learning (DL) solutions to the proposed model-based hybrid beamformers. Numerical experiments demonstrate that both techniques outperform the conventional approaches in terms of spectral efficiency and radar beampatterns, as well as exhibiting less hardware cost and computation time.

While considering beamforming at (0.1–1 THz) THz, following unique characteristics differentiate the THz-band (low THz band: 0.1–1 THz) from mm-Wave (24–77 GHz):

- The path loss in THz channels includes both spreading loss and molecular absorption. The latter is more significant at THz than mm-Wave. The severe path loss is compensated for by deploying much larger antenna arrays that require more creative choices for subarray geometries.

 DOI: 10.1201/9781003499480-12

- Both line-of-sight (LoS) and non-LoS (NLoS) paths are significant at mm-Wave. However, at THz, the NLoS paths have insignificant contribution to the received power. This leads to LoS-dominant and NLoS-assisted communications scenario at THz.
- Significant attenuation implies shorter ranges for THz systems than mm-Wave. Additionally, the THz-specific molecular absorption effect leads to the channel bandwidth varying with the range. This requires stricter limitations on the deployment and coverage of THz communications and radar.
- THz channels are extremely sparse in the angular domain and have smaller angular spread than mm-Wave. Therefore, it is feasible to adopt subarray models such as array-of-subarrays (AoSA) and GoSA to bring down the high-frequency hardware and computational complexities. These subarray structures have been shown to overcome the limited communications range while also retaining a reasonable spatial multiplexing gain.
- THz channel exhibits peculiarities such as misalignment and phase uncertainties in phase-shifters. The frequency-independent analog beamformers largely used in the broadband mm-Wave communications may lead to beam split effect in THz channels: the generated beams split into different physical directions at each subcarrier due to ultra-wide bandwidth and large number of antennas. This phenomenon has also been called beam squint in mm-Wave works. While both beam squint and beam split pertain to a similar phenomenon, the latter has more severe achievable rate degradation. In particular, the main lobes of the array gain corresponding to the lowest and highest subcarrier frequencies do not overlap at THz at all, while there is a relatively small deviation in mm-Wave band. While the beam squint depends on the array size, the beam split is a function of both wide bandwidth and large arrays.
- The THz-band has several other propagation and scattering effects. The specular scattering is less dominant, and partially diffused scattering is a more appropriate model for THz. The THz channel coherence time may be smaller than the symbol time and the channel may no longer be considered time-invariant. Further, the molecular absorption is range-dependent leading to very different noise models than mm-Wave. At much shorter distances, the THz bandwidth is also range-dependent and spherical wave propagation must be accounted for.

In this work, we consider the THz hybrid beamforming problem by incorporating the above-mentioned unique THz features except the last scattering characteristic, which is beyond the scope of this chapter. Contrary to prior works, we focus on the THz wideband hybrid beamforming for an ultra-Massive MIMO Joint Radar Communications (JRC) configuration. To this end, we develop both model-based and model-free techniques that rely on both channel state information (CSI) and channel covariance matrix. To reduce the hardware complexity, we propose a GoSAs structure, in which the antenna elements in the same subarray are connected to the same phase-shifter. In other words, GoSA forms an array of subarray-of-subarrays, which is different from the prior AoSA structures. Thus, the proposed structure employs even fewer phase-shifters than that of fully connected arrays or partially connected AoSA structures, while providing satisfactory radar and communications performance in terms of the beampattern and the spectral efficiency, respectively. To improve the radar performance, the higher degrees of freedom (DoF) are provided by using partially connected GoSAs. Nevertheless, partially connected structure has poor spectral efficiency performance compared to the fully connected array. Hence, we suggest a partially connected with overlapped (PCO) GoSA structure for performance improvement. To design the hybrid beamformers based on the PCO structure, we propose a modified version of the manifold optimization (MO)-based alternating minimization (AltMin) technique, which is originally suggested to solve the beamformer design problem in fully connected arrays. Our numerical experiments show that the proposed approach has much lower hardware complexity than the state-of-the-art techniques, while maintaining satisfactory radar and communications performance. The main contributions are:

- THz ultra-Massive MIMO JRC. Our proposed JRC approach based on ultra-Massive MIMO is inspired by recent advancements in THz technologies and is, therefore, closer to practical feasibility. It is particularly helpful for short range vehicular applications, wherein the ego vehicle simultaneously communicates with the user equipment and detect/track the radar targets with higher angular resolution due to high beamforming gain of using ultra-Massive number of antennas.
- Model-based THz hybrid beamforming. Previous research examined THz hybrid beamforming without ultra-Massive MIMO. Our optimization-based hybrid beamforming for ultra-Massive MIMO relies on both CSI and channel covariance matrix. While the former provides higher spectral efficiency, the latter has lower channel overhead at the cost of slight performance loss.
- Novel GoSA structure. We propose GoSA structure to lower the hardware cost which could be high for THz systems due to use of many antennas. GoSA allows us to employ fewer number of phase-shifters as compared to AoSA. We analyze the performance of GoSA with respect to several design parameters, such as the number of antennas and the antenna spacing. To provide a trade-off between the hardware complexity and the spectral efficiency, PCO-based analog precoder is proposed based on modified MO method.
- Beam split correction. We present a hardware-efficient approach to correct the beam split effect in THz channels arising from their ultra-wide bandwidth. While prior works consider an additional time-delay network for this operation, the proposed approach effectively mitigates the beam split effect without requiring such a complex structure.
- DL solutions. We design two learning models using convolutional neural networks (CNNs), one of which is employed to estimate the direction of the radar targets, whereas the other is used to design the hybrid beamformers. While DL-based beamforming techniques have been proposed earlier, THz JRC hybrid beamformer design remains unexamined in prior literature.
- The Kronecker and elementwise Hadamard product are denoted by $\otimes$ and $\odot$, respectively. The notation expressing a convolutional layer with N filters/channels of size $D \times D$ is given by $N @ D \times D$.

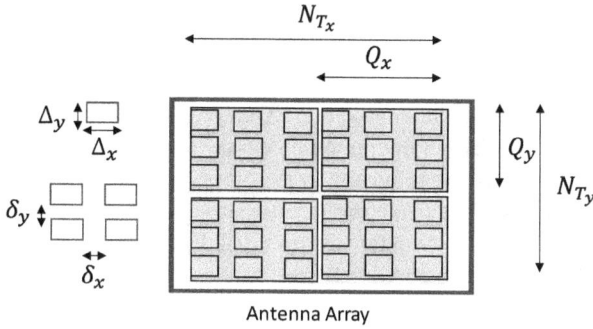

Antenna Array

FIGURE 12.1 A radar communications system for a vehicle-to-vehicle (V2V) and vehicle-to-device (V2D) scenario, wherein a single THz radar communications unit, with a $N_T = N_{Tx} \times N_{Ty}$ antenna array, is mounted onto a vehicle to simultaneously transmit toward both communications receiver and vehicular targets.

SYSTEM MODEL

We consider a wideband ultra-Massive MIMO architecture in the context of a JRC system for a vehicle-to-vehicle (V2V) and vehicle-to-device (V2D) scenario, in which the transmitter (TX) senses the environment via probing waveforms to the targets and communicates with the receiver (RX), as illustrated in Figure 12.1. The antenna arrays at the TX and the RX employ graphene-based plasmonic nano-antennas, which are placed on a metallic surface layer, with a dielectric layer between them. The antennas form GoSA structure as each subarray consists of $Q_x \times Q_y$ uniform rectangular arrays (URAs) with $Q = Q_x Q_y$ antennas, as shown in Figure 12.1.

Also, there are $N_T = N_{Tx} N_{Ty}$ and $N_R = N_{Rx} N_{Ry}$ subarrays of size Q at the TX and RX, respectively, which form an $N_T Q \times N_R Q$ ultra-Massive MIMO transceiver architecture. In each $Q_x \times Q_y$ subarray, the antenna spacing along the x-and y-axis are δ_x, δ_y and the distance between each subarray are Δ_x, Δ_y respectively.

COMMUNICATIONS MODEL

In the downlink, the TX with N_T subarrays, each of which has Q antenna elements, aims to transmit N_S data streams toward the RX in the form of:

$$s[m] = \left[s_1[m], \ldots, s_{N_S}[m] \right]^T$$

by using hybrid analog and digital beamformers with NRF RF chains, where:

$$E\left\{ s[m] s^H[m] \right\} \text{ and } N_S \leq N_{RF}$$
$$m \in \mathcal{M} = \{1, \ldots, M\}$$

M is the number of subcarriers. Due to beamforming at subarray level, each subarray of size Q generates a single beam. This is done by connecting the Q antennas in each subarray to a single phase-shifter to lower the hardware complexity. Thus, the TX first applies subcarrier-dependent $N_{RF} \times N_S$ RF baseband precoder $F_{BB}[m]$. The signal is, then, transformed to the time-domain via M-point inverse fast Fourier transform (IFFT). After adding the cyclic prefix, the TX employs a subcarrier-independent RF precoder $F_{RF} \in \mathbb{C}^{N_T \times N_{RF}}$ by employing N_T phase-shifters.

Hybrid beamforming-based transmitter structures for (a) fully connected, (b) partially connected AoSAs, and (c) partially connected GoSAs architectures. While all the architectures employ $N_T Q$ antennas with N_{RF} RF chains, each antenna is connected to each RF chain via combiners in the fully connected model with $N_T Q N_{RF}$ phase-shifters. In partially connected AoSA, the same RF chain is connected to $\bar{N}Q$ $\left(\bar{N} = \dfrac{N_T}{N_{RF}} \right)$ antennas with $N_T Q$ phase-shifters totally. In partially connected GoSA model, each RF chain is connected to $\bar{N}Q$ antennas while each phase-shifter is connected only Q antennas, introducing N_T GoSAs with only N_T phase-shifters.

Also, each group consists of $\bar{N}$ subarrays of size Q. The main difference between AoSA and GoSA is that each RF chain is connected to $\bar{N}Q$ phase-shifters in the former, while each RF chain is connected to only $\bar{N}$ phase-shifters in the latter. Hence, the number of phase-shifters in GoSA is Q times lower than that of AoSA. In GoSA, we assume that the antennas in each subarray are fed with the same phase-shift to reduce the hardware complexity and power consumption, which is critical in THz systems.

In the proposed GoSA model, the RF precoder has unit-modulus constrain:

$$\left| \left[F_{RF} \right]_{i,j} \right| = \frac{1}{\sqrt{N_T}}, \quad \begin{array}{l} i \in \left\{ 1, \ldots, N_T \right\} \\ j \in \left\{ 1, \ldots, N_{RF} \right\} \end{array}$$

since F_{RF} is constructed by using phase-shifters. Furthermore, we have power constrained:

$$\sum_{m \in M} \left\| F_{RF} F_{BB} \left[m \right] \right\|_{\mathcal{F}} = MN_S$$

Thus, the $N_T \times 1$ transmitted signal from the TX is given by $x[m] = F_{RF} F_{BB}[m]s[m]$. Assuming frequency-selective fading over multicarrier transmission between the TX and RX, the received signal at the RX is given by:

$$y \left[m \right] = \sqrt{\rho} H \left[m \right] F_{RF} F_{BB} \left[m \right] s \left[m \right] + n \left[m \right]$$

where $y[m] \in \mathbb{C}^{N_R}$ is the output of N_R subarrayed antennas at the RX, ρ is the received power and $n[m] \in \mathbb{C}^{N_R}$ denotes the additive white Gaussian noise (AWGN) vector with:

$$n \left[m \right] \sim \mathbb{CN} \left(0, \sigma_n^2 I_{N_R} \right)$$

$H[m] \in \mathbb{C}^{N_R \times N_T}$ denotes the THz channel matrix between the TX and the RX.

In THz transmission, the wireless channel $H[m]$ can be represented by a single dominant LoS path with assisting a few NLoS paths, which are small due to large reflection losses, scattering, and refraction. Channel modeling at THz-band has been a challenge largely because of lack of realistic measurements. Very recently, measurement campaigns at 140 GHz have been reported. While the delay/angular spread at 140 GHz and lower frequencies are comparable, the correlation distance of shadow fading at the former is much shorter. The same study mentions multiple NLoS and dominant LoS paths at low-THz (0.1–1 THz).

However, we are not focused on only 140 GHz, and, therefore, employs the assumptions theorized for the entire upper-mm-Wave/low-THz region. While the ray-tracing techniques assume the channel to be sparse and dominated by the LoS component for the graphene nano-transceivers, the other channel models such as the 3GPP model are also popular for THz beamforming.

In this work, we adopt the Saleh-Valenzuela (SV) THz channel model channel, wherein $H[m]$ is constructed by the superposition of a single LoS path and the contribution of N_{clu} cluster of N_{ray} NLoS paths. Then, the $N_R \times N_T$ THz ultra-Massive MIMO channel matrix is given by:

$$H\left[m\right] = \gamma \left(\alpha_{1,m} A_R^m \left(\Theta_1\right) A_T^{m^H} \left(\Psi_1\right) + \sum_{l=2}^{L} \alpha_{l,m} A_R^m \left(\Theta_l\right) A_T^{m^H} \left(\Psi_l\right) \right)$$

where $\gamma = \sqrt{\dfrac{N_T N_R}{L}}$ and $L = 1 + N_{\text{clu}} + N_{\text{ray}}$ denotes the total number of LoS and NLoS paths. Moreover, $\alpha_{l,m}$ represents the channel gain of the lth path for mth subcarrier, and we have:

$$\alpha_{l,m} = \begin{cases} \alpha_m^{\text{LoS}}, & l = 1 \\ \alpha_{rc,m}^{\text{NLoS}}, & l = \left(N_{\text{ray}} - 1\right)c + r + 1 \end{cases}$$

for which $l \in \{1, ..., L\}$, $r \in \{1, ..., N_{\text{ray}}\}$ and $c \in \{1, ..., N_{\text{clu}}\}$. $\alpha_m^{\text{LoS}} \in C$ denotes the channel gain of the LoS path and it is defined as:

$$\alpha_m^{\text{LoS}} = \left(\frac{c_0}{4\pi f_m \bar{d}} \right)^{\frac{\bar{\gamma}}{2}} e^{-\frac{1}{2}\bar{\kappa}\left(f_m\right)\bar{d}}$$

c_0 is the speed of light, $\bar{\kappa}\left(f_m\right)$ is the frequency-dependent molecular absorption coefficient, and d is the distance between the TX and RX. In mm-Wave channels (24–776 GHz), the path loss exponent $\bar{\gamma}$ is around 2, while it has typical values between 3 and 4.5 in THz channels (0.1–1THz) for dense urban environments.

$\alpha_{m,rc}^{\text{NLoS}} = \left|\alpha_{m,rc}^{\text{NLoS}}\right| e^{j\bar{\beta}_{m,rc}}$ corresponds to the channel gain of the rth NLoS path in the cth cluster, and $\bar{\beta}_{m,rc}$ is an independent uniformly distributed phase-shift over $[0, 2\pi)$.

The notations $\Theta_1 = \{\phi_l, \theta_l\}$ and $\Psi_l = \{\varphi_l, \vartheta_l\}$ denote the azimuth/elevation angle-of-arrival (AoA) and angle-of-departure (AoD) of the received/transmitted paths at the RX and the TX, respectively.

The matrices $A_R^m(\Theta_l) \in \mathbb{C}^{N_R \times Q}$ and $A_T^m(\Psi_l) \in \mathbb{C}^{N_T \times Q}$ are the steering matrices corresponding to the AoAs and AoDs of the GoSAs, respectively, and they are defined as:

$$A_R^m(\Theta_l) = \begin{bmatrix} a_{R,1}^{m^T}(\Theta_l) \\ \cdot \\ \cdot \\ a_{R,N_R}^{m^T}(\Theta_l) \end{bmatrix}$$

$$A_T^m(\Psi_l) = \begin{bmatrix} a_{T,1}^{m^T}(\Psi_l) \\ \cdot \\ \cdot \\ a_{T,N_T}^{m^T}(\Psi_l) \end{bmatrix}$$

where $a_{R,n_R}^{m^T}(\Theta_l)$ and $a_{T,n_T}^{m^T}(\Psi_l)$ is $Q \times 1$ steering vector corresponding to the antennas in the $n_R{}^{th}$ and $n_T{}^{th}$ subarray for $n_R \in \{1, ..., N_R\}$ and $n_T \in \{1, ..., N_T\}$, respectively.

The ith of the transmit steering vector $a_{T,n_T}^{m^T}(\Psi_l)$ is given by:

$$\left| a_{T,n_T}^m(\Psi_l) \right|_i = \frac{1}{\sqrt{N_T}} \exp\left\{ -j \frac{2\pi}{\lambda_m} \kappa_{n_T,i}^T \Omega_l \right\}$$

where

- $\lambda_m = \frac{c_0}{f_m}$ is the wavelength for the subcarrier m with frequency of $f_m = f_c + \frac{B}{M}\left(m - 1 - \frac{M-1}{2} \right)$.
- B denotes the bandwidth.
- $\kappa_{n_T,i} = [x_{nT,i}, y_{nT,i}, z_{nT,i}]^T$ denotes the position of the ith antenna of the n_Tth subarray in Cartesian coordinate system.
- $\Omega_l = [\cos \varphi_l \sin \vartheta_l, \sin \varphi_l \sin \vartheta_l, \cos \vartheta_l]^T$ denotes the direction-independent parameter.
- the structure of $a_R^m(\Theta_l)$ is like that of $a_T^m(\Psi_l)$.

Without loss of generality, we assume that the antennas are perfectly calibrated against mutual coupling and gain/phase mismatches. Finally, by exploiting the GoSA structure, the $(n_R, n_T)^{th}$ element of $H[m] \in N_R \times N_T$ is given as:

$$\left[H[m] \right]_{n_R,n_T} = \gamma \sum_{l=1}^{L} \alpha_{l,m} a_{R,n_R}^m \odot a_{T,n_T}^{m^*}(\Psi_l)$$

By connecting the Q antennas in the subarrays to a single phase-shifter, we can construct an $N_R \times N_{RF}$, (instead of $N_T Q \times N_{RF}$) RF precoder. Using partially connected GoSA, the associated RF precoder has the form of:

$$
F_{RF} = \begin{bmatrix} u_1 & 0 & \dots & 0 \\ 0 & u_2 & \dots & 0 \\ \dots & 0 & \dots & 0 \\ 0 & 0 & \dots & u_{N_{RF}} \end{bmatrix} \in \mathbb{C}^{N_T \times N_{RF}}
$$

where $u_i \in \mathbb{C}^{\bar{N}}$ represents a portion of $N_T \times 1$ phase-shifter values with indices $\{(i-1)\bar{N}+1,\dots,i\bar{N}\}$ for $i \in \{1, \dots N_{RF}\}$, where $\bar{N} = \dfrac{N_T}{N_{RF}}$. Each entry of u_i is then applied Q antennas in N_T subarrays to steer the transmitted beams so that a total of $N_T Q$ antennas are fed.

To address the performance degradation due to GoSA, the columns of F_{RF} is designed with overlapping terms. Assume $\bar{u}_i \in \mathbb{C}^{\bar{M}}$ to include the overlapped phase-shifter terms, where $\bar{M} \in \left[\bar{N}, N_T - N_{RF} + 1 \right]$ provides non-overlapped partially connected structure), while $\bar{M} = N_T - N_{RF} + 1$ provides maximum overlap among the phase-shifters. In this case, the performance improvement is at the cost of using more phase-shifters. Nevertheless, it still has a lower number of phase-shifters as compared to the partially non-overlapped case in conventional AoSA. The use of partially connected/PCO GoSA structure provides higher DoF as compared to the simple phased-array MIMO radar structure, for which $N_{RF} = 1$, and we have a fully connected MIMO structure when $N_{RF} = N_T$.

While MIMO radar outperforms the phased array in terms of angular resolution and DoF for parameter estimation and parameter identification, phase-array provides higher coherent processing gain and lower computation and hardware complexity. This complexity is further reduced by using the GoSA structure by feeding each of Q antennas with the same phase-shift. Thus, the partially connected GoSA provides a trade-off between the DoF and the hardware complexity, both of which increase as $N_{RF} \rightarrow N_T$.

In communications-only systems, the aim is to design the hybrid precoders such that the spectral efficiency at the TX is maximized, while there are also other related performance metrics, such as energy efficiency and minimum mean-squared-error (MMSE). By decoupling the beamformer design problem at the TX and the RX, the mutual information at the TX is maximized instead of spectral efficiency, for which a perfect combiner is assumed at the receiver. Once the transmitter is designed, the receive beamforming design is done by using the MMSE as performance metric. Then the mutual information of the communications system is given by:

$$
\mathcal{X} = \frac{1}{M} \sum_{m=1}^{M} \mathcal{X}\left(F_{RF}, F_{BB}[m] \right)
$$

where

$$\mathcal{X}\left(F_{\mathrm{RF}}, F_{\mathrm{BB}}\left[m\right]\right) = \log_2 \left| I_{N_R} + \frac{\rho}{N_S \sigma_n^2} H\left[m\right] F_{\mathrm{RF}} F_{\mathrm{BB}}\left[m\right] \times F_{\mathrm{BB}}^H\left[m\right] F_{\mathrm{RF}}^H H^H\left[m\right] \right|$$

corresponds to the mutual information for subcarrier m.

We note here that the maximization is provided by exploiting the similarity between the hybrid beamformer $F_{RF}F_{BB}[m]$ and the optimal unconstrained beamformer $F_C[m] \in \mathbb{C}^{N_T \times N_S}$. The latter is obtained from the right singular matrix of the channel matrix $H[m]$. The singular value decomposition of the channel matrix is $H\left[m\right] = U_H\left[m\right] \prod V_H^H\left[m\right]$, where $U_H[m] \in \mathbb{C}^{N_R \times \mathrm{rank}(H[m])}$ and $V_H[m] \in \mathbb{C}^{N_T \times \mathrm{rank}(H[m])}$ are the left and the right singular value matrices of the channel matrix, respectively, and $\prod[m]$ is rank $(H[m]) \times$ rank $(H[m])$ matrix composed of the singular values of $H[m]$ in descending order.

By decomposing $\prod[m]$ and $V_H[m]$ as follows: $\prod\left[m\right] = \mathrm{diag}\left\{\widetilde{\prod\left[m\right]}, \overline{\prod\left[m\right]}\right\}$ and $V_H\left[m\right] = \tilde{V}_H\left[m\right], \overline{V}_H\left[m\right]$, where $\tilde{V}_H\left[m\right] \in \mathbb{C}^{N_T \times N_S}$, the unconstrained precoder is readily obtained as $F_C\left[m\right] = \tilde{V}_H\left[m\right]$. Then, the maximization is achieved by minimizing the Euclidean distance between and $F_C[m]$ and $F_{RF}F_{BB}[m]$ as follows:

$$\min_{F_{\mathrm{RF}}, \{F_{\mathrm{BB}}\left[m\right]\}_{m \in M}} \frac{1}{M} \sum_{m \in M} F_{\mathrm{RF}} F_{\mathrm{BB}}\left[m\right] - F_C\left[m\right]_{\mathcal{F}}$$

$$\text{s.t.:} \sum_{m \in M} \left\| F_{\mathrm{RF}} F_{\mathrm{BB}}\left[m\right] \right\|_{\mathcal{F}} = M N_S, \quad \left| \left[F_{\mathrm{RF}} \right]_{i,j} \right| = \frac{1}{\sqrt{N_T}}, \forall i, j$$

RADAR MODEL

The goal of radar processing is to achieve the highest possible SNR gain toward the direction of interest. The radar first transmits an omni-directional waveforms to detect the unknown targets within the angular space of interest in the search phase, then it generates directional beams toward to the targets for tracking purposes. We assume a subarrayed MIMO radar architecture with GoSAs, wherein each GoSA is used to coherently transmit waveforms that are orthogonal to the ones generated by other GoSAs, thereby coherent processing gain is achieved. To this end, the transmit waveform of the kth GoSA ($k \in \{1, ..., K\}$) is designed as $w_k(t) = W(t)e^{j2\pi k \Delta f t}$, $0 < t < T_0$, where $W(t)$ is the pulse shape with duration of T_0, so that the orthogonality of $w_k(t)$ is satisfied for a variety of time delays and Doppler shifts if the frequency increment among the GoSA waveforms satisfies:

$$\Delta_f = \left| f_{k+1} - f_K \right| \gg \frac{1}{T_0}$$

Denote $\{\Phi_1, ..., \phi_K\}$ as the set of target directions $\left\{\phi_k = \left(\overline{\varphi}_k, \overline{\vartheta}_k\right)\right\}$, then, the $N_T \times K$ GoSA-MIMO radar-only beamformer is modeled as:

$$F_R = \mathrm{blkdiag}\left\{v_1, ..., v_K\right\}$$

similar to above, where $v_k \in \mathbb{C}^{\bar{K}}$ denotes the values of the transmit steering vector $a_T(\phi_K) \in \mathbb{C}^{N_T}$ with indices $\{(k-1)\bar{K}+1,...,k\bar{K}\}$ for $\{k = 1, ..., K\}$ and $\bar{K} = \dfrac{N_T}{K}$. It is possible to construct F_R via overlapped GoSA with $\bar{v}_k \in \mathbb{C}^{N_T-K+1}$ for $\{k = 1, ..., K\}$.

The estimation of the target directions $\{\phi_K\}_{k=1}^{K}$ is performed in the search phase of the radar. This is achieved via both: model-based methods, such as multiple signal classification (MUSIC) algorithm, and model-free techniques based on DL. In this work, we assume that search operation is completed, and the direction information of the targets is acquired prior to the beamformer design. The beampattern of the radar with GoSA structure is:

$$B\left(\tilde{\Phi}.m\right) = \text{Trace}\left\{A_T^H\left(\tilde{\Phi}\right)R[m]A_T\left(\tilde{\Phi}\right)\right\}$$

where $R[m] \in \mathbb{C}^{N_T \times N_T}$ is the covariance matrix of the transmitted signal, then the design of the radar beampattern is equivalent to the design of the covariance matrix of the radar probing signals subject to the hybrid architecture of the beamformers. In case of a single-target scenario, the optimal beamformer is known to be conventional nonadaptive beamformer, i.e., steering vector corresponding to the direction of interest. When there are multiple targets, the covariance matrix of the transmitted signal is utilized. In case of multiple targets in radar-only scenario with hybrid beamforming, we define the covariance matrix of the transmitted signal $x[m]$ as:

$$R[m] = E\left\{x[m]x^H[m]\right\} = E\left\{F_{RF}F_{BB}[m]s[m]s^H(m)F_{BB}^H[m]F_{RF}^H\right\}$$

$$= F_{RF}F_{BB}[m]s[m]E\left\{s[m]s^H(m)\right\}F_{BB}^H[m]F_{RF}^H = F_{RF}F_{BB}[m]F_{BB}^H[m]F_{RF}^H$$

which requires the design of hybrid beamformers F_{RF} and $F_{BB}[m]$. The hybrid beamformer design problem for radar-only system is solved by minimizing the Euclidean distance between $F_{RF}F_{BB}[m]$ and $F_R R[m]$ as:

$$\min_{F_{RF},\{F_{BB}[m],P[m]\}_{m\in M}} \frac{1}{M}\sum_{m\in M}\left\|F_{RF}F_{BB}[m] - F_R P[m]\right\|_{\mathcal{F}}$$

$$\text{s.t.}: \sum_{m\in M}\left\|F_{RF}F_{BB}[m]\right\|_{\mathcal{F}} = MN_S, \left|\left[F_{RF}\right]_{i,j}\right| = \frac{1}{\sqrt{N_T}}, \forall i,j, \ P[m]P^H[m] = I_{N_S}$$

where the unitary matrix $P[m] \in \mathbb{C}^{K \times N_S}$ is an auxiliary variable to provide a change of dimension between $F_{RF}F_{BB}[m]$ and F_R, which have different dimensions (i.e., $N_T \times N_S$ and $N_T \times K$, respectively), without causing any distortion in the radar beampattern and $P[m]P^H[m] = I_K$.

MODEL-BASED HYBRID BEAMFORMING

The analytical models of these two items have not been included here for the sake of brevity. It is noteworthy that Deep Multiple Signal Classification (Deep MUSIC) and Deep Beam Former (Deep BF) AI/ML/DL models as depicted in Figure 12.2 have been used.

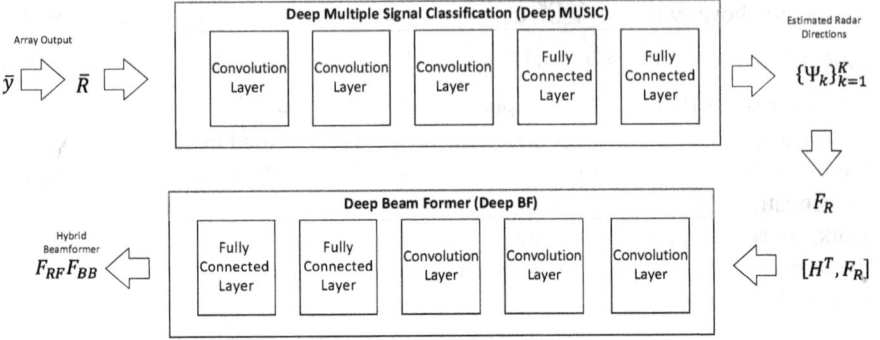

FIGURE 12.2 Model-free hybrid beamforming framework, in which Deep Multiple Signal Classification (Deep MUSIC) and Deep Beam Former (Deep BF) are employed to predict radar target directions and hybrid beamformer weights, respectively.

The interested readers might see [1] for detail mathematical models for Deep MUSIC and Deep BF. However, the simulation results have been discussed.

SIMULATION RESULTS

We evaluate the performance of the proposed hybrid beamforming approach for different array structures. The communications performance of the algorithms is evaluated in terms of spectral efficiency, while the radar performance is presented with the beampattern analysis of the hybrid beamformers. Furthermore, we analyze the trade-off between both tasks by sweeping η for [0, 1]. The hybrid beamformers are designed for fully connected, partially connected, and PCO array structures. The proposed MMO-based approach is used to design PCO array. Then, it is compared with the partially connected and fully connected arrays, which employ the MO-AltMin and Triple AltMin (TAltMin) approach, respectively, while the fully digital unconstrained beamformers are used as a benchmark.

In the simulations, unless stated otherwise, we select the operating frequency as $f_c = 300$ GHz with $M = 64$ and $B = 15$ GHz bandwidth, which is in low-THz band (100 GHz–1 THz) and applicable for long-range radar (LRR) (~150 m). We also select $\Delta = \Delta_x = \Delta_y = \lambda/2$ and $\delta = \delta_x = \delta_y = \lambda/4$, where λ denotes the wavelength corresponding to the carrier frequency. At the TX and RX, $N_{Tx} = N_{Ty} = 32$ ($N_T = 1024$) and $N_{Rx} \times N_{Ry} = 9$ ($N_R = 81$) subarrays are used, respectively, with $Q_x = Q_y = 3$ ($Q = 9$). Thus, the resultant architecture forms a 729×9216 ultra-Massive MIMO transceiver. We assume that $N_{RF} = 16$ RF chains are used at the TX to transmit $N_S = 4$ data streams to the RX via the THz channel which is assumed to include one LoS and four NLoS (i.e., $L = 5$) paths, where $\phi_l, \varphi_l \in [-150°, 150°]$ and $\theta_l, \vartheta_l \in [70°, 90°]$. The TX simultaneously generates beams toward both RX and $K = 3$ radar targets located at {(60°, 70°), (110°, 75°), (140°, 80°)}.

For model-free approach, we consider 1-D scenario, i.e., the elevation angles of the targets are 90° for simplicity. The learning model DeepBF is realized as a CNN with 11 layers. The first layer is the input layer of size $N_T \times (N_R + K) \times 2$. The second, fourth, and sixth layers are convolutional layers with 256@3 ×3 filters. After the first two convolutional layers, there is a max pooling to reduce dimension by 2.

The seventh and ninth layers are fully connected layers with 1,024 units. The eighth and tenth layers are dropout layers with 50% rate. Finally, the last layer is a regression layer of size $2N_TN_S$. Let $\mathcal{D}_i = \left(\mathcal{G}_i, \mathcal{E}_i\right)$ be the ith input-output tuple of the training dataset for $i = 1, ..., D$, where $D = |\mathcal{D}|$ denotes the number of samples in the dataset. To generate the training dataset, we consider $Z_C = 10^2$ channel realizations with the channel statistics and $K = 3$ radar target directions, which are generated uniform randomly from the interval $[-50°, 50°]$ with $1°$ resolution for $Z_R = 10^4$ realizations. Once the input data $\mathcal{G}_i$ is prepared, the optimization problem is solved for each input data, then the corresponding output label, i.e., $\mathcal{E}_i$ is computed for $i = 1, ...,$ D in an offline manner. As a result, the resulting dataset is comprised of samples of $D = Z_CZ_R = 10^6$ size $N_T \times (N_R + K) \times 2$. The cost function for Deep BF is the MSE between $\mathcal{L}\left(\mathcal{G}_i, \theta\right)$ and $\mathcal{E}_i$. The Deep MUSIC model is constructed. Then, the learning models are realized in MATLAB on a PC with 2304 GPU cores. We use the stochastic gradient descent (SGD) algorithm with momentum of 0.9 and update the network parameters with learning rate 0.001 when the mini-batch size is 64.

The simulation result has been plotted showing the number of phase-shifters with respect to N_T and Q for different array structures, i.e., AoSA and GoSA, respectively. The fully connected structures employ N_TQN_{RF} and N_TN_{RF} phase-shifters for AoSA and GoSA, respectively, while the partially connected structures are more efficient since only N_TQ and N_T phase-shifters are used for AoSA and GoSA. Compared to AoSA, the proposed GoSA structure employs much less phase-shifters than that of AoSA for $Q \geq N_{RF}$, and they become equal if $Q = 1$. Thus, GoSA is much more energy efficient than AoSA. While GoSA provides lower hardware complexity, it has slightly poorer spectral efficiency performance, which is ameliorated via the PCO structure by increasing the number of phase-shifters from N_T (non-overlapped) up to $N_{RF}(N_T - N_{RF} + 1)$ (fully overlapped). Nevertheless, the fully overlapped or fully connected GoSAs still have lower phase-shifters than that of AoSA with partially connected structure.

Another simulation has been plotted the spectral efficiency with respect to SNR for CSI-based hybrid beamforming when $\eta = 0.5$. We observe that GoSA performs slightly lower than AoSA structure while using $Q = 9$ times less phase-shifters, which significantly lowers the hardware complexity of ultra-Massive MIMO system. While partially connected structures have the lowest hardware complexities, they perform the worst as compared to the fully connected case. The GoSA with PCO improves the spectral efficiency by employing relatively more phase-shifters which still less than that of AoSA. The gap between the unconstrained (fully digital) and hybrid beamformers is large due to the trade-off between radar and communications tasks with $\eta = 0.5$.

Also, the spectral efficiency has been plotted with respect to η, wherein we note that as $\eta \rightarrow 1$, the spectral efficiency for the fully connected, partially connected, and PCO approaches to the performance of unconstrained beamformer, i.e., $F_C[m]$. When $\eta \rightarrow 0$, then the RF precoder FRF generates the beams toward the radar targets only; thus, the spectral efficiency is reduced. As a result, the selection of η is critical. In practice, η is increased if the communications task is more critical than tracking the targets or when there is no target. Conversely, lower η is selected if the radar task demands more resources, e.g., more transmit power is required depending on the range of the radar targets.

We have also illustrated that the beampattern of the designed hybrid beamformers for 1-D and 2-D angle distributions, respectively. It is seen that the beampatterns are presented for $\eta = 0$, $\eta = 0.5$ and $\eta = 1$, where we assume that all the paths have the elevation angle of $90°$. The ideal beampatterns correspond to the radar-only beamformer F_R for AoSA and GoSA structures. We note that for $\eta = 0$ ($\eta = 1$) all the beams are generated toward the radar targets (the RX), respectively, while F_{RF} generates $K + 1$ beams toward both targets and the RX when $\eta = 0.5$. The proposed GoSA PCO structure provides lower-side lobes and narrower beams toward both RX and radar targets as compared to the other algorithms. The 2-D angular distribution has been illustrated [1] for $\eta = 0.5$, where we present the beampattern corresponding to the GoSA with PCO. We observe that the proposed hybrid beamforming approach accurately generates beams toward both targets and RX paths in 2-D angular space. This is provided with the 2-D structure of the antenna array in two dimensions.

While the design of the antenna array is straightforward in AoSA case by selecting the antenna spacing as $\lambda/2$, the selection of δ is critical for the GoSA structure has been illustrated [1]. The spectral efficiency performance has been illustrated with respect to $\bar{d}$, where $\bar{d} = \dfrac{\lambda}{\delta}$. As $\bar{d}$ increases, we reduce the antenna element spacing in the subarrays of GoSA. Specifically, when $\bar{d} = \infty$, we have $\delta = 0$; thus, the $Q_x \times Q_y$ antennas in each subarray become co-located. From this, we infer that as the performance of AoSA approaches to that of GoSA as $\bar{d}$ increases. While slight performance loss is observed from AoSA with partially connected array, the performance of the fully connected structure and the fully digital beamformers significantly reduce as $\bar{d}$ increases. As a result, this figure is helpful when designing the GoSA because of the improvement in the spectral efficiency by changing $\bar{d}$ while the lower limit is $\bar{d} = 2$ (i.e., $\delta = \lambda/2$) to avoid spatial aliasing among the antennas.

In another study, it shows the spectral efficiency of the competing algorithms for both CSI- and channel covariance matrix-based beamforming when $\eta = 0.5$. A slight performance loss is observed for all channel covariance matrix-based approaches due to loss of precision in the angle and path gain information, while less channel overhead is involved in channel covariance matrix-based beamforming. In the CSI-based approach, the RX should feedback $N_R \times N_T$ (81×1024) channel matrix, whereas only the angle and path information need to be sent to the TX in the channel covariance matrix so that $C[m]$ is constructed.

The effect of frequency has been studied on the generated beams for (a) sub-6 GHz, (b) mm-Wave and (c) THz MIMO systems. We observe that different beams point to very close physical directions at low frequencies while the beam split occurs at THz, wherein the main lobes corresponding to the lowest/highest and center subcarrier frequencies do not overlap. We also present the effect of beam split on spectral efficiency with respect to bandwidth in another illustration. A severe loss in the spectral efficiency is observed when the bandwidth is large (i.e., > 20 GHz for $f_c = 300$ GHz). This arises from the use of frequency-independent analog beamformer, which causes a misalignment of the generated beams at different subcarriers, hence degrades the spectral efficiency. This loss can be effectively mitigated by the proposed beam split correction technique, which tunes the phase mismatches in the analog beamformer due to the use of a single frequency, i.e., f_c.

The performance of the hybrid beamformers in terms of radar target direction estimation has been studied together with the cost function for communications- and radar-only beamformers, respectively. It shows that SNR is swept for $[-20, 30]$dB and the corresponding direction root-mean-squared error (RMSE) and the spectral efficiency are computed. Both radar (direction RMSE) and communication (spectral efficiency) performances improve proportionally as SNR increases, while the partially connected arrays perform poorer than the fully connected ones for both AoSA and GoSA structures. Nevertheless, the proposed GoSA PCO array exhibits satisfactory performance for both radar and communications.

The spectral efficiency comparison of model-based and model-free techniques has been performed when GoSA structure is used. The simulations are averaged over 500 Monte Carlo trials, each of which is conducted for different realization of radar target angles. Note that we considered narrowband scenario in this simulation due to the memory limitations of the computation platform used for model training, while the results can be generalized for wideband scenario. While a slight loss is observed for the model-free techniques compared to CSI-based beamforming, they have close performance to the channel covariance matrix-based methods. Another advantage of the model-free approach is computational complexity thanks to its implementation via parallel processing units, such as GPUs. The performance of the learning model depends on the size of the dataset, which should cover a large portion of the whole input space. In case of smaller datasets, transfer learning-based approaches may be used. Based on simulations for the TX-RX settings, the computation time for MO, triple AltMin (TAltMin), DeepMUSIC, and DeepBF are, respectively, 2.124, 0.68, 0.0036, and 0.0058 seconds, which shows the advantage of model-free techniques.

SUMMARY

In this chapter, we analyzed the performance of MIMO radar for a moving target detection problem. We developed signal models for the MIMO radar and a phased-array radar to compare their performance in detecting moving targets in colored Gaussian noise-plus-clutter. We derived GLRT moving target detectors for centralized MIMO, distributed MIMO, and a phased-array radar system, respectively. We showed that the MIMO radar approach obtains a probability of detection for moving targets with small radial velocities that cannot be obtained by the conventional phased-array radar. We quantified the degradation by employing distributed MIMO radar in place of the more computationally demanding centralized MIMO radar implementation. The importance of proper antenna placement was discussed, and an adaptive version of the MIMO radar moving target detector was proposed. This adaptive detector was shown to provide constant false alarm probability for homogeneous clutter.

We have introduced a THz ultra-Massive MIMO JRC architecture and investigated model-based and model-free hybrid beamforming techniques. To lower the hardware complexity critical in THz systems, we proposed GoSA ultra-Massive MIMO architecture. We developed hybrid beamforming via PCO structures to provide a trade-off between higher spectral efficiency and hardware complexity in terms of the number of phase-shifters. The hybrid beamformers for THz JRC system are designed relying on both CSI and channel covariance matrix of the wireless channel information between the TX and the RX. The computation times for beamformer design could be prohibitively high for ultra-Massive MIMO THz systems.

We have addressed this by suggesting a model-free DL-based approach. We evaluated the performance of the proposed methods in terms of spectral efficiency and radar beampattern. We demonstrated that GoSA provides less hardware complexity compared to full-array and AoSA structures. To mitigate the beam split effect, we also introduce a hardware-efficient approach by correcting the phases of the frequency-independent beamformers. Compared to CSI-based beamforming, the channel covariance matrix-based approach has a slight performance loss, while the latter enjoys less channel overhead. The model-free method is advantageous in terms of computational complexity and exhibits approximately 500 times lower computation time as compared to the MO-based approaches, while maintaining spectral efficiency performance close to that of channel covariance matrix-based technique.

REFERENCE

1. Elbir, A. M. et al., "Terahertz-Band Joint Ultra-Massive MIMO Radar-Communications: Model-Based and Model-Free Hybrid Beamforming," arXiv:2103.00328v2 [eess.SP] 29 Sep 2021.

13 Device-to-Device Communications with Mesh Networking

SYSTEM ARCHITECTURE

The internet of things (IoT) is a global industrial movement that brings together new types of network communications by combing people, processes, data, and objects. As we see today, the IoT is evolving quickly, and with billions of devices connected to the internet in different areas and many devices connected to each other, challenges are emerging. One of the biggest challenges in IoT is the scale and range of such communications. Considering these challenges, the technologies proposed in low-power wide area network (LPWAN) are among the communication requirements in IoT. Various technologies and standards have been proposed for LPWAN networks based on the IoT requirements. Narrowband internet of things (NB-IoT) is one of the technologies related to cellular telecommunication introduced by 3GPP standardization organization with the aim of creating a long-range and low- consumption network. In addition to these two characteristics, the low cost and reuse of existing infrastructure for cellular and long-term evolution (LTE) technologies make the importance of using NB-IoT on the IoT even more evident. Long-term evolution–advanced (LTE-A) telecommunication technologies include device-to-device (D2D) communication (Figure 13.1), which transmits information directly between the user and the base station (BS).

It improves power consumption, spectral efficiency, and reduces network latency. In this research, we consider a heterogeneous network where the NB-IOT system is located. In this case, the communication link between the user and the BS cannot meet the quality of service (QoS) required to transmit sensitive and important data. To solve this problem, we will use D2D communication as a routing method. This method for the NB system-IoT allows the connection between the user and the BS to communicate through D2D relays with dual-jump routing. This mechanism also include optimization issues and finding an algorithm to reduce latency, increase EDR, and improve performance. The proposed scheme shows a stable double-hop delay even for a low transmission power. This D2D communication has the following features:

- Improve customer experience through the delivery of relevant, actionable, and location-based information and marketing in retail, stadiums, events, and hospitality.
- Deliver operational efficiency for enterprise networks in healthcare, smart buildings, and manufacturing by combining D2D smartphone communication networks and IoT devices into a single mesh network.

DOI: 10.1201/9781003499480-13

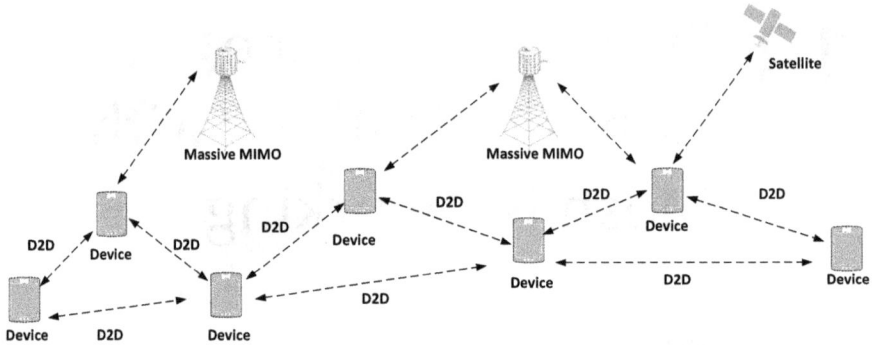

FIGURE 13.1 Device-to-device mesh communications.

- Build location aware networks to support the "new normal" COVID-19 era of social distancing monitoring and alerting.
- Support business in general, education, and health facilities and protect employees, students, and visitors.
- Build hybrid mesh topology to eliminate a single point of failure.
- Create D2D mesh using P2P Wi-Fi and Bluetooth protocols and multihop technologies.
- Map the future road to support LTE-A Direct and 5G D2D.
- Use Multicast domain system (mDNS)/DNS-service discovery (SD) for SD with algorithm based on evolutionary game theory.
- Harden underground cable for self-healing and fast reconnection.
- Create WebSocket-based bidirectional streaming data and control messages.

A highly scalable, fast-presence server [1] (delivers session tokens for anonymity. Data is cached on device until gateway is connected. Wyld Mesh utilizes a global positioning system (GPS), beacons, hotspots as pseudo beacons for location reckoning. It uses collective positioning (Patent GB2001939.4). It securely exchanges location information between device and using received signal strength indicator (RSSI) to perform trilateration to improve the accuracy of a GPS.

The following is designed to keep personal identification information (PII) secure and standards based: elliptic-curve Diffie-Hellman (ECDH) keys, advanced encryption standard (AES) global system for mobile (GSM) 256-bit encryption, signatures, etc.. Policy-based options are stored in a secure dictionary (digitally signed schemas). A keychain is created on registration and designed to rotate keys monthly. Keys are exchanged between meshed devices through the discovery process. Sensitive data never leaves the phone and is encrypted at rest. Data from smartphone mesh network and IoT devices are collected. Create visualization of relevant user data to inform on marketing outcomes, sales outcomes, and activity of actionable content and of IoT sensor data to make informed decision and actions. Support monetization of date, time, and location-aware footfall, buying habits, response to notification, and likes.

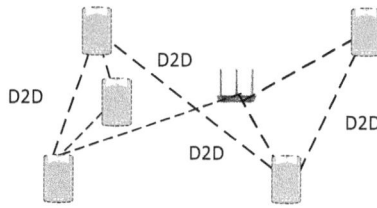

FIGURE 13.2 D2D and hybrid communications.

The mesh network is a collection of smartphones and IoT devices that are connected directly to each other in a device-to-device (D2D) mode in a decentralized topology. As such, the network is created without the need for traffic to be passed through a centralized set of nodes as in a traditional wireless network comprising of BSs, Wi-Fi access points, and core networks. The traffic hops instead from smartphone directly to other smartphones and IoT devices within a certain proximity of each other.

This means that the mesh network can carry traffic in areas where traditional networks are congested or there is no coverage. Each smartphone can connect with up to six other devices within range using the inherent peer-to-peer (P2P) capability of smartphones using Wi-Fi Direct or P2P Wi-Fi (Figure 13.2). Each of those six other devices can connect to another six devices, creating a robust and self-healing resilient mesh network.

To join the mesh network, the smartphone user needs to download a mobile App, whether it be a Dallas Cowboys or a Waitrose Loyalty Application with a Wyld software agent embedded into that application. Once inside the geo-zone, the smartphones are connected; and once the user leaves the geo-zone, the smartphone is disconnected.

The mesh can create a mass expansive network through multihop routing and ensures that as a device leaves the mesh, routing is automatically redirected through the most optimal route to ensure integrity. The Wyld Mesh technology routing algorithm uses evolutionary game theory modeling to intelligently select the routing based on optimal resiliency as well as device battery charge. Theoretically, there is no limit to the number of smartphones that can be connected in the mesh network. IoT devices are integrated into the smartphone mesh network also using Bluetooth or LPWAN protocols to create a people-to-people and a people-to-things network.

RESILIENCE AND ROBUSTNESS

- Hybrid mesh topology to eliminate a single point of failure
- D2) mesh created using P2P Wi-Fi and Wi-Fi Direct and multihop technologies
- Future road mapped to support LTE-A Direct and 5G D2D
- Uses mDNS/DNS-SD for SD with algorithm based on evolutionary game theory (Patent GB2001939.4)
- Hardened on London Underground for self-healing and fast reconnection

RELEVANT: USER LOCATION CONTEXT

- Wyld Mesh utilizes a GPS, beacons, hotspots as pseudo beacons for location, reckoning collective positioning (Patent GB2001939.4)
- Securely exchanges location information between devices and uses RSSI to perform trilateration to improve the accuracy of a GPS

SECURITY AND PRIVACY

- Designed to keep PII secure and standards based:
  - ECDH keys
  - AES GSM communications 256-bit encryption
  - Elliptic curve digital signature algorithm (ECDSA) signatures
- Policy-based options stored in a secure dictionary (digitally signed schemas)
- A keychain created on registration and designed to rotate keys monthly; keys exchanged between meshed devices through discovery process
- Sensitive data never leaves the phone and is encrypted at rest

SUMMARY

We have described the D2D communications in mesh networking environment. Since D2D communications are low-power communications, we cannot use the traditional P2P protocol. Fortunately, Patent GB2001939 provides the mechanisms for low-power D2D communications, even for mesh networking environment. Security features include PII protections and many other security features, including ECDH, AES, ECDSA, and others.

REFERENCE

1. Patent GB2001939.4. Mobile Mesh & Content Delivery Technology.

14 Vehicle-to-Vehicle and Vehicle-to-Everything Communications

OVERVIEW

Vehicle-to-everything (V2X) communication exchanges real-time and seamless data between vehicles, networks, and even autonomous vehicles, thus revolutionizing the intelligent transportation system (ITS). In 6G V2X systems, vehicle users have the characteristics of high-speed movement. Robust channel estimation in time-varying channels is used to guarantee the quality of communication services, especially for V2X scenarios. To improve the channel estimation accuracy and reduce the pilot overhead, multi-input multi-output (MIMO) radar is deployed to assist millimeter wave (mm-Wave) channel estimation.

The vehicle-to-vehicle (V2V) communication is a subset of V2X communications. V2V communication enables vehicles to wirelessly exchange information about their speed, location, and heading. The technology behind V2V communication allows vehicles to broadcast and receive omnidirectional messages (up to 10 times per second), creating a 360-degree "awareness" of other vehicles in proximity. Vehicles equipped with appropriate software (or safety applications) can use the messages from surrounding vehicles to determine potential crash threats as they develop. The technology can then employ visual, tactile, and audible alerts – or a combination of these alerts – to warn drivers. These alerts allow drivers the ability to take action to avoid crashes. To provide stable and reliable wireless link quality and to improve channel gain are the key areas for research.

The V2V communication messages have a range of more than 300 meters and can detect dangers obscured by traffic, terrain, or weather. V2V communication extends and enhances currently available crash avoidance systems that use radars and cameras to detect collision threats. This new technology doesn't just help drivers survive a crash – it helps them avoid the crash altogether. Vehicles that could use V2V communication technology range from cars and trucks to buses and motorcycles. Even bicycles and pedestrians may one day leverage V2V communication technology to enhance their visibility to motorists. Additionally, vehicle information communicated does not identify the driver or vehicle, and technical controls are available to deter vehicle tracking and tampering with the system. V2V communication technology can increase the performance of vehicle safety systems and help save lives. Connected vehicle technologies will provide drivers with the tools they need to anticipate potential crashes and significantly reduce the number of lives lost each year. In this section, we have provided only the introduction of V2V and V2X communications with some salient features. Moreover, Sections 15, 16, 17, 18, and 19 have described V2V, P2P

DOI: 10.1201/9781003499480-14

V2V, V2V with massive MIMO channels, V2X with Ultra-massive MIMO communications, and V2X millimeter wave MIMO radar, respectively.

VEHICLE-TO-VEHICLE COMMUNICATION

The vehicle communications sector is becoming so broad that it is often subdivided into smaller sectors. The broadest is "vehicle to everything" (V2X), which is generic for the whole sector, while "vehicle to vehicle" (V2V) and "vehicle to infrastructure" (V2I) are the most common, and where most of the development is taking place. Some others are emerging, including "vehicle to people" (V2P), where vehicles and pedestrians will be able to have a two-way communication. V2V would be based around a peer-to-peer mesh network where each element of the network (a vehicle) is able to generate, receive, and relay messages. With this approach, an expansive network can be created in populated areas without the need for expensive infrastructure. Typically, each vehicle would be able to transmit information about their speed, direction, location, braking, and turning intent – although this list is likely to expand over time. Alongside V2V will be V2I, where the vehicle is able to interact bidirectionally with fixed infrastructure such as stop signals, road construction sites, train crossings, and more.

Initially, the systems will provide a warning to the driver of the vehicle, but once systems are more mature, they may be able to control the vehicle by braking or steering around obstacles. Obstacle detection is one of the primary areas for vehicle vision systems, and, although great advances have been made, there are limitations. For example, a vision system would struggle to detect a hazard such as a broken-down vehicle or queue of traffic around a bend, but a V2V implementation would not be constrained by line-of-sight and could reliably detect such situations. The inputs from vision systems and V2V are likely to be merged in "sensor fusion" for confirmation and to eliminate false notifications.

V2V would form a mesh network, and dedicated short-range communication (DSRC) is one technology being proposed by organizations such as the Federal Communications Commission (FCC) and International Standard Organization (ISO). This is like Wi-Fi, as it operates at 5.9 GHz and has a range of approximately 300 meters – which equates to around 10 seconds on a highway. However, with up to 10 "hops" on the mesh the "visibility" of the V2V system extends to around a mile, which gives plenty of warning on busy roads where the mesh can extend this far. With systems such as automated emergency braking (AEB) becoming more prevalent, it won't be long before vehicles are able to bring themselves to a safe halt based upon an alert from the V2V system.

The challenges for designers broadly fall into three key areas:

- Developing a system that, within congested modern airwaves, can provide clear interruption-free communication to and from a vehicle that has several sources of significant interference including high current switching, inverters, and Bluetooth/Wi-Fi devices
- Providing safe and secure connectivity to integrate the V2V/V2X system into the rest of the increasingly complex vehicle systems

The critical role of passive components and connectivity:

- While semiconductor solutions will be integral to the communications standards battle, passive components, including radio frequency (RF) components such as antennas, electromagnetic compatibility (EMC) components, and transient protection devices, will be required in the development of every V2V system, now and in the future.
- One of the most fundamental components is the antenna. To assist designers, devices such as standard-off-the-shelf ceramic patch antennas are available that eliminate the design costs and cycle time associated with bespoke solutions. These high-performance single-chip solutions are optimized for DSRC and offer proven high performance in a package measuring just 13 mm × 13 mm × 4 mm.
- Complete antenna solutions are also available, either optimized for the DSRC band or configured for multi-band (2.4–6 GHz) operation. These are not generally polychlorinated biphenyl (PCB) mounting, but they include a suitable RF connector on a flying lead allowing remote placement of the antenna for optimum performance. With vehicle styling becoming increasingly important, "Pulse Shark Fin" antennas are available that support multiple applications, thereby consolidating multiple system needs in a single stylish solution [1–3].

VEHICLE-TO-EVERYTHING COMMUNICATIONS

OVERVIEW

Vehicle-to-everything (V2X) communication exchanges real-time and seamless data between vehicles, networks, and even autonomous vehicles, thus revolutionizing the Intelligent Transportation System (ITS). In 6G V2X systems, vehicle users have the characteristics of high-speed movement. Robust channel estimation in time-varying channels is used to guarantee the quality of communication services, especially for Vehicle-to-Everything (V2X) scenarios. To improve the channel estimation accuracy and reduce the pilot overhead, multi-input multi-output (MIMO) radar is deployed to assist millimeter wave (mm-Wave) channel estimation [1–3].

The vehicle-to-vehicle (V2V) communication is a subset of V2X communications. V2V communication enables vehicles to wirelessly exchange information about their speed, location, and heading. The technology behind V2V communication allows vehicles to broadcast and receive omni-directional messages (up to 10 times per second), creating a 360-degree "awareness" of other vehicles in proximity. Vehicles equipped with appropriate software (or safety applications) can use messages from surrounding vehicles to determine potential crash threats as they develop. The technology can then employ visual, tactile, and audible alerts—or a combination of these alerts—to warn drivers. These alerts allow drivers the ability to take action to avoid crashes. Providing stable and reliable wireless link quality and improved channel gain are the key areas for research.

The V2V communication messages have a range of more than 300 meters and can detect dangers obscured by traffic, terrain, or weather. V2V communication extends and enhances currently available crash avoidance systems that use radars and cameras to detect collision threats. This new technology doesn't just help drivers survive a crash—it helps them avoid the crash altogether. Vehicles that could use V2V communication technology range from cars and trucks to buses and motorcycles. Even bicycles and pedestrians may one day leverage V2V communication technology to enhance their visibility to motorists. Additionally, vehicle information communicated does not identify the driver or vehicle, and technical controls are available to deter vehicle tracking and tampering with the system. V2V communication technology can increase the performance of vehicle safety systems and help save lives. Connected vehicle technologies will provide drivers with the tools they need to anticipate potential crashes and significantly reduce the number of lives lost each year.

Vehicle-to-Everything (V2X) Communications

We are on the verge of a new age of linked autonomous cars with unheard-of user experiences, dramatically improved air quality and road safety, extremely varied transportation settings, and a plethora of cutting-edge apps. A substantially improved Vehicle-to-Everything (V2X) communication network that can simultaneously support massive hyper-fast, ultra-reliable, and low-latency information exchange is necessary to achieve this ambitious goal. These needs of the upcoming V2X are expected to be satisfied by the Sixth Generation (6G) communication system. Vehicle-to-Everything (V2X) communications have the potential to drastically reduce the number of vehicle collisions and as a result the number of deaths.

Brain Controlled V2X

Instead of a physical link between the driver and the vehicle, a Brain-Controlled Vehicle (BCV) is driven by the driver's thoughts. BCV may considerably boost freedom for people with disabilities by providing them with an additional interface for controlling automobiles. The brain-vehicle interface has the potential to improve manual driving by anticipating a driver's actions and sensing discomfort. While the aim is fully automated vehicles, managing the unpredictability and intricacies of autonomous driving will necessitate human adaptability. A brain-vehicle interface is expected to mitigate the challenges of autonomous driving in challenging and uncertain situations, like rural and less organized areas, by maintaining human involvement. While services connected to brain-machine interactions would demand concurrent ultrahigh dependability, ultralow latency, ultrahigh data rate connectivity, and ultrahigh-speed computing, current wireless communications (e.g., 5G) and compute technologies are unable to actualize BCV. For instance, a rough estimate of the full brain recording requirement puts it at roughly 100 Gb/s, which present wireless technology cannot handle. However, to assist learning and adapt to the behavior of human drivers, 6G-V2X must use a fully phased brain-vehicle interface and artificial intelligence/machine learning/deep learning (AI/ML/DL) approaches.

UAV/Satellite-Assisted V2X

Due to their extensive aerial coverage, UAVs have the potential to act as airborne wireless access points within the 6G-V2X network. These unmanned aerial vehicles can offer various services to mobile customers, such as relaying data, storage, and computational capabilities. UAVs can work together with base stations and other components of the immobile network infrastructure, especially in highly congested vehicle environments, to manage the wireless network and improve the user experience. Unmanned Aerial Vehicles (UAVs) offer numerous distinctive V2X applications as airborne agents, benefiting from their nearly unrestricted three-dimensional mobility. These applications include:

- Offering early reports of road accidents prior to the arrival of rescue teams.
- Surveillance of traffic violations to support law enforcement agencies.
- Disseminating warnings regarding road hazards in regions lacking pre-installed Roadside Units (RSUs).

Another possible airborne communication vehicle for 6G-V2X communications is satellites. Satellites are currently utilized in the V2X specifications for localization. It is important to note that satellite transmission data rates have increased significantly recently. Multibeam satellites, for example, have been widely employed in satellite communication networks due to their ability to boost wireless transmission throughput. As a result, 6G-V2X may use satellite communication as a method to help contact between a car and a distant data center in a situation where there is no terrestrial service. Like UAV-based V2X communication, a satellite can also perform network and computing tasks.

Integrated Control V2X

A key component of 6G will be integrated communication and management, which may also contribute to the advancement of sophisticated and autonomous V2X services. One use of integrated communication and management is vehicle platooning, in which a group of automobiles travels closely together in a car-coordinated motion without any mechanical connection. The capacity of the road is increased, fuel economy improves, and pleasant road trips are three of the main advantages of car platooning. To synchronize their acceleration and slowdown, each vehicle in the platoon needs to know the relative distance and velocity of its nearby vehicles.

Block-Based V2X

The wide adoption of V2X networks is highly dependent on authentication and message propagation for large-scale vehicle traffic. This concept sets new restrictions on how resources may be distributed in V2X networks. To combat potential malicious assaults or jamming, mission-critical communications, for instance, should have ultra-resilient security, but multimedia data services may simply need lightweight protection owing to the volume of data. Different frame structures, routing, power, and spectrum allocation techniques result from these two different kinds of security needs. 6G-V2X might use a blockchain system, which is considered a disruptive technology for safe multiparty decentralized transactions. The utilization of blockchain technology presents a range of enhanced security and privacy services, distinct from traditional security solutions, as it eliminates the need for intermediaries. Blockchain's inherent distributed ledger technology enables decentralized security management, offloading tasks to mobile cloud/edge/fog computing, and caching content within 6G-V2X communication.

In the context of 6G-V2X, a blockchain-based security solution, such as a smart contract or consensus mechanism, is expected to not only verify message authenticity but also safeguard the sender's privacy. Blockchains also play a pivotal role in managing unlicensed spectrum, enabling multiple users to share the same spectrum. 6G-V2X may adopt a blockchain-based approach to spectrum sharing, potentially delivering a more secure, intelligent, cost-effective, and highly efficient decentralized spectrum sharing solution. Although there have been various attempts to implement a blockchain-based communication network, the dynamic nature of the V2X communication situation and the need for real-time data processing preclude a straightforward adoption of existing blockchain technology.

The technology itself suffers from excessive latency, despite blockchain's enormous promise to provide improved security and network management. Consequently, the development of new blockchain algorithms with ultra-low latency is imperative before their integration into 6G-V2X can be realized. Some significant outstanding issues with existing blockchain technology that require in-depth research include its limited throughput and scalability.

INTELLIGENT REFLECTING SURFACE V2X

The applications of upcoming 6G-empowered vehicle-to-everything (V2X) communications depend heavily on large-scale data exchange with high throughput and ultra-low latency to ensure system reliability and passenger safety. However, in urban and suburban areas, signals can be easily blocked by various objects. Moreover, the propagation of signals with ultra-high frequencies such as millimeter waves and terahertz communication are severely affected by obstacles. To address these issues, the Intelligent Reflecting Surface (IRS), which consists of nearly passive elements, has gained popularity because of its ability to intelligently reconfigure signal propagation in an energy-efficient manner. Due to the promise of ease of deployment and low cost, IRS has been widely acknowledged as a key technology for both terrestrial and non-terrestrial networks to improve signal strength, physical layer security, positioning accuracy, and reduce latency. The integration of IRS into V2X communications can significantly improve the system performance. These improvements can be achieved in the form of security, capacity, energy efficiency and coverage extension. In addition, IRS enhanced vehicle-to-infrastructure (V2I), V2V, vehicle-to-drone (V2D), and vehicle-to-satellite communications are some examples of use cases.

TACTILE COMMUNICATIONS V2X

Tactile communication is a revolutionary technology, which enables a paradigm shift from the current digital content-oriented communications to steer/control-oriented communications by allowing real-time transmission of haptic or sensual information (i.e., touch, motion, vibration, and surface texture). By integrating human sensual information, tactile communication in 6G-V2X is expected to provide a truly immersive experience for onboard vehicle users. In addition to traditional applications of multimedia communications (e.g., onboard meetings/demonstrations and infotainment), tactile communication will enhance vehicular specific applications, such as remote driving, vehicle platooning, and driver training by enabling fast and reliable transfer of sensor data along with the haptic information related to driving experience and trajectories. Several haptic-based warning signals (e.g., waking up drowsy drivers or catching distracted drivers' attention) have been developed and tested for automotive applications to improve driving safety. On the other hand, tactile-based V2X can be extremely helpful to VRUs by providing them with appropriate haptic signals that will enhance their safety and activity. For instance, some researchers used haptic signals to combine cycling with cooperative driving while supporting cyclists moving in a platoon.

It is observed that the proposed system enhanced cycling behavior without negatively impacting concentration levels. Despite tactile communication's immense potential, there are still many challenges. For example, tactile communication requires extremely high-speed and extremely low-latency communication to ensure reliable and real-time exchange of large volumes of haptic information. These stringent connectivity constraints are very difficult to meet in high-mobility vehicular environments. This is because they require higher frequencies (e.g., mm-Wave or even THz) to meet their data demand.

However, those higher frequencies are not very reliable, particularly in mobile environments. For example, it is shown that, even in an indoor environment, THz networks may not be able to provide highly reliable high-rate communications. This, in turn, motivates research to develop a new breed of services called highly reliable high rate low-latency communications (HRLLCs) that can provide a combination of traditional 5G services for example, enhanced mobile broadband (eMBB) services that ignore reliability and URLLC services that ignore rate). Apart from the above challenges, tactile communication poses several fundamental challenges, including the design of application-specific control and communication protocols, the development of human-to-machine interfaces for wireless haptic interactions, and the design of suitable haptic codecs to capture and represent the haptic data and exact reconstruction of received haptic data.

Hybrid RF-VLC V2X

In 6G-V2X, it is expected that the vehicle and its occupants will be served at extraordinarily high data rates and with extremely low latency. However, this feature may not be feasible with standalone radio frequency (RF)-based V2X communication as conventional RF-based vehicular communication often suffers from high interference, large latency, and low packet delivery rates in highly dense scenarios. One alternative approach may be the combination of RF and visible light communication (VLC)-based V2X communications, where, along with radio waves, visible light can be used as a medium of communication in vehicular networks.

The ultrahigh data rate (potentially up to 100 Gb/s) achieved by light-emitting diode (LED) or laser diode (LD)-based VLC and its inherent features (such as low power consumption, enhanced security, and anti-electromagnetic interference) make VLC technology an ideal candidate for future ITS. Moreover, a VLC-based V2X communication system will require minimum setup cost as VLC-based V2X can be implemented by using the existing LEDs/LDs in vehicle headlights or preinstalled street/traffic lights.

In V2X networks, VLC can be mainly used in the following three scenarios: V2V communication through headlights/backlights, V2X communication through traffic lights, and V2X communication through streetlights. Note that the traffic/streetlights can be used to establish backhaul links with one another by using free-space coherent optical communications. In addition to enhancing the data rate, VLC can boost the performance of V2X networks by eliminating the limitations of traditional RF-based V2X communications. For example, in the presence of big vehicle shadowing, RF-based V2V communication suffers from severe packet drop due to high path-loss and packet collision. In this scenario, the transmitting vehicle can communicate with the big vehicle through VLC, and then, the big vehicle can relay the messages to the vehicles in the shadow region. Similarly, using VLC, traffic/streetlights can also be used in the urban intersections to relay the messages to facilitate communication between vehicles from perpendicular streets, where traditional RF-based V2V communication often suffers from severe packet loss. Note that, while RF-based solutions (e.g., big vehicle or roadside unit (RSU) relaying) to the above problems are studied in the literature, such solutions can cause severe interference in the high-density scenarios due to the RF-based re-transmissions.

Although extensive research has been carried out on VLC-based V2X communication in the past decade, VLC has not been included in the 5G-V2X standard. Several open issues still need to be solved for enabling hybrid RF-VLC V2X. These include interoperability between VLC and RF technologies, and deployment issues. In an outdoor environment, the performance of VLC degrades due to the interference caused by natural and artificial light sources. On the other hand, the received signal strength in VLC may dramatically vary due to the vehicles' mobility. Hence, ambient lighting-induced interference and mobility-induced channel variations need to be properly addressed before deploying VLC in 6G-V2X systems.

TERAHERTZ-ASSISTED V2X

THz communication, which operates at THz bands (0.1–10 THz), is envisioned as a promising approach to alleviate increasingly congested spectrum at lower frequencies. Leveraging the availability of ultrawide bandwidth, THz communication will be able to provide transmission rates ranging from hundreds of gigabits per second to several terabits per second. Such an extremely high throughput will enable a plethora of new V2X application scenarios, such as ultrafast massive data transfer between vehicles and haptic communications. Since THz communication is able to provide fiber-like data rates without the need for wires between multiple devices at a distance of a few meters, it may also be used in onboard use cases, such as the BCV scenario, where extremely high throughput and low-latency wireless communication is required.

While the THz spectrum brings several unique benefits, there are many major challenges to be addressed, such as transceiver architectures, materials, antenna design, propagation measurement, channel modeling, and new waveforms. It is essential to characterize and understand THz radio propagation in different V2X scenarios, such as highway, urban, and in-vehicle. One of the main challenges in THz-assisted 6G-V2X will be the effective use of traditional cellular and new THz bands. As such, suitable dynamic resource scheduling is required to exploit their unique benefits. For example, while THz communication offers very high data rates, it is only suitable for short-range V2X communications. In this case, resources may be allocated in THz bands to those transmitters with receivers within a short range. Note that appropriately designed relaying or IRS techniques can be potential solutions to extend the coverage of THz-based V2X communications.

QUANTUM COMPUTING-AIDED V2X

Quantum computing is considered one of the revolutionary technologies for generic 6G wireless communications in several seminal works. However, the development of practical quantum computing and communication systems is in its infancy, and practical solutions may be quite some time off. Therefore, quantum computing and communications may potentially play a role toward the end of the 6G development or even beyond 6G+ technologies. Nevertheless, once some form of quantum computing is available for 6G communications, it can be expected to make its way into V2X applications as well.

If available, we can envision that quantum computing will offer enhanced security in V2X communications. Note that security in V2X communications is significantly more important than in traditional communications since, for example, a security breach in an autonomous vehicle can cause fatal accidents. As the wireless spectrum is shared between vehicles and other types of cellular users (e.g., pedestrians), V2X communications may be vulnerable to malicious attacks, and traditional encryption strategies may not be adequate. Quantum computing has the inherent security feature of quantum entanglement that cannot be cloned or accessed without tampering with it, making it an appropriate technology to enhance 6G-V2X communications security. Moreover, quantum domain security is based on the quantum key distribution (QKD) framework that allows detecting any malicious eavesdropping attempt. For example, the use of quantum federated learning to securely execute learning tasks among vehicles can be an important use case.

In addition to the enhanced security feature, the advent of quantum computing promises a radically enhanced computational capability offering to significantly enhance and optimize 6G-V2X services through the fast execution of extremely complex and currently time-consuming optimization algorithms. For example, the implementation of advanced ML algorithms that require big data processing and massive training (e.g., finding an optimum geographic route with multiple objectives) is a very challenging task. In such scenarios, traditional computing often sacrifices optimality, while quantum computing can efficiently achieve optimality with reduced complexity.

Although quantum computing is seen as a promising technology, much more research is required to turn it into a widely usable technology to exploit its potential.

For example, current quantum computer chips can only operate at extremely low temperatures (close to zero Kelvin), which makes them at best only usable on the vehicular infrastructure side. To use them in vehicles, significant research is needed on the thermal stability of quantum computer chips. Other fundamental challenges include the development of large-scale quantum computing, the design of quantum security architectures, and the characterization of entanglement distribution.

AI-ENABLED V2X

Recent developments in artificial intelligence, machine learning, and deep learning (AI/ML/DL) studies have allowed the development of new technologies such as self-driving cars and speech assistants due to the availability of large data sets, storage and computing capacity. Given this context, ML has become more and more crucial to the highly independent and clever functioning of tomorrow's 6G vehicular networks. Traditional wireless communication system design is heavily dependent on model-based methodologies, in which different communication system building elements are carefully built based on analysis of measurement data. Although these model-based methods have proven effective in designing conventional communication systems, there may be some 6G-V2X situations where precise modeling (such as an accurate interference model and channel estimation) is unlikely. In the previously mentioned scenarios, where traditional communication system design might encounter discrepancies in model compatibility, AI/ML/DL proves to be a potent instrument due to its ability to discern features and detect relationships, even those concealed deeply, between input and output data. Furthermore, the data-driven essence of machine learning can facilitate predictions and projections regarding user behavior, network traffic, application needs, security threats, and channel dynamics. This, in turn, leads to improved resource allocation and enhanced network performance.

Recent advancements in AI/ML/DL methods are significantly contributing to the progress of autonomous vehicles, while also playing a crucial role in enhancing the overall driving experience and road safety. Take, for instance, the wealth of data streams originating from sources like cameras, Light Detection and Ranging (LiDAR) sensors, GPS units, and various sensors. These data streams can be efficiently processed, allowing for the application of modular perception, planning, action, or end-to-end learning techniques in automated driving, enabling data-informed intelligent decision-making. In the realm of AI/ML/DL-based vision, there's ongoing exploration of multimodal reasoning. This involves the integration of camera frames and LiDAR scans to enhance object detection, a vital element in automated driving safety. Our focus in this paper is on the network perspective, highlighting the influence of AI/ML/DL within 6G-V2X networks. However, it's worth noting that numerous AI/ML/DL-driven applications for intelligent driving are anticipated in the future. We also introduce one of the most cutting-edge AI/ML/DL techniques, federated learning.

The merger of various forms of communication and the strict data delivery requirements for 6G-V2X will worsen security problems. The inherent broadcast nature of vehicular communication exposes it to potential security vulnerabilities and malicious attacks, although 6G-V2X seeks to provide smooth access between infrastructure sites and cars. A vehicular network could be the subject of numerous malicious attacks, including approved and sanctioned attacks, data forgeries, and distributions. A new user identification and verification method must be developed to maintain safe and legal access to data, services, and systems because private user information, such as user identity or trajectory, is shared over wireless links in a V2X system. Supervised learning with categorization capacity is suggested as a useful tool to detect abnormal driving behavior in cars. It is stated that because training and detection processes rely on previously labeled data, supervised learning may not be able to identify new or undiscovered assaults. Unsupervised learning, which can aggregate data without the need for labeled information, is being considered for real-time recognition. For vehicle networks, intruder detection using K-means clustering is specifically suggested. Anomaly detection using unsupervised learning is researched to combat assaults that can occur spontaneously in real-time.

Nonetheless, these approaches exclusively focus on either misuse detection or anomaly detection, which may not effectively address real-world scenarios where both known and undiscovered attacks could occur simultaneously. Additionally, to reduce transmission costs, reactive detection is considered primarily in current detection methods. Nonetheless, it is anticipated that proactive exploration-based security methods will be beneficial for elevating the security standards within a 6G-V2X network, particularly in scenarios where communication resources are relatively abundant. For example, some researchers have adopted a proactive anomaly detection strategy for linked vehicles to avoid cyber threats.

Typically, cryptographic techniques are used in the higher layers of the protocol stack to handle security problems in wireless transmission. However, in diverse and dynamic V2X networks where cars may haphazardly enter or exit the network at any moment, the administration and exchange of private keys will be difficult. In this sense, Physical Layer Security (PLS) solutions can be used to complement conventional encryption methods. While PLS techniques leverage the physical attributes and stochastic behavior of wireless channels to mitigate eavesdropping, they can still be influenced by the accuracy of channel modeling. In a V2X context marked by high mobility and frequent channel variations, the use of machine learning becomes advantageous for accurate channel prediction and monitoring. This, in turn, has the potential to enhance the effectiveness of PLS-based methods. Furthermore, various security levels are anticipated according to the situations and services. Consider, for instance, two cars that track one another either at a busy junction or on a deserted road. Due to the mobility of cars, the latter has a greater number of variables that can influence decision making, leading to strict security requirements. In the latter scenario, AI/ML/DL may be used to dynamically determine the necessary security degree and the best PLS answer. ML can also play a role in enhancing the management and communication systems designed to safeguard against data injection attacks within specific vehicle platoons. When AI/ML/DL is used to improve security, the end-to-end network efficiency of the ML-based system must be validated. As stated above, AI/ML/DL can be applied to functional components at various network levels. To ensure that all interactions are secure, AI/ML/DL usage should be coordinated across the network.

Federated Learning for 6G-V2X is also very useful. Training AI/ML/DL models, which can be used at base sites or in vehicles, is a key problem for effective use of

AI/ML/DL. Training a large AI/ML/DL model in distant clouds is an apparent answer, but it may take some time. One issue is that a sluggish reaction to external changes could result from rapidly changing vehicle network and communication conditions, which would affect performance. Furthermore, given that most training data is generated at network edges, such as base stations and mobile vehicles, the expense and latency associated with transmitting this data to a remote cloud can be substantial. In view of this, conducting local training of machine learning models within 6G-V2X networks emerges as a more favorable choice. To increase the accuracy and generalization of the performance of AI/ML/DL models, joint training samples are a possible solution.

This is because each base station or car may only be able to store a limited number of training samples. While base stations and vehicles may have concerns about compromising their privacy by sharing training data, privacy remains a significant concern in collaborative training. Federated learning, a relatively recent approach designed to address privacy and transmission overhead issues associated with ML model development, has garnered considerable research attention to enhance wireless networks. There are numerous technological challenges to address when implementing federated learning applications, which are regarded as a promising machine learning approach for enhancing the efficiency of 6G-V2X networks.

Supervised learning is considered primarily in the study that has already been conducted on shared learning in wireless networks. As the use of Reinforcement Learning (RL) models increases, a flexible federated RL system is required that can accommodate a variety of 6G-V2X use cases. Furthermore, initiating federated RL from a blank slate is often infeasible since many V2X applications are mission-critical, and such an approach could lead to an unstable initial phase during the learning process. The limited connection between vehicles presents another difficulty for federated learning involving cars.

Vehicles may not be in contact with base units or other federated learning-related vehicles. Therefore, while they are stopped, the cars might need to participate in federated learning. Lastly, a more comprehensive study is needed to determine how the wireless route affects the effectiveness of federated learning. Wireless delays and errors can affect the precision of federated learning, as demonstrated. The high-speed mobility of vehicles and the dynamic nature of channels within a mobile V2X network could exacerbate this effect. Further research is essential to explore the integrated development of wireless and learning processes for V2X applications.

Vehicle-to-Vehicle (V2V) Communication

The vehicle communications sector is becoming so broad that it is often sub-divided into smaller sectors. The broadest is 'vehicle to everything' (V2X) which is generic for the whole sector while 'vehicle to vehicle' (V2V) and 'vehicle to infrastructure' (V2I) are the most common, and where most of the development is taking place. Others are emerging including 'vehicle to people' (V2P) where vehicles and pedestrians will be able to have two-way communication. V2V would be based around a peer-to-peer mesh network where each element of the network (a vehicle) is able to generate, receive and relay messages. With this approach, an expansive network can be created in populated areas without the need for expensive infrastructure. Typically, each vehicle would be able to transmit information about their speed, direction, location, braking and turning intent – although this list is likely to expand over time. Alongside V2V will be V2I where the vehicle is able to interact bi-directionally with fixed infrastructure such as stop signals, road construction sites, train crossings and more.

Initially, the systems will provide a warning to the driver of the vehicle, but once systems are more mature, they may be able to control the vehicle by braking or steering around obstacles. Obstacle detection is one of the primary areas for vehicle vision systems and, although great advances have been made, there are limitations. For example, a vision system would struggle to detect a hazard such as a broken-down vehicle or queue of traffic around a bend, but a V2V implementation would not be constrained by line-of-sight and could reliably detect such situations. The inputs from vision systems and V2V are likely to be merged in 'sensor fusion' for confirmation and to eliminate false notifications.

V2V would form a mesh network and dedicated short-range communications (DSRC) is one technology being proposed by organizations such as the Federal Communications Commission (FCC) and International Standard Organization (ISO). This is like Wi-Fi, as it operates at 5.9 GHz and has a range of approximately 300 meters – which equates to around 10 seconds on a highway. However, with up to 10 'hops' on the mesh the 'visibility' of the V2V system extends to around a mile, which gives plenty of warning on busy roads where the mesh can extend this far. With systems such as automated emergency braking (AEB) becoming more prevalent, it won't be long before vehicles are able to bring themselves to a safe halt based upon an alert from the V2V system.

The challenges for designers broadly fall into three key areas:

- Developing a system that, within congested modern airwaves, can provide clear interruption-free communication to and from a vehicle that has several sources of significant interference including high current switching, inverters, and Bluetooth/Wi-Fi devices.
- Providing safe and secure connectivity to integrate the V2V/V2X system into the rest of the increasingly complex vehicle systems.

The critical role of passive components and connectivity:

- While semiconductor solutions will be integral to the communications standards battle, passive components including radio frequency (RF) components such as antennas, electromagnetic compatibility (EMC) components, and transient protection devices will be required in the development of every V2V system, now and in the future.
- One of the most fundamental components is the antenna. To assist designers, devices such as standard-off-the-shelf ceramic patch antennas are available that eliminate the design costs and cycle time associated with bespoke solutions. These high-performance single-chip solutions are optimized for DSRC and offer proven high performance in a package measuring just 13mm × 13mm × 4mm.

- Complete antenna solutions are also available, either optimized for the DSRC band or configured for multi-band (2.4GHz to 6GHz) operation. These are not generally polychlorinated biphenyl (PCB) mounting but they include a suitable RF connector on a flying lead allowing remote placement of the antenna for optimum performance. With vehicle styling becoming increasingly important, 'Pulse Shark Fin' antennas are available that support multiple applications, thereby consolidating multiple system needs in a single stylish solution.

REFERENCES

1. Noor-A-Rahim, M. et al., "6G for Vehicle-to-Everything (V2X) Communications: Enabling Technologies, Challenges, and Opportunities," *Proceedings of the IEEE*, Vol. 110, No.6, June 2022.
2. Khan, W. U., et al., "Opportunities for Intelligent Reflecting Surfaces in 6G-Empowered V2X Communications," URL: arXiv:2210.00494v1 [eess.SP] 2 Oct 2022.
3. Wang, D. et al., "A Short Overview of 6G V2X Communication Standards," URL: arXiv: 2311.16810v1 [eess.SP] 28 Nov 2023.

15 V2V Communications Network

SELF-DRIVING V2V COMMUNICATIONS

Self-driving vehicles (SDVs) might seem to be futuristic, and these vehicles are expected to be the norm soon. They will provide safer, cheaper, and less congested transportation solutions for everyone, everywhere. A core component of SDVs is their ability to perceive the world [1]. From sensor data, the SDV needs to reason about the scene in 3D, identify the other agents, and forecast how their futures might play out. These tasks are commonly referred to as perception and motion forecasting. Both strong perception and motion forecasting are critical for the SDV to plan and maneuver through traffic to get from one point to another safely. SDVs require to improve the perception (e.g., sensor data, representations of messages, and others) and motion forecasting performance. By intelligently aggregating the information received from multiple nearby vehicles, one can observe the same scene from different viewpoints. This allows the viewers to see through occlusions (i.e., blockers) and detect actors at long range, where the observations are very sparse or non-existent. Artificial intelligence (AI)-based approach of sending compressed deep feature map activations achieves high accuracy while satisfying communication bandwidth requirements.

The reliability of perception and motion forecasting algorithms has significantly improved [1] in the past few years due to the development of neural network architectures that can reason in 3D and intelligently fuse multisensor data (e.g., images, light detection and ranging (LiDAR), maps). Motion forecasting algorithm performance has been further improved by building good multimodal distributions [that capture diverse actor behavior and by modeling actor interactions. Recently, some papers [1] have proposed approaches that perform joint perception and motion forecasting, dubbed perception and prediction (P&P), further increasing the accuracy while being computationally more efficient than classical two-step pipelines.

In 3D modeling, a point cloud is a set of data points in a 3D coordinate system – commonly known as the XYZ axes. Each point represents a single spatial measurement on the object's surface. Taken together, a point cloud represents the entire external surface of an object. Perspective-n-Point (**PnP**) is the problem of estimating the pose of a calibrated camera given a set of n 3D points in the world and their corresponding 2D projections in the image. The PnP problem – usually referred to as PnP – is the problem of finding the relative pose between an object and a camera from a set of n pairings between 3D points of the object and their corresponding 2D

DOI: 10.1201/9781003499480-15

projections on the focal plane, assuming that a model of the object is available. Some of the salient points for V2V communications are as follows:

- Vehicle-to-vehicle (V2V) communication to improve the perception and motion forecasting performance of SDVs.
- LiDAR is a remote sensing method that uses light in the form of a pulsed laser to measure ranges (variable distances) to the earth.
- By intelligently aggregating the information received from multiple nearby vehicles, we can observe the same scene from different viewpoints.

A novel perception and motion forecasting model is defined, which enables the self-driving vehicle to leverage the fact that several SDVs may be present in the same geographic area. Following the success of joint P&P algorithms, we design our approach as a joint architecture to perform both tasks, which is enhanced to incorporate information received from other vehicles. Specifically, the P&P model is devised to do the following by the SDV with the given sensor data:

- Process this data.
- Broadcast the processed data.
- Incorporate information received from other nearby SDVs.
- Generate final estimates of where all traffic participants are in the 3D space and their predicted future trajectories.

Two key questions arise in the V2V setting:

- What information should each vehicle broadcast to retain all the important information while minimizing the transmission bandwidth required?
- How should each vehicle incorporate the information received from other vehicles to increase the accuracy of its perception and motion forecasting outputs?

WHICH INFORMATION TO BE TRANSMITTED

An SDV can choose to broadcast three types of information: (i) the raw sensor data, (ii) the intermediate representations of its P&P system, or (iii) the output detections and motion forecast trajectories. While all three message types are valuable for improving performance, the message sizes need to be minimized while maximizing P&P accuracy gains. Note that small message sizes are critical because cheap, low-bandwidth, decentralized communication devices are to be leveraged. While sending raw measurements minimizes information loss, they require more bandwidth. Furthermore, the receiving vehicle would need to process all additional sensor data received, which might prevent it from meeting the real-time inference requirements. On the other hand, transmitting the outputs of the P&P system is very good in terms of bandwidth, as only a few numbers need to be broadcasted. However, we may lose valuable scene context and uncertainty information that could be very important to better fuse the information.

By sending the intermediate representations of messages, the P&P network achieves the best of both worlds. First, each vehicle processes its own sensor data and computes its intermediate feature representation. This is compressed and broadcasted to nearby SDVs. Then, each SDV's intermediate representation is updated using the received messages from other SDVs. This is further processed through additional network layers to produce the final perception and motion forecasting outputs. This approach has two advantages: (1) Intermediate representations in deep networks can be easily compressed, while retaining important information for downstream tasks. (2) It has low computation overhead, as the sensor data from other vehicles has already been pre-processed. First, the intermediate representations of messages need to be computed. Second, the computed messages will then need to be compressed. In the following text, we explain how each vehicle should incorporate the received information to increase the accuracy of its P&P outputs (Algorithm 15.1).

Algorithm 15.1 for Cross-Aggregation [1].

ALGORITHM 15.1   Cross-Vehicle Aggression

1. **Input**: Representation $\hat{z}_i$, Relative Pose Δp_i, and time delay $\Delta t_{i \rightarrow k}$ for each SDV_i

2. **for** each vehicle i **do**

3. $h_i^{(0)} = CNN\left(\hat{z}_i, \Delta t_{i \rightarrow k}\right) \big\| 0$ $\rightarrow$ Compensate time delay, initial node state

4. **end for**

5. **for** l iterations **do** $\rightarrow$ Message passing

6.   **for** each vehicle i **do** $\rightarrow$ Process in parallel

7. $m_{i \rightarrow k}^{(l)} = CNN\left(\left(h_i^{(l)}, \xi_{i \rightarrow k}\right), h_k^{(l)}\right).M_{i \rightarrow k}$ $\rightarrow$ Spatially transform message

8. $h_k^{(l+1)} = ConvGRU\left(h_i^{(l)}, \phi_M\left(\left[\forall_{j \in N(i)}, m_{j \rightarrow i}^{(l)}\right]\right)\right)$ $\rightarrow$ Node state update

9.   **end for**

10. **end for**

11. $z_i^{(L)} = MLP\left(h_i^{(L)}\right)$ $\rightarrow$ Output updated intermediate representation

---

LEVERAGING MULTIPLE VEHICLES

V2VNet has three main stages: (1) a convolutional network block that processes raw sensor data and creates a compressible intermediate representation, (2) a cross-vehicle aggregation stage, which aggregates information received from multiple vehicles with the vehicle's internal state (computed from its own sensor data) to compute an updated intermediate representation, and (3) an output network that computes the

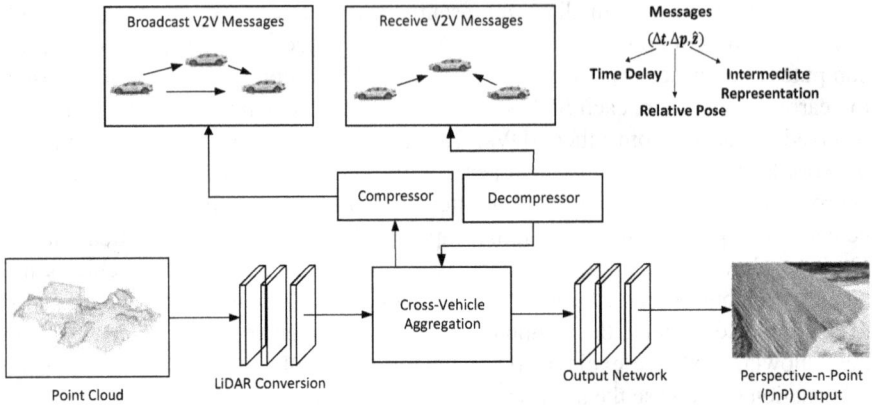

FIGURE 15.1   V2V network architecture.

final P&P outputs. We now describe these steps in more details. We refer the reader to Figure 19.1 for our V2VNet architecture, while Figure 15.1 shows different processes of V2V Network Architecture for the self-driving V2V communications.

Voxelization is the process of converting a data structure that stores geometric information in a continuous domain (such as a 3D triangular mesh) into a rasterized image (a discrete grid).

LiDAR Convolution Block

We extract features from LiDAR data and transform them into bird's-eye-view (BEV). Note that voxelization is the process of converting a data structure that stores geometric information in a continuous domain (such as a 3D triangular mesh) into a rasterized image (a discrete grid). Specifically, we voxelize the past five LiDAR point cloud sweeps into 15.6 cm^3 voxels, apply several convolutional layers, and output feature maps of shape $H \times W \times C$, where $H \times W$ denotes the scene range in BEV and C is the number of feature channels. We use three layers of 3×3 convolution filters (with strides of 2, 1, 2) to produce four times the down-sampled spatial feature map. This is the intermediate representation that we then compress and broadcast to other nearby SDVs.

Compression

We now describe how each vehicle compresses its intermediate representations prior to transmission. We adapt Balle et al.'s variational image compression algorithm [1] to compress our intermediate representations; a convolutional network learns to compress our representations with the help of a learned hyperprior. The latent representation is then quantized and encoded lossless with very few bits via entropy encoding. Note that our compression module is differentiable and therefore trainable, allowing our approach to learn how to preserve the feature map information while minimizing bandwidth.

CROSS-VEHICLE AGGREGATION

After the SDV computes its intermediate representation and transmits its compressed bitstream, it decodes the representation received from other vehicles. Specifically, we apply entropy decoding to the bit stream and apply a decoder CNN to extract the decompressed feature map. We then aggregate the received information from other vehicles to produce an updated intermediate representation. Our aggregation module must handle the fact that different SDVs are located at different spatial locations and see the actors at different timestamps due to the rolling shutter of the LiDAR sensor and the different triggering per vehicle of the sensors. This is important as the inter-mediate feature representations are spatially aware.

Toward this goal, each vehicle uses a fully connected graph neural network (GNN) as the aggregation module, where each node in the GNN is the state representation of an SDV in the scene, including itself (see Figure 15.2). Each SDV maintains its own local graph based on which SDVs are within range (i.e., 70 meters). GNNs are a natural choice as they handle dynamic graph topologies, which arise in the V2V setting.

GNNs are deep-learning models tailored to graph structured data: each node maintains a state representation, and for a fix number of iterations, messages are sent between nodes and the node states are updated based on the aggregated received information using a neural network. Note that the GNN messages are different from the messages transmitted/received by the SDVs: the GNN computation is done locally by the SDV. We design our GNN to temporally warp and spatially transform the received messages to the receiver's coordinate system. We now describe the aggregation process that the receiving vehicle performs. We refer the reader to Algorithm 15.1 for pseudocode.

We first compensate for the time delay between the vehicles to create an initial state for each node in the graph. Specifically, for each node, we apply a convolutional neural network (CNN) that takes as input the received intermediate representation $\hat{z}_i$, the relative 6 degrees of freedom (DoF) pose Δp_i between the receiving and transmitting SDVs and the time delay $\Delta t_{i \to k}$ with respect to the receiving vehicle sensor time. Note that for the node representing the receiving car, $\hat{z}$ is directly its intermediate representation. The time delay is computed as the time difference between the sweep

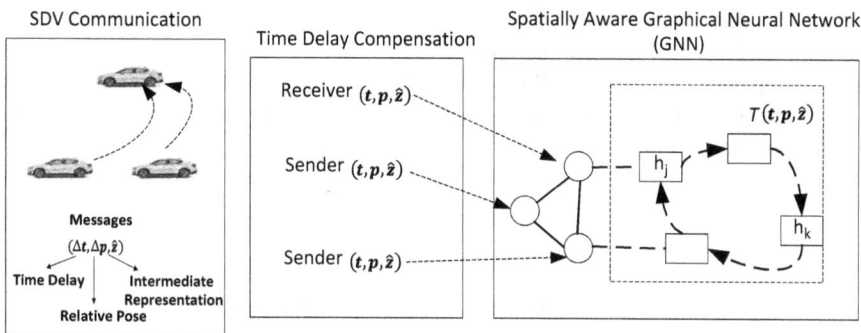

FIGURE 15.2 Self-driving vehicles (SDV) and spatially aware graphical neural network (GNN).

start times of each vehicle, based on universal GPS time. We then take the time delay-compensated representation and concatenate with zeros to augment the capacity of the node state to aggregate the information received from other vehicles after propagation (line 3 in Algorithm 15.1).

Next, we perform GNN message passing. The key insight is that because the other SDVs are in the same local area, the node representations will have overlapping fields of view. If we intelligently transform the representations and share information between nodes where the fields of view overlap, we can enhance the SDV's understanding of the scene and produce better output P&P. Figure 15.2 visually depicts our spatial aggregation module. We first apply a relative spatial transformation $\xi_{i \to k}$ to warp the intermediate state of the ith node to send a GNN message to the kth node. We then perform joint reasoning on the spatially aligned feature maps of both nodes using a CNN. The final modified message is computed as in Algorithm 15.1 line 7, where T applies the spatial transformation and resampling of the feature state via bilinear-interpolation, and $M_{i \to k}$ masks out nonoverlapping areas between the fields of view. Note that with this design our messages maintain the spatial awareness.

We next aggregate at each node the received messages via a mask-aware permutation-invariant function M and update the node state with a convolutional gated recurrent unit (ConvGRU) (Algorithm 15.1 line 8), where $j \in N(i)$ are the neighboring nodes in the network for node i and ϕ_M is the mean operator. The mask-aware accumulation operator ensures only overlapping fields of view are considered. In addition, the gating mechanism in the node update enables information selection for the accumulated received messages based on the current belief of the receiving SDV. After the final iteration, a multilayer perceptron outputs the updated intermediate representation (Algorithm 15.1 line 11). We repeat this message propagation scheme for a fixed number of iterations.

OUTPUT NETWORK

After performing message passing, we apply a set of four inception-like convolutional blocks to capture multiscale context efficiently, which is important for prediction. Finally, we take the feature map and exploit two network branches to output detection and motion forecasting estimates, respectively. The detection output is(x, y, ω, h, θ), denoting the position, size, and orientation of each object. The output of the motion forecast branch is parameterized as (x_i, y_i), which denotes the object's location at future time step t. We forecast the motion of the actors for the next 3 seconds at 0.5 s intervals. Please see supplementary for additional architecture and implementation details.

LEARNING

We first pretrain the LiDAR backbone and output headers, bypassing the cross-vehicle aggregation stage. Our loss function is cross-entropy on the vehicle classification output and smooth on the bounding box parameters. We apply hard-negative mining to improve performance. We then fine-tune jointly the Li-DAR backbone, cross-vehicle aggregation, and output header modules on our novel V2V dataset with synchronized inputs (no time delay) using the same loss function. We do not use the temporal warping function at this stage. During training, for every example in the mini batch, we randomly sample the number of connected vehicles uniformly on [0; min (c; 6)], where c is the number of candidate vehicles available. This is to make sure that V2VNet can handle arbitrary graph connectivity while also making sure the fraction of vehicles on the V2V network remains within the GPU memory constraints. Finally, the temporal warping function is trained to compensate for time delay with asynchronous inputs, where all other parts of the network are fixed. We uniformly sample time delay between 0.0 s and 0.1 s (time of one 10 Hz LiDAR sweep). We then train the compression module with the main network (backbone, aggregation, output header) fixed. We use a rate-distortion objective, which aims to maximize the bit rate in transmission while minimizing the distortion between uncompressed and decompressed data. We define the rate objective as the entropy of the transmitted code, and the distortion objective as the reconstruction loss (between the decompressed and uncompressed feature maps).

V2V Simulation – Dataset for V2V Communication

No realistic dataset for V2V communication exists in the literature. Some approaches simulate the V2V setting by using different frames from the KITTI datasets [1] to emulate multiple vehicles. However, this is unrealistic since sensor measurements are at different timestamps, so moving objects may be at completely different locations (e.g., a 1 s time difference can cause 20 m change in position). Other approaches utilize a platoon strategy for data collection [1], where each vehicle follows behind the previous one closely. While more realistic than using KITTI, this data collection is biased: the perspectives of different vehicles are highly correlated with each other, and the data does not provide the richness of different V2V scenarios. For example, we will never see SDVs coming in the opposite direction, or SDVs turning from other lanes at intersections.

To address these deficiencies, we use a high-fidelity LiDAR simulator, LiDAR-sim [1], to generate our large-scale V2V communication dataset, which we call V2V-Sim. LiDAR-sim is a simulation system that uses a large catalog of 3D static scenes and dynamic objects that are built from real-world data collections to simulate new scenarios. Given a scenario (i.e., scene, vehicle assets and their trajectories), LiDAR-sim applies ray casting followed by a deep neural network to generate a realistic LiDAR point cloud for each frame in the scenario.

We leverage traffic scenarios captured in the real world ATG4D dataset [1] to generate our simulations. We recreate the snippets in LiDAR-sim's virtual world using the ground-truth 3D tracks provided in ATG4D. By using the same scenario layouts and agent trajectories recorded from the real world, we can replicate realistic traffic. At each timestep, we place the actor 3D assets into the virtual scene according to the real-world labels and generate the simulated LiDAR point cloud seen from the different candidate vehicles. We define the candidate vehicles to be nonparked vehicles that are within the 70 meter broadcast range of the vehicle that recorded the real-world snippet. We generate 5,500 25 s snippets collected from multiple cities. We subsample the frames in the snippets to produce our final 46,796/4,404 frames for train/test splits for the V2V-Sim dataset. V2V-Sim has on average ten candidate vehicles that could be in the V2V network per sample, with a maximum of 63 and a variance of 7, demonstrating the traffic diversity. The fraction of vehicles that are candidates increases linearly w.r.t broadcast range.

EXPERIMENTAL EVALUATION

In this section, we showcase the performance of our approach compared to other transmission and aggregation strategies as well as single-vehicle P&P. Metrics: We evaluate both detection and motion forecasting around the ego vehicle with a range of: $x \in [-100, 100]$ m, $y \in [-40, 40]$ m. We include completely occluded objects (0 LiDAR points hit the object), making the task much more challenging and realistic than standard benchmarks. For object detection, we compute average precision (AP) and precision-recall (PR) curve at intersection-over-union (IoU) threshold of 0.7. For motion forecasting, we compute absolute l_2 displacement error of the object center's location at future timestamps (3s prediction horizon with 0.5 s interval) on true positive detections.

We set the IoU threshold to 0.5 and recall to 0.9 (we pick the highest recall if 0.9 cannot be reached) to obtain the true positives. These values were chosen such that we retrieve most objects, which is critical for safety in self-driving. Note that most self-driving systems adopt this high recall as the operating point. We also compute trajectory collision rate (TCR), defined as the collision rate between the predicted trajectories of detected objects, where collision occurs when two cars overlap with each other more than a specific IoU (i.e., collision threshold τ). This metric evaluates whether the predictions are consistent with each other. We exclude the other SDVs during evaluation, as those can be trivially predicted.

BASELINES

We evaluate the single-vehicle setting, dubbed No Fusion, which consists of LiDAR backbone network and output headers only, without V2V communication. We also introduce two baselines for V2V communication: LiDAR Fusion and Output Fusion. LiDAR Fusion warps all received LiDAR sweeps from other vehicles to the coordinate frame of the receiver via the relative transformation between vehicles (which is known, as all SDVs are assumed to be localized) and performs direct aggregation. We use the state-of-the-art LiDAR compression algorithm Draco [1] to compress LiDAR Fusion messages. For Output Fusion, each vehicle sends post-processed outputs, i.e., bounding boxes with confidence scores, and predicted future trajectories after nonmaximum suppression (NMS). At the receiver end, all bounding boxes and future trajectories are first transformed to the ego-vehicle coordinate system and then aggregated across vehicles. NMS is then applied again to produce the results.

EXPERIMENTAL DETAILS

For all analysis, we set the maximum number of SDVs per scene to be 7 (except for an ablation study measuring how the number of SDVs affect V2V performance). All models are trained with Adam [1], which is a method for stochastic optimization.

TABLE 15.1

Detection Average Precision (AP) at Intersection-over-Union (IoU) = {0.5, 0.7}, Prediction with l_2 Error at Recall 0.9 at Different Timestamps, and Trajectory Collision Rate (TCR)

Method	AP@IoU↑		l_2 Error (m)↓			TCR↓
	0.5	0.7	1.0 s	2.0 s	3.0 s	$\tau = 0.1$
No Fusion	77.3	68.5	0.43	0.67	0.98	2.84
Output Fusion	90.8	86.3	**0.29**	**0.50**	0.80	3.00
LiDAR Fusion	92.2	88.5	**0.29**	**0.50**	0.79	2.31
V2V Network	**93.1**	**89.9**	**0.29**	**0.50**	**0.78**	**2.25**

COMPARISON TO EXISTING APPROACHES

As shown in Table 15.1, V2V-based models significantly outperform No Fusion on detection (~20% at IoU 0.7) and prediction (~0.2 m l_2 error reduction at 3). LiDAR Fusion and V2VNet also show strong reduction (20% at 0.01 collision threshold) in TCR. These results demonstrate that all types of V2V communication provide substantial performance gains. Among all V2V approaches, V2VNet either is on-par with LiDAR Fusion (which has no information loss) or achieves the best performance.

V2VNet's slight performance gain over LiDAR Fusion may come from using the GNN in the cross-vehicle aggregation stage to reason about different vehicles' feature maps more intelligently than naive aggregation. Output Fusion's drop in performance for TCR is due to the large number of false positives relative to other V2V methods (at recall > 0.6). It is also seen that the percentage of objects with an l_2 error at 3 s smaller than a constant. This metric shows similar trends consistent with Table 15.1. The fusion algorithms results that are same as those V2V Network are shown in "bold."

COMPRESSION

Simulation results show the trade-off between transmission bandwidth and accuracy for different V2V methods with and without compression. Draco [1] achieves 33× compression for LiDAR Fusion, while our compressed intermediate representations achieved a 417× compression rate. Note that compression marginally affects the performance. This shows that the intermediate P&P representations are much easier to compress than LiDAR. Given the message size for one timestamp with a sensor capture rate of 10 Hz, we compute the transmission delay based on V2V communication protocol. At the broadcast range 120 meters, the data rate is roughly 25 Mbps. This means sending V2VNet messages may induce roughly 9 ms delay, which is very low.

SDV DENSITY

We now investigate how V2V performance changes as a function of % of SDVs in the scene. To make this setting like the real world, for a given 25 s snippet, we choose a fraction of candidate vehicles in the scene to be SDVs for the whole snippet. It is also shown that V2V performance increases linearly with the % of SDVs in both detection and prediction.

Number of LiDAR Points and Velocity

V2V methods boost the performance on completely and mostly occluded objects (0 and 1 ~ 6 LiDAR points) by over 60% in AP. This is an extremely exciting result, since the main challenges of perception and motion forecasting are objects with very sparse observations. Simulations result shows performance on objects with different velocities. While other V2V methods drop in detection performance as object velocity increases, V2VNet has consistent performance gains over No Fusion on fast-moving objects. Output Fusion and LiDAR Fusion may have deteriorated due to the rolling shutter of the moving SDV and the motion blur of moving agents during the temporal sweep of the LiDAR sensor. These effects are more severe in the V2V setting, where SDVs may be moving in opposite directions at high speeds while recording moving actors. Although not explicitly tackling such issue, V2VNet performs contextual and iterative reasoning on information from different vehicles, which may indirectly handle rolling shutter inconsistencies.

Imperfect Localization

We have simulated inaccurate pose estimates by introducing different levels of Gaussian $\sigma = 0.4\ m$ and von Mises $\left(\sigma = 4^\circ\right); \frac{1}{\kappa} = 4.873 \times 10^{-3}$ noise to position and heading of the transmitting SDVs. It is also seen, on both noise types, V2VNet outperforms LiDAR Fusion and Output Fusion in P&P performance. The only exception is Output Fusion l_2 error with heading noise larger than 3°. We hypothesize that Output Fusion's performance is better at this setting due to its low-recall (fewer true positives) relative to V2VNet (0.62 vs. 0.73 at 4° noise). Fewer true positives can cause lower l_2 error relative to higher recall methods. Degradation from heading noise is more severe than position noise, as subtle rotation in the ego-view will cause substantial misalignment for far-off objects; a vehicle bounding box (5 m × 2 m) rotated by 1° with respect to a pivot 70 meters away generates an IoU of 0.39 with the original.

Asynchronous Propagation

We have simulated random time delay by delaying the messages of other vehicles at random from $u(0,t)$, where $t = 0.1$. We apply a piece-wise linear velocity model (computed via finite differences) in Output Fusion to compensate for time delay. We do not adjust for LiDAR Fusion as it is nontrivial. It is seen that V2VNet demonstrates robustness across different time delays. Output Fusion does not perform well at high time delays as the piece-wise linear model used is sensitive to velocity estimates.

Mixed Fleet

We have also investigated the case that the SDV may receive different types of perception messages (i.e., sensor data, intermediate representation, and P&P outputs). We analyze the setting where every SDV (other than the receiving vehicle) has one-third chance to broadcast each measurement type. We then perform Sensor Fusion, V2VNet, Output Fusion for the relevant set of messages to generate the final output. The result is in between the three V2V approaches: 88.6 AP at IoU = 0.7 for detection, 0.79 m error at 3.0 s prediction, and 2.63 TCR.

Qualitative Results

Form simulation results, it is seen that V2VNet can handle occlusion. For example, from the simulation results, we have perceived, and motion forecast, a high-speed vehicle in our right lane, which can give the downstream planning system more information to better plan a safe maneuver for a lane change. V2V-Net also detects many more vehicles in the scene that were originally not detected by No Fusion.

SUMMARY

We have proposed a V2V approach for P&P that transmits compressed intermediate representations [1] of the P&P neural network, achieving the best compromise between accuracy improvements and bandwidth requirements. To demonstrate the effectiveness of our approach, we have created a novel V2V-Sim dataset that realistically simulates the world when SDVs will be ubiquitous. We hope that our findings will inspire future work in V2V perception and motion forecasting strategies for safer self-driving cars. This allows us to see through occlusions and detect actors at long range, where the observations are very sparse or nonexistent. We also show that our approach of sending compressed deep feature map activations achieves high accuracy while satisfying communication bandwidth requirements.

REFERENCE

1. Wang, T. H. et al., "V2VNet: Vehicle-to-Vehicle Communication for Joint Perception and Prediction," Ubert-ATG and University of Toronto, arXiv:2008.0751v1 [cs.CV] 17 Aug 2020.

16 Peer-to-Peer V2V Communication

OVERVIEW

The existing vehicle-to-vehicle networks that are based on central structures are prone to single point of failures and may consequently result in system paralysis. The peer-to-peer (P2P) network communications architecture that does not have any single point of failure will enable the V2V network to be fault tolerance and maintain the stability of the network system. Moreover, P2P network does not need the centralized network operation and maintenance. Consequently, the base station and the roadside units will no longer be necessary facilities [1]. The vehicles can be used as relays to directly participate in the exchange and transfer of information. The relay nodes (vehicles) are selected based on the degree distribution and the consensus algorithm. The real-time capability, efficiency, and cost-effectiveness of the proposed P2P model is verified through experimental results.

Vehicles interconnect with each other through various vehicular sensors, mobile communication devices, navigation systems, and other communication modules that can be termed as the "internet of things" (IoT). V2V communication can help broadcast possible traffic accidents, congestion, and dynamic obstacle information, to improve travel efficiency [1]. The traditional V2V communication mainly requires on-board units (OBUs) and roadside units (RSUs). RSUs unify configuration and management of resources as the central nodes within a certain region. This centralized structure has defects such as poor scalability and low bandwidth utilization. Besides, it is prone to single failures. If a central node in a region fails, the main information publisher would be paralyzed, yielding serious system failure in this region. Besides, the implantation cost would be extremely high with the centralized V2V communication structure, as it requires numerous base stations.

P2P V2V network is a self-organizing system of equal and autonomous entities (peers) which aims for the shared usage of distributed resources in a networked environment avoiding central services. There are a few routing protocols for P2P vehicular communication which can be grouped into three categories: unicast approach, broadcast approach, and geocast approach. Unicast protocols provide information delivery between two nodes via multiple wireless hops. Unicast is the most straightforward way to implement; however, it suffers a low packet delivery ratio and a large delay of packet transmission since the routing paths are not optimal.

A broadcast is an effective approach for safety-related information exchange such as emergency accident, traffic information services, announcements, and advertisement, to achieve cooperative driving in vehicle ad hoc network (VANET). However, it has disadvantages such as message redundancy and link unreliability that seriously

DOI: 10.1201/9781003499480-16

degrade the network efficiency [1]. Geocast routing is essentially a location-based routing, to broadcast the packet from a source node to all other nodes within a specified geographical region (zone of relevance – ZOR). Geocast reduces network congestion by restricting the transmission area. However, it does not consider the untrusted environment [1].

P2P NETWORK

In the network layer, high security and good communication quality are the basic requirements of V2V communication. Besides, construction of RSUs and upgrading existing base station equipment usually means a huge cost. P2P is a decentralized network structure, which possesses the following advantages compared to the traditional centralized network. The P2P network has a good robustness. The service is distributed among the nodes. When some nodes fail in a P2P network, other peers can automatically adjust the overall topology and maintain network connectivity and information transmission. This effectively avoids the occurrence of single-point failures. P2P network has high flexibility and reliability. All participants may serve as relays, thus greatly improving the flexibility and reliability of anonymous communication. P2P network has the advantage of the balanced load. Since each node is both a server and a client, it can reduce the computing amount of traditional client/server (C/S) structure and storage requirements, to achieve the entire network load balance.

DISTRIBUTED CONSCIOUS

There are a few key problems that the P2P network structure needs to solve, e.g. self-organization of nodes and the disagreement among nodes. The blockchain (BC) technology has intrinsic similarities with the vehicle networking. They both have many nodes with a large amount of information broadcasting in between them. The BC technology exploits distributed consensus algorithms to solve the problem of trusted communication in an untrusted network environment. V2V network can also realize self-maintenance of vehicle networking through consensus algorithms.

There are some common consensus algorithms in the realms of BC technology, such as proof of work (POW), proof of stake (POS), delegated POS (DPOS), and so on. But they all demand enormous computing power and computing time. Besides, these algorithms could not adapt to the constantly changing topology of vehicle networking. There are also some traditional consensus algorithms, such as Paxos and the more concise version, i.e., Raft [1] algorithm. Here, we choose the Raft algorithm to establish a self-maintaining vehicle network.

P2P V2V NETWORK COMMUNICATIONS NETWORK

A V2V vehicle network architecture is described based on the P2P network. In this network, each vehicle node undertakes the data/information release and provision at the same time. The vehicle nodes can join in or exit as peers at will, not affecting the

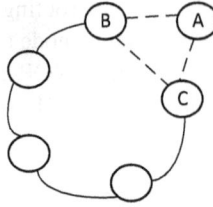

FIGURE 16.1   A schematic view of an example P2P network.

overall performance of the network. We model the vehicle network topology with an undirected graph $g = (V, E)$, representing the connected graph of a vehicle network. There are V nodes in g. E represents the number of edges in the graph. Each vehicle in the vehicle network is modeled as a vertex in the undirected graph. Each path of V2V communication is modeled as an edge in the undirected graph. Distributed and balanced load of the P2P network help enhance the reliability of the system. At the same time, the decentralization topology has good robustness and high security, avoiding the single-point failure. In a P2P vehicle network, any node can join in and exit the network without affecting the system performance. Figure 16.1 illustrates when node A exits the network, node B and node C can quickly connect through a new path.

SELF-ORGANIZATION OF VEHICLE NETWORKING

The average number of hops from one node to another node in the vehicle network is defined as the transmission length $L(g)$, as shown below:

$$L(g) = \frac{1}{(N/2)} \sum_{i,j \in V} d(i,j)$$

where $N/2$ means half numbers of the nodes, $d(i, j)$ refers to the shortest path between node i and node j, and $(i, j \in V)$. The adjacency matrix of a vehicle network is another important concept. In the vertex set V, the set of the adjacency list of elements is defined as the adjacency matrix D_{con}, where the element a_{ij} represents the connection state between the ith node and the jth node. In graph theory and computer science, an adjacency matrix is a square matrix used to represent a finite graph. The elements of the matrix indicate whether pairs of vertices are adjacent or not in the graph. In the special case of a finite simple graph, the adjacency matrix is a $(0, 1)$-matrix with zeros on its diagonal as shown in Figure 16.2 and its symmetric adjacent matrix is provided here.

$$\text{Adjacent Matrix} = \begin{bmatrix} 0 & 0 & 0 & 1 \\ 0 & 0 & 0 & 1 \\ 0 & 0 & 0 & 1 \\ 1 & 1 & 1 & 0 \end{bmatrix}$$

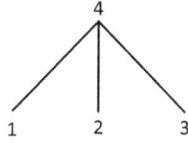

FIGURE 16.2 A simple graph with four vertexes and three links.

The adjacency matrix is a symmetric matrix with all diagonal elements being zero, as shown below:

$$D_{con} = \begin{bmatrix} 0 & a_{12} & \cdots & a_{1j} \\ a_{21} & 0 & \cdots & a_{2j} \\ a_{31} & a_{32} & \cdots & a_{3j} \\ a_{i1} & a_{i2} & \cdots & 0 \end{bmatrix}$$

Each column of the adjacency matrix is the adjacency table of the node. In a P2P vehicle network, the adjacency table p can be used as the routing table of vehicles in practice. Take the first column of D_{con} as an example; it represents the connection between the first car and the other i vehicle nodes:

$$p = \begin{bmatrix} 0, a_{21}, a_{31}, \ldots, a_{i1} \end{bmatrix}^T$$

The degree of the vehicle network K_v is defined as the number of edges at which the vertex of a vehicle node is v and the number of edges connected to the vertex at the nth moment. The degree of each node in the P2P vehicle network can reflect the topology of the network. The above equation illustrates the transformation relation between K_v and p_n.

$$K_v = \sum_1^n p_n$$

The adjacency matrix at the vehicle node cannot be obtained directly. We use the distance matrix instead, and transform the distance matrix into the adjacency matrix:

$$M_{dis} = \begin{bmatrix} d_{11} & d_{12} & \cdots & d_{1t} \\ d_{21} & d_{22} & \cdots & d_{2t} \\ \cdots & \cdots & \cdots & a_{3j} \\ d_{t1} & d_{t2} & \cdots & d_{t \times t} \end{bmatrix}$$

Each element d is obtained through the global positioning satellite (GPS) information of the vehicles. Differential GPS positioning technology ensures the positioning accuracy at the centimeter level, thus satisfying the requirement of V2V. Suppose an OBU obtains the GPS information of the surrounding t vehicles at one moment, yielding M_{dis} with the size of $t \times t$. For example, in a real scenario, M_{dis} may exist as follows: (in a unit of meter):

$$M_{dis} = \begin{bmatrix} 0 & 99.71 & \dots & 151.88 \\ 99.71 & 0 & \dots & 269.09 \\ \dots & \dots & \dots & \dots \\ 151.88 & 269.0 & \dots & 0 \end{bmatrix}$$

Suppose the communication distance of OBUs is d. When $d = 250$ m, we have

$$d_{ij} = \begin{cases} 1 & (d_{txt} \geq 250) \\ 0 & (d_{txt} < 250 \cup i = j) \end{cases}$$

The distance matrix is further simplified to an adjacency matrix with only 0 or 1 as explained earlier. Then by calculating the sum of each column of the adjacency matrix, the degree of the node can be obtained:

$$D_{con} = \begin{bmatrix} 0 & 1 & \dots & 1 \\ 1 & 0 & \dots & 0 \\ \dots & \dots & \dots & \dots \\ 1 & 0 & \dots & 0 \end{bmatrix}$$

where the element a_{ij} still represents the connection state between the K_v ith node and the jth node. When $a_{ij} = 0$, it means there is no routing demand between the i and j nodes, and vice versa. To implement a self-organized vehicle network, selection of the appropriate relay node is important. When each vehicle node chooses its own relay node for V2V communication, it is necessary to reach a consensus. In this part, we use the Raft consensus algorithm to elect a relay node. Raft ensures that each

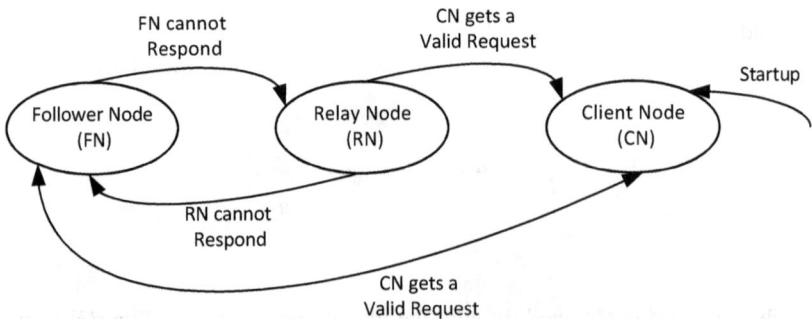

FIGURE 16.3   State transition at relaying [1].

node in the cluster agrees on the same series of state transitions. The State Machine
Safety is guaranteed by a restriction on the election process.

ALGORITHM 1   Pseudo Code for Messages Relaying among Vehicles

Input: Request from vehicles to transfer information
Output: The vehicle receives useful information by relaying.
Initialize: CN (Client Node), FN (Follower Node), RN (Relay Node), relay_message_
    num = 0,
FN_useful_message = 0;
*While (CN = 0) do:
            CN broadcast;
            If (FN_useful_message = 1)
            FN = 1;
            FN broadcast, CN = 1;
            Else if
            FN = 0, FN broadcast;
            CN compare {min(destination)and max (Kv)}
            RN = 1;
            End if
            relay_message_num + +;
    End

In our P2P network, each vehicle node maintains its own routing table about the GPS
information of surrounding vehicles, through periodic handshaking. When the node
hopes to know the traffic information of the destination, it enquires the surrounding
vehicles in the routing table p for the information. If the surrounding vehicles in the
routing table p have no relevant information, the node can extend the request through
the next relay node.

All messages are sent in broadcast packets. The nodes in the vehicle network may
be used for three rolls: the client node (CN), the relay node (RN) and the follower
node (FN). The transition between the three rolls is shown in Figure 16.3. The entire
process is for a vehicle to send a request and then get an effective response. The proposed vehicle relay algorithm is illustrated in Algorithm 1. If a client gets a valid
reply, let CN = 1, otherwise CN = 0. If in the following mode, the vehicle responds
to the request, let FN = 1, otherwise FN = 0. If the vehicle node is selected as a relay
node, let RN = 1, otherwise RN = 0. After each relay process, the count of relay messages (relay_message_num) is accumulated by 1.

The client node sends requests to all nodes within the communication scope.
When nodes receive the request, the surrounding vehicle nodes transit to the corresponding rolls. If a surrounding vehicle carries useful information, then *FN_useful_
message* = 1. At this time, the surrounding vehicles can directly reply to the client
node without relaying. If FN_useful_message = 0, the following node need replies to
their destination to the client node, and the client node selects the node whose destination is similar to itself and has higher K_v value as the relay node.

The client node selects the node whose destination is like itself and has higher K_v value as the relay node. In summary, the relay nodes should have the similar destination as the requesting node. From the network topology, when the node has a higher degree K_v, it can quickly share the request and broadcast it to the surrounding nodes. This helps improve the efficiency of information transmission in the P2P network.

SIMULATION OF V2V COMMUNICATION AT A CROSSROAD

Interactions are often traffic-intensive area and most prone to traffic accidents. We choose interactions as the experiment scenario. The parameters are shown in Table 16.1 for an example scenario; the sampling frequency is once per minute.

The vehicle network is constructed based on the P2P model, with no central node. The P2P communication is directly through OBUs, within the broad coverage radius d of the OBU. If beyond the communication coverage radius d, information will be transferred through vehicle relays. As a comparison, a traditional centralized vehicle network is also simulated. In traditional scenario, where the central node CENTRE is located at the coordinate of (250, 250). The CENTRE node refers to the base station. The information sent by the CN needs to go first to the resource allocation of CENTRE and then reach the destination node. All nodes communicate directly through the CENTRE node. The coverage of the CENTRE is set to be 800 m.

We measure the average path length of the traditional vehicle network and the P2P network, to demonstrate the information transfer efficiency. As shown in Table 16.2, when the number of nodes varies from 50 to 150, the average path length of the P2P vehicle network is much shorter than the traditional vehicle network.

TABLE 16.1
Traffic Parameters at a Crossroad

Parameters	Value
Average speed of vehicle	36 km/hour
Number of vehicles in each street	50
Total number of vehicles	200
Street width	36 m
Regional area	500×500 m^2
Radius d of OBU communication	250 m

TABLE 16.2
Average Path Length for Traditional and P2P Network

Number of Nodes	P2P	Tradition
50	~1.22	~1.23
100	~1.23	~1.29
150	~1.24	~1.32
200	~1.33	~1.34

Due to the impact of the distribution of vehicles at the interactions, when the number of nodes increases up to 200, the P2P network has a similar average path length to the traditional centralized vehicle network.

EXPERIMENTAL RESULTS

Based on the above simulation results, which verify the topological properties of proposed P2P vehicle network, we demonstrate the experiment on actual vehicles. In the experiment, we implement a P2P network with 11 nodes on the 3G network and Wi-Fi network, respectively. The schematic diagram of the node distribution is shown in Figure 16.4.

In the experiment, we measured round-trip time (RTT), the number of sent acknowledgement (ACK) packets, and the number of received ACK packets. The experimental result of RTT is shown in Table 16.3.

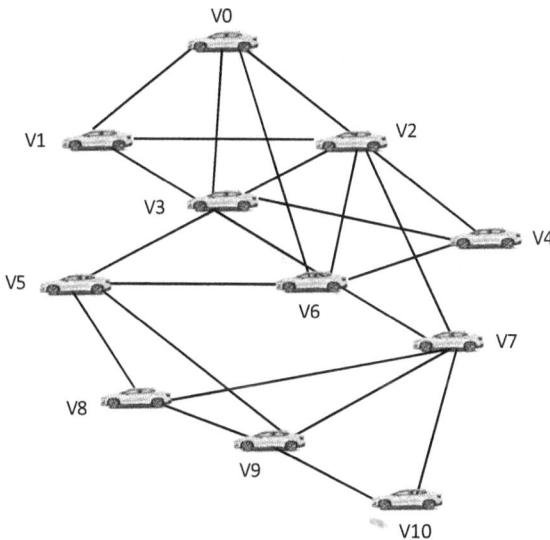

FIGURE 16.4 Example P2P node (vehicle) distribution.

TABLE 16.3
Round-Trip Time Measurement

Sent ACK	Receive ACK	RTT (milliseconds)
12	12	13.450
13	13	11.992
14	14	10.780
16	16	9.284

TABLE 16.4

Round-Trip Time Measurement with Changes in Vehicle Speeds

Speed of Vehicle (km/h)	RTT (milliseconds)
0	~11.8
20	~9
40	~9
60	~9.2
80	~9.5

After 75 trials, the average RTT is about 12 milliseconds. The results show that the proposed V2V P2P networks satisfy the real-time requirement. Table 16.4 shows that RTT changes little when vehicle speed changes.

Information transmission delay (i.e., measured as RTT) is the critical criterion for the vehicle network. If the communication delay between nodes is as low as in the scale of milliseconds, the P2P vehicle network is considered feasible. The total delay is provided by:

$$d_{sum} = d_{queue} + d_{trans} + d_{proc} + d_{prop}$$

where the total delay d_{sum} is divided into four parts: transmission delay (d_{trans}, almost same as RTT), propagation delay (d_{prop}), queuing delay (d_{queue}), and processing delay (d_{proc}). Because the propagation speed of the electromagnetic wave is close to the speed of light, the propagation delay (d_{prop}) is usually ignored. Table 16.5 lists the environment parameters of IEEE 802.11p. The following results can be obtained after substituting:

$$d_{sum} = d_{queue} + d_{trans} + d_{proc} = 2.1\,\text{ms}$$

TABLE 16.5

Environmental Parameters in IEEE802.11p

Parameters	Value
Channel model	Two ray ground*
Packet length L_{data}	1543 Bytes
Data rate R_{link}	6 Mbps
Slot length T timeslots	13 μs
Short interframe interval T_{sifs}	32 μs
Decentralized interframe space T_{difs}	58 μs
Propagation delay d_{prop}	2 μs

* The two-rays ground-reflection model is a multipath radio propagation model, which predicts the path losses between a transmitting antenna and a receiving antenna when they are in line of sight (LOS).

TABLE 16.6
Environment Parameters in LTE

Parameters	Value
User equipment (UE) processing delay	1.0 ms
Frame adjustment	0.5 m
TTI for uplink (UL) data packet	1.0 ms
eNB processing delay	1.0 ms

Similarly, in the long-term evolution (LTE) environment, only the user interface delay is estimated, and the required parameters are as Table 16.6. The total delay can be calculated as follows:

$$d_{\text{sum}} = d_{\text{queue}} + d_{\text{trans}} + d_{\text{proc}} = 3.5\,\text{ms}$$

The total delay of the node is related to hybrid automatic repeat request (HARQ) and the number of relay nodes. Generally, in the prescheduling mode of LTE, the value of HARQ is 0–30% [1]. When the delay of the propagation delay is neglected, the computation of the user interface delay T_d is:

$$T_d = d_{\text{queue}} + d_{\text{trans}} + d_{\text{proc}} = s\left(d_{\text{sum}} + 5bn\right) + m$$

In the above equation, T_d means user's interface delay in relay mode, m means the number of queuing nodes, n means the number of retransmissions, s means the number of relay nodes, and b means the probability of an error occurring during the first retransmission. Therefore, when $s = 2$, $n = 2$, $m = 3$, $p = 30\,\%$, $m = 2$, $P = 30\,\%$., $T_d = 16$ ms.

In general, the response time of the driver T_e should be at the time of early warning of the 1–2 s [1]. The time of V2V communication should be in millisecond time delay and should be left out of sufficient T_e to respond to the driver, or to brake in time in unmanned driving. The total delay time of V2V communication T_d is 16 ms. Since $T_d = 16$ ms $\ll T_e = 1.197$ s, we conclude that with either DSRC or LTE, millisecond delay can satisfy the P2P vehicle networks.

SELECTION OF RELAY FOR MULTIHOP TERMINATION CONDITIONS

Earlier, we proposed that the vehicle nodes with higher K_v value are selected as relay nodes. To broadcast real-time information, it is also necessary to determine the threshold of the relay multihops. In this section, we will discuss the selection of a relay node for multiple hops. For example, if the request issued by CN lacks the necessary termination conditions, information flooding will occur. It will cause congestion of the network and reduce the throughput of the entire vehicle network system.

TABLE 16.7

Number of Nodes (Vehicles) vs. Path Length

Number of Nodes	Average Path Length (Number of Hop)
100	~2.78
200	~2.68
300	~2.65
400	~2.60
500	~2.59
600	~2.62

We simulate the network connectivity of multiple interactions; as shown in Figure 16.4, there are four interactions, and each crossroad is spaced with 500 meters. When the number of vehicles varies from 200 to 600, the average path length of P2P vehicle network is within three hops, as shown in Table 16.7.

Therefore, the threshold of maximum three relay multihops is chosen as the convergence condition of vehicle flooding routing. If a client node still cannot get valid information after three relays, it abandons the request and starts a new round of request.

IMPLEMENTATION COST

Compared to the traditional centralized network, the cost of the P2P vehicle network is lower. Take Shenzhen as an example; as of November 2016, the number of cars (N_{cars}) in Shenzhen was 3.4 million. Within city limits, it is assumed that a base station could accommodate 250 vehicle nodes $A_{capacity}$ simultaneously online. The number of base stations N_{eNB} required is 1.36×10^4.

$$N_{eNB} = \frac{N_{cars}}{A_{capacity}}$$

The cost of each base station (M_{eNB}) is about \$120,000, and each base station has an annual repair cost (M_{repair}) about \$12,000. eNB refers to the number of years used by base stations. The base station cost C_{eNB} can be calculated as:

$$C_{eNB} = N_{eNB}\left(M_{eNB} + M_{repair}y_{eNB}\right)$$

Suppose each vehicle in the vehicle network is equipped with an OBU to communicate. The cost of traditional vehicle networks is the sum of base stations and OBU costs. But the cost of the P2P network only needs to pay the price of OBU):

$$C_{tradition} = C_{eNB} + C_{OBU}M_{OBU}$$

$$C_{P2P} = C_{OBU}M_{OBU}$$

TABLE 16.8
P2P and Traditional Cost

Number of Vehicles × 10⁴	Traditional (Centralized) Cost $ × 10⁶	P2P (Distributed) Cost $ × 10⁶
1	0.37	0.30
10	3.72	3.00
100	37.15	29.96
200	74.30	59.92
300	111.45	89.88

Table 16.8 shows the comparison between the cost of traditional methods and the P2P vehicle networks in five years, with the number of vehicles ranging from 10,000 to 3 million.

The cost of the decentralized vehicle network is far lower than that of the traditional vehicle network. For example, with a city containing 3 million vehicles, the decentralized vehicle network can save $2.16 billion for the V2V network construction.

SUMMARY

Traditional vehicle networks are usually centralized structures, which are prone to single-point failures. A P2P vehicle network is proposed in this chapter to increase the network fault tolerance and reduce the maintenance costs. The relay nodes (vehicles) are selected based on the degree distribution and the consensus algorithm. The real-time capability, efficiency, and cost-effectiveness of the proposed P2P vehicle network is verified through simulation and experimental results. Moreover, the superiority of cost-effectiveness to the P2P vehicle network is proven by the cost analysis. With a city containing 3 million vehicles, the decentralized vehicle network can save up to $2.16 billion for the V2V network construction.

REFERENCE

1. Yang, L. and Li, H., "Vehicle-to-Vehicle Communication Based on a Peer-to-Peer Network with Graph Theory and Consensus Algorithm," October 2018. https://doi.org/10.1049/iet-its.2018.5014, www.ietdl.org

17 V2V Massive MIMO Channels

Vehicle-to-vehicle (V2V) communication is enjoying substantial research attention as a benefit of its compelling applications. However, the ever-increasing tele-traffic is expected to result in overcrowding of the available band. As a first resortV2V communication [Chen] with multiple input multiple output (MIMO) antennas can be utilized to enhance the attainable bandwidth efficiency or link reliability. However, in hostile V2V wireless propagation environments, the achievable multiple-antenna gain is eroded by the channel correlation. As a promising MIMO technique, spatial modulation (SM) only activates a single transmit antenna (TA) in any symbol interval and, hence, completely avoids the inter-antenna interference, hence showing robustness against channel correlation [1]. As a further powerful solution, non-orthogonal multiple access (NOMA) has been used for improving the bandwidth efficiency and robustness. Inspired by the robustness of SM against channel correlation and the benefits of NOMA, this capability has been intrinsically amalgamated into NOMA-SM in order to deal with the deleterious effects of wireless V2V environments as well as to support improved bandwidth efficiency.

Moreover, the bandwidth efficiency of NOMA-SM is further boosted with the aid of a massive TA configuration. Specifically, a spatiotemporally correlated Rician channel is considered for a V2V scenario. We investigate the bit error ratio (BER) performance of NOMA-SM via Monte Carlo simulations, where the impact of the Rician K-factor, spatial correlation of the antenna array, time varying effect of the V2V channel, and the power allocation factor is discussed. Furthermore, we also analyze the capacity of NOMA-SM. By analyzing the capacity and deriving closed form upper bounds on the capacity, a pair of power allocation optimization schemes are formulated. The optimal solutions are demonstrated to be achievable with the aid of our proposed algorithm. Again, instead of simply invoking a pair of popular techniques, we intrinsically amalgamate SM and NOMA to conceive a new system component exhibiting distinct benefits in the V2V scenarios considered.

Compared to NOMA relying on Vertical-Bell Laboratories Layered Space-Time (VBLAST), NOMA-SM has been demonstrated to exhibit improved robustness against the spatial and temporal effects of the V2V channel. By analyzing the capacity and deriving analytical upper bounds in closed form, a pair of power allocation optimization schemes have been formulated for NOMA-SM. The optimal solutions have also been shown to be achievable with the aid of the proposed power allocation algorithm. The numerical results have verified that with the aid of an appropriate power allocation, NOMA-SM can satisfy the QoS support of a low-priority flow, whilst maximizing the throughput of the high-priority flow. In summary, NOMA-SM has been demonstrated to cooperatively improve the link reliability and bandwidth efficiency of V2V

DOI: 10.1201/9781003499480-17

transmissions. Interested readers might see all details of performance analysis in [1]. However, we are providing some performance analysis results here.

BER RESULTS

The BER performance of the NOMA-SM scheme is compared to NOMA relying on the popular VBLAST technique, where NOMA-VBLAST is used as a reference. The effects of the Rician K-factor, adjacent antenna correlation coefficient, temporal correlation, and power allocation factor are all taken into consideration. The Rician K-factors are configured as $K = 2.186$ and $K = 0.2$ for low and high vehicular traffic density, respectively. More specifically, QPSK and 16QAM are applied for NOMA-SM and NOMA-VBLAST, respectively.

The MIMO configuration of the reference is the same as that of NOMA-SM except for using number of transmitters $N_t = 2$. Thus, the following BER comparisons are carried out for the same bandwidth efficiency of 8 bits per channel use (bpcu). All simulation results of this subsection are obtained through a Monte Carlo method.

We have observed the BER performance for different Rician K-factor. It is seen that NOMA-SM significantly outperforms the benchmark. Additionally, the increase of K imposes a more dominant degradation on NOMA-VBLAST, which relies more vitally on the presence of rich non–line-of-sight (non-LoS) scattering. This phenomenon can be explained as follows. The higher Rician factor K represents a stronger LoS component, which increases the spatial correlation among the adjacent channel paths. For NOMA-VBLAST, the multiple-stream information is conveyed with the aid of multiple degrees of freedom (DoFs). By contrast, for NOMA-SM, although the more severe spatial correlation of the LoS scenario makes it difficult to determine the index of the TA, the remaining information related to the amplitude phase modulation (APM) signal-domain is transmitted over a single DoF; hence, it is less susceptible to spatial correlation.

We have investigated the BER results associated with different adjacent TA-correlation coefficients. NOMA-SM is less susceptible to spatial correlation. This phenomenon can be interpreted similarly to the trend as stated above. We investigate the impact of the V2V channel's time varying nature. The BER has been substantially degraded compared to that of the non–time-varying channel. Although a perfect channel estimation procedure is assumed for the receivers, the estimated channel coefficients used for ML detection becomes partially outdated due to the channel's time-varying nature, hence resulting in a degraded BER performance. Nevertheless, the NOMA-SM scheme maintains its advantage over the reference, regardless of the grade of temporal correlation. More importantly, we observe that NOMA-SM consistently outperforms NOMA-VBLAST. By jointly considering the above observations, we conclude that NOMA-SM constitutes a potent amalgam.

CAPACITY RESULTS

Compared to open mobile alliance (OMA)-SM, NOMA-SM provides substantial capacity gains both for the collaboration-aided vehicle and for the in-car user, and accordingly obtains a significant sum capacity enhancement. Specifically, the capacity of the in-car user has been beneficially boosted by the proposed scheme, about twice as high as that of OMA-SM. Since the APM signal-domain of the proposed scheme is combined with a NOMA strategy, each user accesses the channel resources via power domain multiplexing.

SUMMARY

The BER performance of the NOMA-SM transmission strategy has been investigated [1] with the impact of the Rician K-factor, spatial correlation of antenna array, time-varying effect of the V2V channel, and the power allocation factor being discussed. Compared to NOMA relying on VBLAST, NOMA-SM has been demonstrated to exhibit improved robustness against the spatial and temporal effects of the V2V channel. By analyzing the capacity and deriving analytical upper bounds in closed form, a pair of power allocation optimization schemes have been formulated for NOMA-SM. The optimal solutions have also been shown to be achievable with the aid of the proposed power allocation algorithm. Our numerical results have verified that with the aid of an appropriate power allocation, NOMA-SM can satisfy the QoS support of a low-priority flow, whilst maximizing the throughput of the high-priority flow. In summary, NOMA-SM has been demonstrated to cooperatively improve the link reliability and bandwidth efficiency of V2V transmissions.

REFERENCE

1. Chen, Y. et al., "Performance Analysis of NOMA-SM in Vehicle-to-Vehicle Massive MIMO Channels," *IEEE Journal on Selected Areas in Communications* December 2017, 35(12), 2653–2666.

18 V2X Communications Network

V2X ULTRA-MASSIVE MIMO COMMUNICATIONS

6G vehicle-to-everything (V2X) communication will be combined with vehicle automatic driving technology and play an important role in automatic driving. However, in 6G V2X systems, vehicle users have the characteristics of high-speed movement. Therefore, how to provide stable and reliable wireless link quality and improve channel gain has become a problem that must be solved. To solve this problem, a new multiuser scheduling (MUS) algorithm based on block diagonalization (BD) precoding for 6G ultra-massive multiple-input multiple-output (MIMO) systems is described here [1]. The algorithm takes advantage of the sensitive nature of BD precoding to channel correlation, uses the Pearson coefficient after matrix vectorization to measure the channel correlation between users, defines the scheduling factor to measure the channel quality according to the user noise enhancement factor, and jointly considers the influence of the correlation between user channels and channel quality, ensuring the selection of high-quality channels while minimizing channel correlation.

We consider a typical 6G V2X communication scenario and the ultra-Massive MIMO downlink system in the 6G V2X autonomous scenario shown in Figure 18.1 The system consists of a base station (BS), a roadside unit (RSU), and vehicles connected to the BS. BS and RSU are connected by optical fiber. The RSU (transmitting end) is equipped with N_t antennas, and the set of vehicles is $N = \{1, 2, ..., k\}$, where each vehicle user is equipped with N_r receiving antennas and multiple sensors. The system can schedule $R = \{1, 2, ..., R\}$, $R \subset N$ vehicles each time. In the downlink of the system with the assumption that the channel state information (CSI) is available, the maximum number of transmission data streams that each user can support is N_s, the total number of data streams sent by the BS is RN_s.

It can be seen from Figure 18.1 that the original signal is transmitted through the precoder at the antenna end and received by the receiving antenna of the receiving end through the channel. The received signal passes through the combiner to eliminate the interference between the user sub-channels, and finally undergoes signal processing to obtain the transmitted signal vector:

$$s = \sum_{k=1}^{R} F_k s_k$$

where $s \in \mathbb{C}^{(R \times N_s) \times 1}$ denotes the original transmission symbol vector and is such that $E\{ss^H\} = I_{RNs}$, $s_k \in \mathbb{C}^{N_s \times 1}$ is the original vector sent by the RSU to use k with $E\{s_k s_k^H\}$,

DOI: 10.1201/9781003499480-18

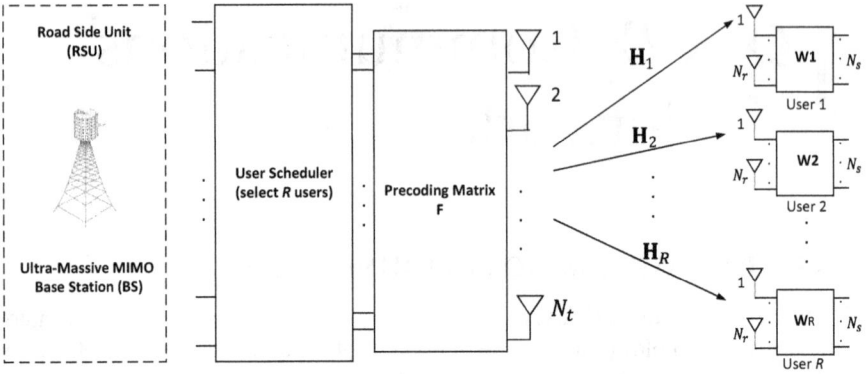

FIGURE 18.1 Multiuser scheduling for ultra-Massive MIMO base station with multiantenna mobile users.

and $F_k \in \mathbb{C}^{N_t \times N_s}$ is the precoding matrix of user k with $\|F_k\|^2 = 1$, which is the unitary matrix. If the vehicle has perfect time and frequency synchronization for each time slot t, the received signal of user k after passing through the channel is expressed as:

$$y_k(t) = H_k(t)F_k s_k + \sum_{i=1, i \neq k}^{R} H_k(t)F_i s_i + n_k(t), \quad k = 1, 2, \ldots, R$$

where $H_k(t)F_k s_k$ is the kth user useful data, $H_k(t) \in \mathbb{C}^{N_r \times N_t}$ is the channel matrix between the RSU and user k in the tth time slot, and each element of the matrix obeys a Gaussian distribution with zero mean and unit variance. $F_i s_i$ denotes interference from other user data to the given user data. $n_k(t) \in \mathbb{C}^{N_r \times 1}$ is the additive white Gaussian noise of $\mathbb{C}N(0, \sigma_k^2 I)$. The received signal obtained after the combiner is given by:

$$\tilde{y}_k(t) = W_k^H H_k(t)F_k s_k + \sum_{i=1, i \neq k}^{R} W_k^H H_k(t)F_i s_i + W_k^H n_k(t), \quad k = 1, 2, \ldots, R$$

where W_k is the combiner of the kth user. If proper precoding is used, the data interference between users can be eliminated at the receiving end, and the above equation is expressed as follows:

$$\tilde{y}_k(t) = W_k^H H_k(t)F_k s_k + W_k^H H_k(t)F_i s_i + W_k^H n_k(t), \quad k = 1, 2, \ldots, R$$

Assuming that BD precoding algorithm is used at the transmitting end for M numbers of users and the power P is equally distributed, the channel capacity of user k having bandwidth B is provided by:

$$C_k = B \log_2 \det\left(I_{kN} + \frac{P}{M\sigma^2} H_i^H(t) H_j^H(t) \right)$$

The use set channel capacity C can be expressed as:

$$C\left(\left[H_k(t)H_k^H(t)\right]^H\right) = B\log_2 \det\left(I_N + \frac{P}{M\sigma^2}\left[H_k(t)H_k^H(t)\right]^H\left[H_k(t)H_k^H(t)\right]\right)$$

SYSTEM PERFORMANCE COMPARISON

To verify the feasibility and effectiveness of the algorithm proposed in this chapter, and to study the impact of vehicles on the algorithm performance in different mobile environments, this chapter uses MATLAB simulation software to build a multiuser ultra-Massive MIMO system to simulate the proposed algorithm and the comparison algorithm, and finally to compare the simulation results. In the simulation, it is assumed that the receiving end and the transmitting end can obtain perfect channel state information. There are M users, the number of each user RF chain is $N_t^{RF} = N_t^{RF} \times M$. Other main simulation parameters are summarized in Table 18.1.

The simulation result mainly compares the bit error rate (BER) performance and spectrum efficiency of the proposed algorithm and the algorithm based on subspace correlation (SC), the geometric angle (GA) algorithm, and the condition number algorithm. The vehicle speed is 80 km/h, and the vehicle density is determined by the speed of the vehicle.

Tables 18.2 and 18.3, respectively, show the BER and spectral efficiency comparison diagrams of the proposed algorithm and the comparison algorithm when 256 transmitting antennas are configured at the BS and two vehicle users are scheduled. We can see from Tables 18.2 and 18.3 that the BER and channel capacity of the four

---

TABLE 18.1
Simulation Parameters

Simulation Parameters	Values
Carrier frequency	28 GHz
Number of transmit antennas, N_t	8
Number of single-user receiving antennas, N_s	2
Number of user, M	2/4
Channel model	Time varying geometric model
Number of clusters, N_c	3
Number of path in each cluster, N_{ray}	7
Azimuth mean distribution	$[0, 2\pi]$
Mean elevation angle distribution	$[-2/\pi, 2/\pi]$
Antenna array structure	Uniform linear array (ULA)
Antenna spacing, d	0.5λ
Vehicle antenna height	1.5 m
Vehicle density	0.0025/0.005/0.0083 vehicle/m
Vehicle moving direction	Move in a straight line along the road
Channel estimation	Idle

TABLE 18.2

SNR of Power and BER for Different Algorithms

	BER				
SNR/dB	Multiuser Scheduling Algorithm Based on Diagonalization Precoding for UM-MIMO System	Subspace Correlation Algorithm	Geometric Angle Algorithm	Conditional Number (CN) Algorithm	Remarks
−40	~4×10^{-1}	~4×10^{-1}	~4×10^{-1}	~4×10^{-1}	Number of
−35	~3.5×10^{-1}	~4×10^{-1}	~4×10^{-1}	~4×10^{-1}	user: $M = 2$
−30	~3.0×10^{-1}	~3.2×10^{-1}	~3.3×10^{-1}	~3.4×10^{-1}	Number of
−25	~2.0×10^{-1}	~3.0×10^{-1}	~4.0×10^{-1}	~4.0×10^{-1}	transmit antenna:
−20	~9.0×10^{-2}	~1.3×10^{-1}	~1.6×10^{-1}	~1.8×10^{-1}	$N_t = 256$
−15	~2.0×10^{-2}	~3.5×10^{-2}	~5.0×10^{-2}	~8.0×10^{-2}	
−10	~2.8×10^{-3}	~5×10^{-3}	~1.2×10^{-2}	~1.9×10^{-3}	

TABLE 18.3

Signal-to-Noise Power of Power and Spectral Efficiency for Different Algorithms

	Spectral Efficiency (Bits Per Second/Hz)				
SNR/dB	Multiuser Scheduling Algorithm Based on Diagonalization Precoding for UM-MIMO System	Subspace Correlation Algorithm	Geometric Angle Algorithm	Conditional Number (CN) Algorithm	Remarks
−40	~0.5	~0.5	~0.5	~0.5	Number of
−35	~1.0	~0.9	~0.85	~0.84	user: $M = 2$
−30	~2.0	~1.9	~1.8	~1.7	Number of
−25	~4.5	~4.3	~4.2	~4.1	transmit antenna:
−20	~7.8	~7.6	~7.5	~7.1	$N_t = 256$
−15	~13.5	~12.0	~11.5	~11.0	
−10	~19.5	~17.8	~17.0	~16.5	

algorithms are relatively close at low signal-to-noise ratio (SNR). With the increase in SNR, the difference of system BER and total system capacity realized by different scheduling algorithms gradually increases. In general, the proposed algorithm is better than the comparison algorithm in terms of BER and spectral efficiency. This is mainly because the conditions of the channel itself have a great impact on the information transmission effect. In general, the proposed MUS algorithm with 256 transmitting antennas is better than the comparison algorithm in terms of BER and spectral efficiency. This is mainly because the conditions of the channel itself have a great impact on the information transmission effect.

There will be a better transmission effect when the conditions of the channel itself are good. The comparison algorithm only considers the correlation of the channel matrix and ignores the influence of the conditions of the channel itself. The algorithm in this chapter considers not only the channel correlation but also the conditions of the channel itself. Adding a scheduling factor to measure the conditions of the channel can effectively eliminate the system interference and improve the system performance. The simulation results (no table included) also show that the spectrum efficiency achieved by several algorithms in the scenario where the number of vehicle users is four is greater than that of the same algorithm in the scenario where the number of vehicle users is two. This is because with the increase in number of scheduling users, the system capacity increases, and the transmitter can higher diversity gain, so the spectral efficiency performance increases.

Table 18.4 shows the comparison of BER and spectrum efficiency performance of the above algorithms when the number of scheduled users is two when 128 antennas are configured at the BS. Compared with Tables 18.2 and 18.3, the frequency efficiency of the system is also improved when the number of transmitting antennas is increased with the number of scheduling users fixed and the system resources limited. This is because increasing the number of antennas at the transmitting end can improve the antenna array gain and effectively improve the signal-to-interference-plus-noise ratio (SINR) at the receiving end of the system, to improve the spectrum efficiency of the system. However, with the increase in the number of transmitting antennas, the BER performance of the system decreases. The main reason is that when the number of transmitting antennas increases, the interference between channels also increases, which reduces the effective transmission signal of users. It can also be seen from Table 18.4 that the performance of the algorithm proposed in this chapter is the best regardless of BER or spectral efficiency. This is because the

TABLE 18.4
Speed of Vehicles and Bit Error Rate for Different Algorithms

Speed (km/h)	Bit Error Rate				
	Multiuser Scheduling Algorithm Based on Diagonalization Precoding for UM-MIMO System	Subspace Correlation Algorithm	Geometric Angle Algorithm	Conditional Number (CN) Algorithm	Remarks
90	~1.5×10^{-5}	~7×10^{-5}	~6×10^{-4}	~1.3×10^{-3}	Number of
100	~4×10^{-5}	~3.5×10^{-4}	~2.3×10^{-3}	~5.5×10^{-3}	user: 2
110	~9.0×10^{-5}	~9×10^{-4}	~6×10^{-3}	~1.8×10^{-2}	Number of
120	~0.8×10^{-4}	~1.8×10^{-3}	~0.4×10^{-3}	~4.0×10^{-2}	transmitting
130	~1.2×10^{-4}	~2×10^{-3}	~2.2×10^{-2}	~7×10^{-2}	antenna: 128
140	~2.0×10^{-4}	~4×10^{-3}	~4×10^{-2}	~1.5×10^{-1}	
150	~3×10^{-4}	~4×10^{-3}	~5×10^{-2}	~2×10^{-1}	

TABLE 18.5

Speed of Vehicles and Spectral Efficiency for Different Algorithms

	Spectral Efficiency (Bits Per Second/Hz)				
Speed (km/h)	Multiuser Scheduling Algorithm Based on Block Diagonalization Precoding for UM-MIMO System	Subspace Correlation Algorithm	Geometric Angle Algorithm	Conditional Number (CN) Algorithm	Remarks
90	~14.5	~13	~11.7	~11.5	Number of
100	~14.4	~12.8	~11.6	~11.4	user: 2
110	~14.2	~12.8	~11.4	~11.2	Number of
120	~14	~12.5	~11.2	~11	transmitting
130	~13.6	~12.3	~10.9	~10.7	antenna: 128
140	~13.2	~11.7	~10.5	~10.2	
150	~12.5	~11.2	~10.2	~9.8	

proposed algorithm considers not only the correlation between channels but also the conditions of the channel itself. This can effectively eliminate the interference between selected users, which also proves that the proposed algorithm is useful in the ultra-Massive MIMO system in the 6G V2X scenario.

Tables 18.4 and 18.5, respectively, show the impact of vehicle speed on the proposed algorithm and the performance of the comparison algorithm (including BER and spectral efficiency) under different vehicle speed environments. The vehicle speed is set from 90 to 150 km/h, where SNR = − 10 dB, $M = 2$, $N_t = 128$. It can also be seen that the BER of the proposed MUS algorithm and the comparison algorithm gradually increases, and the spectral efficiency performance of the system gradually decreases with the continuous increase in vehicle speed. This is because the Doppler effect becomes more serious with an increase in vehicle speed, resulting in the decline of system performance. At the same time, the simulation results show that the proposed MUS algorithm has better robustness than the comparison algorithm. This is because the proposed MUS algorithm makes full use of the time dimension information of the time-varying channel. Therefore, compared with the comparison algorithm, the performance of the proposed algorithm does not decline significantly and has high stability when vehicle speed is increasing.

SUMMARY

Simulation results show that compared with the proposed MUS algorithm based on SC, condition number, and GA, the proposed MUS algorithm can obtain higher user channel gain, effectively reduce the system BER, and can be applied to 6G V2X communication. The proposed 6G V2X MUS algorithm is based on BD precoding for ultra-Massive MIMO systems. Aiming at the high-speed movement and millimeter wave characteristics of the vehicle in the 6G V2X scheme, a time-varying geometric channel model is established. On this basis, the BD precoding is sensitive to channel correlation, and the Pearson coefficient after matrix vectorization is used to measure the channel correlation between users, to comprehensively consider the two factors of channel correlation and channel conditions, and to ensure that the channel correlation is minimized while ensuring the selection of high-quality channels to achieve MUS. BD precoding is then used to eliminate data stream interference. The system simulation results show that the proposed algorithm can effectively reduce the BER and improve the system spectrum efficiency and is suitable for 6G V2X communication.

REFERENCE

1. He, S. et al., "Multi-User Scheduling for 6G V2X Ultra-Massive MIMO System," *Sensors* 2021, 21, 6742. https://doi.org/10.3390/s21206742

19 V2X Millimeter Wave MIMO Radar Communications

Robust channel estimation in time-varying channels [1] is used to guarantee the quality of communication services, especially for vehicle-to-everything (V2X) scenarios. To improve the channel estimation accuracy and reduce the pilot overhead, multi-input multi-output (MIMO) radar is deployed to assist millimeter wave (mm-Wave) channel estimation. The MIMO radar-aided channel estimation scheme has been developed using machine learning (ML)/deep learning (DL) for the uplink (UL) mm-Wave multiuser (MU)-MIMO communications depicted in Figure 19.1. To allocate pilot resources reasonably, a transmission frame structure of joint radar module and communication module is developed, which divides the estimation scheme into two stages, i.e., the arrival/departure (AoA/AoDs) estimation stage and the gain estimation stage. In view of the imperfections of array elements in practice, an AoA/AoDs estimation algorithm is developed based on subspace reconstruction in the AoA/AoDs estimation stage named two-step angle estimation (TSAE) algorithm [1]. In the gain estimation stage, a DL-based channel gain estimator is designed. An autoencoder (AE) combined with residual structure named residual denoising autoencoder (RDAE) is proposed to eliminate the noise on wireless signals, which is passed into the least square (LS) estimation module to obtain gains. Simulation results demonstrate that the MIMO radar-aided and the ML/DL-based channel estimator provide an efficient estimation performance of the high-mobility mm-Wave channel with fewer training resources. A novel UL channel estimation scheme for the mm-Wave MU-MIMO system over a time-varying channel is developed. Specially, a novel transmission frame structure including communication and radar modules are developed, where the channel estimation is decomposed into AoA/AoDs estimation with MIMO radar and channel gain estimation. The MIMO radar is equipped for channel angle estimation, where the echo signal is reconstructed, and the reconstructed covariance matrix is used for channel angle estimation with short training resources.

Based on the estimated AoA/AoDs, a UL/DL-based path gain estimator is leveraged to learn the mapping relationship between the raw data and the noised data, which has been leveraged to optimize the performance of gain estimation. The performance of the proposed algorithms is evaluated and analyzed under the single-path MU-MIMO channel case in high-mobility V2X scenarios. The simulation results have demonstrated that this scheme has estimated the time-varying mm-Wave channels effectively when vehicles are moving at high speed. We have not described the complete study here for the sake of brevity. Interested readers will find the details in [1]. However, we have summarized the performances here.

DOI: 10.1201/9781003499480-19

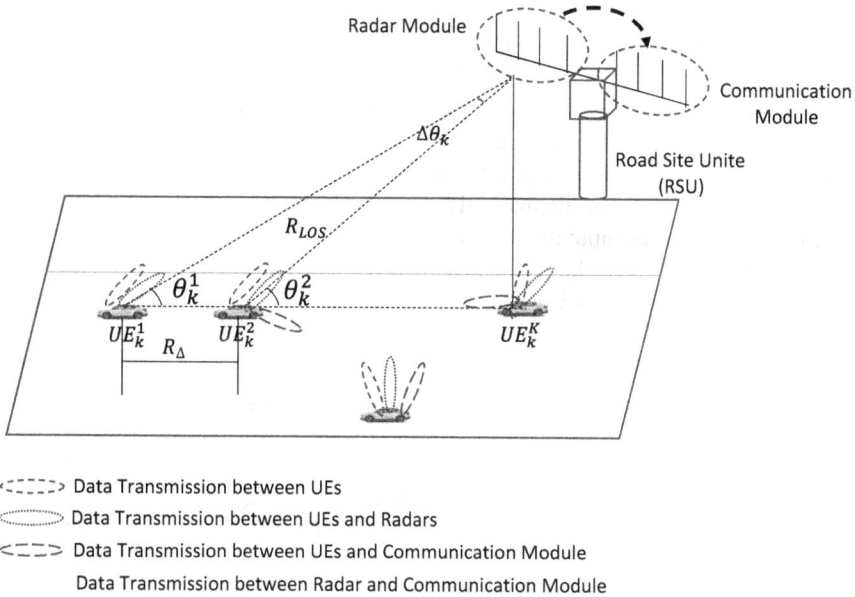

Radar Module

Communication Module

Road Site Unite (RSU)

$\Delta\theta_k$

R_{LOS}

θ_k^1 θ_k^2

UE_k^1 R_Δ UE_k^2 UE_k^K

‹⁻⁻⁻› Data Transmission between UEs

Data Transmission between UEs and Radars

‹═══› Data Transmission between UEs and Communication Module

Data Transmission between Radar and Communication Module

FIGURE 19.1   Radar aided multiuser V2X communications scenario in mm-wave system.

AOA/AODS ESTIMATION PERFORMANCE

We have seen that the performance of subspace reconstruction-based angle estimation algorithm varies with MIMO radar in time-varying mm-Wave channels. The estimation performance fluctuates slightly with the change of SNR, because the signal subspace is reconstructed twice and most of the noise components are removed. Therefore, the increase of SNR performs slight impact on the accuracy of the estimation algorithms. However, the performance differences between the algorithms are significant. Root reduced dimension multiple signal classification (MUSIC) and the proposed scheme have better performance than the other. Especially, this algorithm is more superior because other algorithms have a sharp deterioration in performance due to severe nonorthogonality. Compared with other array errors, the correction of array mutual coupling errors is more complex.

PATH GAINS ESTIMATION PERFORMANCE

For simulations, 35,000 wireless signals are generated, 25,000 for training, 5,000 for validation, and 5,000 for testing. Moreover, the training batch size is set as 256. The proposed RDAE gain estimator is compared with LS estimator, refined estimator [1], and denoising autoencoder (DAE) estimator. The performance comparison without the Doppler frequency shift is illustrated in Figure 19.1. The transmitter and receiver are equipped with 8×16 array elements, respectively. The angle values in the samples are the results of the previous MIMO radar angle estimation scheme, rather than relying on the assumed perfect angle. It is obvious that RDAE achieves the best performance within the testing SNR range. LS is the worst when SNR is lower than 5 dB due to not considering the influence of the noise. For the same problem, different network structures may also cause differences in estimated performance. Clearly, DAE greatly reduces its learning ability, and it basically has no ability to denoise due to the lack of residual learning module.

Compared with the other two comparison methods, the "Refined" has a certain performance improvement, its main idea is that rough channel gains are calculated by LS and then used as input to obtain more accurate gains through further optimization of RDAE. However, the difference between the low-resolution complex gains directly estimated by LS and the more accurate high-resolution gains is not simple noise interference, the learning difficulty of which is undoubtedly higher than the denoising operation of the signal for the DL estimator. The above results can prove that RDAE is more adaptable to high-mobility time-varying channels and has higher estimation accuracy.

There is fast convergence of the proposed RDAE within different training lengths. It is shown that the network has basically completed convergence at the 60th epoch, and the loss of the network at this point is close to zero for all path gain values due to the low difficulty for the network to learn the mapping between the input and the output. From the beginning to the 40th epoch, the loss decreases more rapidly when the path gain is larger, which is indicated that it has more information to guide learning. It is shown that the normalized mean square error (NMSE) increases monotonically with the frequency shift, and this trend becomes more obvious as the SNR increases. However, the curve rises very slowly, which means that our gain estimator is robust to Doppler frequency shift.

CHANNEL ESTIMATION PERFORMANCE

To confirm the performance of our proposed channel estimation scheme based on angle prior, the channel estimation performance curves and Cramer-Rao lower bound (CRLB) have been computed under static conditions. When the SNR is above −5 dB, the user mobility causes the curves to gradually deviate from the CRLB. However, the result indicates that the NMSE deviation is only about 5 dB even when the vehicle is driving at the highest speed, which proves that the proposed scheme is highly adaptable to frequency shift once again.

SUMMARY

In this chapter, a novel UL channel estimation scheme for the mm-Wave MU-MIMO system over a time-varying channel is proposed. Specially, we proposed a novel transmission frame structure including communication and radar modules, where the channel estimation is decomposed into AoA/AoDs estimation with MIMO radar and channel gain estimation. In the first stage, MIMO radar is equipped for channel angle estimation, where the echo signal was reconstructed, and the reconstructed covariance matrix was used for channel angle estimation with short training resources. Based on the estimated AoA/AoDs, a DL-based path gain estimator was leveraged to learn the mapping relationship between the raw data and the noised data, which was leveraged to optimize the performance of gain estimation. The performance of the proposed algorithms is evaluated and analyzed under the single-path MU-MIMO channel case in high-mobility V2X scenarios. The simulation results demonstrated that our proposed scheme estimated the time-varying mm-Wave channels effectively when vehicles are moving at a high speed.

REFERENCE

1. Huang, S. and Gao Y., "MIMO Radar Aided Millimeter-Wave Time-Varying Channel Estimation in MU-MIMO V2X Communications," *IEEE Transaction on Wireless Communications* November 2021, 20(11), 7581–7594.

20 Robotics and Autonomous System Communications

OVERVIEW

Robotics and autonomous systems (RASs) describe systems of platforms, such as automobiles, airplanes, robots, and UAVs, which move and operate in a physical environment for goal-oriented actions. We describe the broad roles that communication formats and technologies have played in enabling multi-robot systems. We approach this field from two perspectives: of robotic applications that need communication capabilities to accomplish tasks and of networking technologies that have enabled newer and more advanced multi-robot systems. Through this review, we identify a dearth of work that holistically tackles the problem of codesign and co-optimization of robots and the networks they employ. We also highlight the role that data-driven and machine learning (ML) approaches play in evolving communication pipelines for multi-robot systems. We refer to recent work that diverges from hand-designed communication patterns and discuss the "sim-to-real" gap in this context. We present a critical view of the way robotic algorithms and their networking systems have evolved and make the case for a more synergistic approach [1]. We also identify some problems as well for RASs.

The use of multiple, connected robots in the place of individually uncommunicative robots provides evident gains by facilitating the inter-robot coordination that allows for work distribution, spatial coverage, and specialization. An increasing variety of applications leverage such networks of robots, including logistics, resource distribution, transport systems, manufacturing, and agriculture. These applications depend on an orchestration of robots over time and space that allows them to jointly work toward common higher-order goals, to deconflict individual actions in shared environments, and to share information in distributed computing schemes. Communication and the mutual exchange of information (state and control) are key to facilitating such interactions.

Early work in the multi-robot domain drew from nature-inspired paradigms, and consequently focused on devising collective behaviors that depended purely on *local* interactions of robots in proximity. A variety of transmission media (e.g., infrared) are used for such near-field communication schemes. Other nature-inspired work built on implicit communication and self-organization, by which robots coordinate indirectly through traces left in the environment. The benefits of such peer-to-peer (P2P) decentralized communication paradigms are manyfold, in particular due to their inherent robustness and scalability. Centralized radio-based communication architectures have become increasingly popular in various instances, especially when the task requires

DOI: 10.1201/9781003499480-20

performance guarantees; representative applications include product pickup and delivery, item retrieval in warehouses, and mobility-on-demand services. Improvements in communication technologies, both hardware and software, have furthered more data-intensive applications, such as cloud robotics. Explicit communication methods generally assume that robots can broadcast information within a local neighborhood that comprises of tens to hundreds of individuals, or that a fixed network infrastructure is available. Yet densely populated workspaces adversely affect communication capabilities because of practical contention over channel bandwidth and airtime. Such networks are additionally burdened by clutter that can induce signal fading, leading to a drastic decrease in the expected communication range.

This problem is compounded by the need for real-time transmission requirements in highly dynamic robot networks. Indeed, topologies and capabilities demanded by robotic applications are practically hostile to radio performance (because these radio networks were not initially designed with robotics in mind). Consequently, most robot applications are designed to merely work around available network technologies and optimize their performance within the given constraints. Our review is motivated by a lack of studies that provide a high-level overview of the interplay between communication networks and their role in robotic applications.

Figure 20.1 graphically demonstrates a typical architecture of a multi-robot control scheme, in which each robotic system is designed strongly around its perception and control strategies. As mentioned prior, robot control algorithms generally do not actively employ the output of the communications network as part of the control loop, and as a result often overlook factors such as network contention.

The result is a wide array of optimizations that work in favor of the network, but often not for autonomy, or vice-versa. Hence, we argue that a better co-optimization

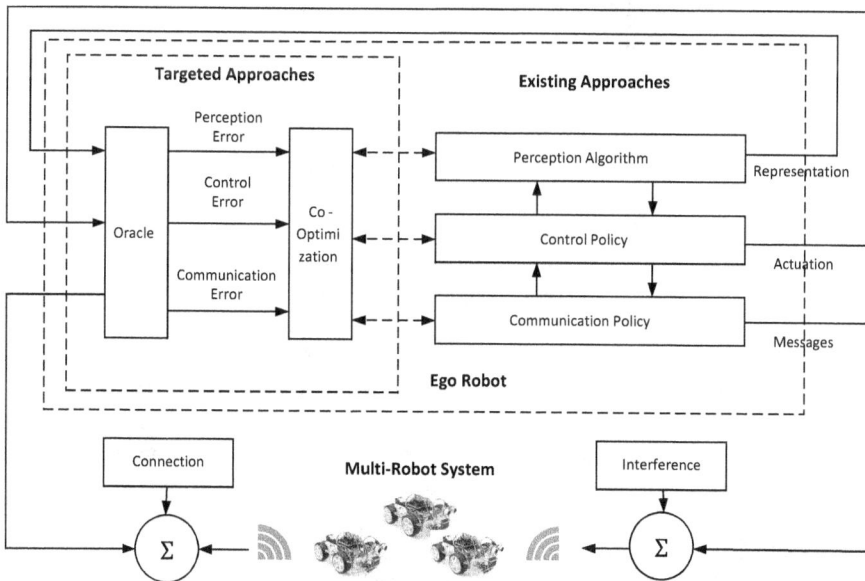

FIGURE 20.1 Typical multi-robot control architecture.

scheme (illustrated on the left in Figure 20.1) would consider all aspects of the architecture simultaneously. The illustration shows this scheme linked via an "oracle," which is a source of error estimation from incomplete information and which facilitates this, i.e., given a hypothetical oracle, we posit that one can codesign the various algorithm layers on the right.

We have captured a variety of network architectures and technologies, and a variety of multi-robot applications that employ them. A careful choice of communications architecture, medium, and algorithm is key to ensuring that a given robot task can be completed. Therefore, we will also explore some of the newer approaches that consider bypassing such handcrafted selections and attempt to model inter-robot communications in a data-driven fashion.

ROBOT NETWORK DESIGN

Choosing an adequate communications architecture, medium, and algorithm is key to ensuring desired robot performance. In the following, we distill the factors that influence the robot network design choices. We elaborate upon them in the following four categories, (a) application, (b) robot, (c) algorithm, and (d) environment, and give an illustrative example for each.

The application: The application defines what the shared information is for, and how the robots need to interact to solve the problem at hand. Examples: real-world applications, such as in environmental monitoring and agriculture, require groups of robots to act over large distances (often operating with robots separated by $\sim 1000 \times$ body lengths). Such sparsely distributed robot systems, hence, necessitate networking capabilities that can span larger spaces. Other applications, such as cooperative driving, formation control, and flocking require uninterrupted, situated, close-range communication for tight inter-robot coordination and control.

The robot: The robot (and the physical hardware) defines local constraints on the frequency and format of information to be transmitted and received. Examples: A quadrotor that uses state information for local stability control requires an update frequency in the order of several hundred Hertz; while on-board inertial measurement units (IMUs) can provide the necessary information for body stabilization, extrinsic pose estimates are still required for tasks, and must be received at relatively high rates (e.g., 100 Hz). Lack of reliable updates naturally poses a significant risk to tasks that require tight coordination, such as outdoor flocking and formation control, while sparse outdoor flight has been demonstrated in a team of 30 drones, there is a dearth of results on dense and agile outdoor flight. Moreover, in GPS-denied environments, robots resort to on-board sensing and, consequently, require dependable inter-robot communication to achieve group behavior.

The algorithm: The algorithm connects the application to the robot, and essentially sets conditions on the nature of information that needs to be received (e.g., global or local) and when (e.g., asynchronously or synchronously, and how often). Examples: In allocation problems, the optimization objective is often global, and to achieve optimality, we deploy centralized algorithms that collect all robot-to-task assignment costs (e.g., expected travel times) to determine the optimal assignment. Similarly, multi-robot path planning has an optimal solution (for both make span and flowtime objectives), but only when the computational unit has access to full system information. In the absence of full observability, robots need to resort to locally available knowledge. In decoupled prioritized path planning, robots communicate to mutually deconflict their path plans in time-space. Each time a robot's plan changes, its robot neighborhood changes, or a new conflict arises, the deconfliction process restarts.

The environment: The environment defines *under what conditions* shared information is delivered. Examples: Are the robots operating indoors or outdoors, or both? Does the workspace afford a fixed (and possibly centralized) communications infrastructure, or must we instead rely on ad hoc networking? Is the environment cluttered with obstacles that interfere with wireless signals? What medium can we use; e.g., are the robots operating in air, under water, or in space? What legal jurisdictions regulate the communication infrastructure? And finally, is the communication channel safe, or can it be spoofed or robots attacked?

MULTI-ROBOT COMMUNICATION SCHEMES

Here we discuss multi-robot communication from the perspective of the underlying communication technologies, focusing upon the challenges, limitations, and optimizations that are relevant in multi-robot system networks (see [1] for more detail).

Synchronicity: Specifying robotic data flows is often the first consideration in discussing the challenges of a wireless data protocol. For example, it is often implicitly assumed that multi-robot control algorithms are executed synchronously by every participant. This introduces a hard timing constraint on the maximum allowable delay in message delivery between those participants. This is, however, a feature that commonly deployed communications protocols are not designed to meet, with "best-effort" message delivery being the standard paradigm.

Dynamic topologies: Hard timing constraints are often exacerbated by highly connected communications topologies that are dynamic, where a robot must communicate its status with many different (or sometimes every) participating robot(s). This can lead to a high degree of contention for radio resources since many messages may need to be sent at every control loop. While there are schemes that aim to minimize redundant data transmissions, it remains true that, as multi-robot networks increase in scale, communication technologies must be selected and designed specifically to manage the dynamics of the application, something that is generally overlooked in robotic networks today.

Message frequency: Bandwidth is often employed as a metric to specify the demands on a communications link. However, this is often an insufficient characterization by itself, since the underlying technology may have significant overheads per message, and robot teams often depend on low-latency messaging as well. This is particularly true for ad hoc networks where there is no central entity enforcing message scheduling, to the extent that many communications protocols will not approach their rated bandwidths in highly connected ad hoc topologies, where overheads such as contention dominate radio resource consumption.

Connectivity: Since a greater connectivity range implies reachability and information exchange with more robots, it has an obvious impact on the overall messaging rate any specific robot must handle. The spatial density of robots must be considered while discussing range, and two key factors of interest emerge consequently. Firstly, as ranges increase, radio-based links are more prone to fading and interference even when transmission power is commensurately increased. Secondly, robot control algorithms often assume a fixed range, which may result in greater interconnectivity, and sometimes increased messaging rates in dense scenarios.

Dynamic routing: The case where the required range of communication exceeds the underlying capabilities of the radio hardware onboard must also be considered, as this implies a mesh-type network where a message must traverse multiple robots (network nodes). This first requires planning the robots' paths before accounting for the computational and protocol over-heads of robots processing messages other than their own. Then, the problem is that of message routing decisions and dynamic topologies. The routing decision problem is generally central to ad hoc mesh networks; this is only made more challenging by the potential for rapid shifts in communications topology, especially in highly mobile or large-scale robotic scenarios. Hard timing guarantees, in the range required by robotic control, are not currently available at nontrivial scales (especially with multihop routing over dynamic topologies), though some attempts have been made in this direction.

Operational environment: Robotic networks will invariably be required to operate in environments with external noise and interference, which cause unpredictable impacts on link quality. This informs the selection of com-munications protocols since some protocols operate in a licensed spectrum with reduced external interference or are otherwise less prone to external noise due to atmospheric attenuation (60 GHz). Doppler shift requires similar consideration because many communication technologies fail at high relative velocities. Protocols that depend upon fine-grained frequency division multiplexing are more prone to Doppler-related errors, and such schemes are often used in high bandwidth techniques.

PEER-TO-PEER COMMUNICATIONS SCHEME

Despite considerable research interest, there are no current wireless data stan-dards explicitly designed for exchanging information between autonomous robots. Currently deployed robot-to-robot networks depend upon more generic wireless data networking standards which are not typically optimized for the challenges dis-cussed above. In the absence of a specific standard, we will discuss the strengths and weaknesses of existing technologies for the multi-robot control application. Ad hoc networks map well to the communications patterns required of decentralized robotic control, and the most relevant for this survey are Mobile Ad Hoc Networks (MANETs), which deal with the problem of facilitating communication between mobile nodes without coordination from infrastructure. More specific forms of interest are Vehicular Ad Hoc Networks (VANETs) and Flying Ad Hoc Networks (FANETS), where the former generally deal with automotive use cases and the latter drones and UAVs, and these are more exposed to dynamic conditions that are expected from robotic ad hoc networks. Local area networking technologies are well suited for ad hoc networking.

In contrast with ad hoc operation, infrastructure networks map more closely to a centralized robotic control, where communications patterns are more like traditional bandwidth-focused networking applications. Despite this, hard latency requirements and rapid robot movement require specific mechanisms at the protocol level, which are not common for either cellular or local area standards.

COMMUNICATION-AWARE ALGORITHMS

Regardless of which underlying communication scheme or protocol is employed, unlimited and unconstrained communication cannot be assumed for any interactive scenario. A significant amount of literature in multi-robot applications, however, has generally focused on designing control schemes that do not explicitly model this dependency. This is reflected in most of the literature in robot flocking. We argue that the problem becomes more pronounced in cases where the robots need to deconflict and replan their motions in tight and constrained spaces. While some consideration for communication a-synchronicity is made in some of the more recent works [1], the challenge is generally far from being solved.

One straightforward approach to handle this is to simply reduce the amount of data (frequency, packet-size etc.) that needs to be communicated between agents. In an exploration problem, this is often done through various novelty metrics that determine whether a new datapoint needs to be communicated. Some [1] have proposed localization estimators that perform well by quantizing the transmitted information to very small packets, thereby tackling severe link constraints.

On the other hand, there is also a sustained research interest in modeling the communication channels between the agents, and factoring that as a constraint into the motion planning problem. This is done primarily to ensure robustness of a control scheme against imperfect and noisy communications. Alternatively, planning schemes have also considered communication as a subtask (almost as if "scheduling" communications at intervals). Finally, there are several approaches that consider a joint optimization scheme, where path planning and communication planning are carried out in tandem. We divide this body of work into these three broad styles.

COMMUNICATION-AWARE PLANNING

As mentioned earlier, planning robot motions or trajectories that consider some model of the underlying communication links is an active area of research. For instance, several authors have considered the task of coverage and formation control by a team of robots. Evidently, these domains require explicit factoring of communication constraints into the planning problem. One way this is approached is by analyzing the stability of a formation under various communication link latencies. This can then be then integrated into the control problem for a more reliable system that also explored that allow agents to maintain some degree of coordination while respecting limited communication ranges of their neighbors. This approach is also sometimes utilized in the context of cooperative target localization. Path planning has also been developed such that connectivity with a subset of base stations or with some agents is maintained. Similar methods that plan for multiple robots (with connectivity constraints) involve using ant colony optimization (ACO)-based planning or genetic algorithms.

PLAN-AWARE COMMUNICATIONS

In heavily constrained spaces, it is often desirable to design a network architecture that considers the planned path and seeks opportunities to communicate therein. Underwater robots, for instance, have very limited communication capacities, and this is an active field of interest. Some [1] have considered scheduling algorithms for underwater robotic sensor networks and show how path planning algorithms depend on these. Since bandwidth and interference constraints are much more severe in these environments, such scheduling algorithms often model the *value* of communicating at a particular timestep. This also plays a role in determining whether communicating has a positive impact on the state of the robot system, and is also studied as an online decision problem, and an optimization problem that considers when/what to communicate. Recent developments in subterranean robots operating under severe communication constraints have also explored the strategy of developing/maintaining communication "backbones" for explorer robots to continue frontiers and of explicitly splitting communication pathways by plan and priority.

JOINT PLANNING

Several of the works listed in the previous subsections may also be seen as jointly optimizing for communication quality as well as path qualities. However, there are other approaches that attempt to explicitly model this optimization problem. For instance, some [1] have proposed a scheme that alternates between the two optimization problems sequentially. The nature of this scheme often makes it difficult to prove hard guarantees regarding optimality; however, a more hybrid approach in which the two controllers interact can offer more guarantees on network integrity available data rates. A joint optimization scheme, on the other hand, can formulate this problem well; for instance, using a linear quadratic (LQ) form can additionally offer robustness guarantees as well. Yet another means of joint optimization is to consider the system as a cyber-physical system (CPS), where the "cyber" controller handles the communications domain, and the "physical" controller handles the kinematics of the robot. Such models allow designers to factor various other elements of a CPS system, such as dynamically adapting one of the subsystems (communication capacity) while still maintaining the coupling with the other.

AI/ML/DL FOR ROBOTS AND AUTONOMOUS SYSTEMS

Leveraging ML methods [2] is a promising new avenue to tackle some of these challenges. Artificial intelligence (AI), ML, and deep learning (DL) are increasingly used in RASs that attempt to mimic the adaptive and smart problems-solving capabilities of humans. Such systems promise a smarter and safer world where, e.g., self-driving vehicles can reduce the number of road accidents, medical robots perform intricate and digital surgeries. Pilots participate in crew flight-operations. In addition, internet of things (IoT) technology could inspire wider applications of RAS. However, many RASs are currently released into the world without full prior analysis of potential inappropriate operations and thus may accomplish things which were not foreseen by their human designers or owners. If not managed and understood properly, there is a risk that system autonomy could devalue human work or give rise to hostile attitudes toward advanced technology. Thus, developing trust in RAS right from the start is paramount on the progression toward fully autonomous solutions.

AI/ML/DL-BASED COMMUNICATION MECHANISMS

Message routing decisions in robotic mesh networks are complicated by highly dynamic topologies. While many routing mechanisms exist in ad hoc networks, these generally depend upon relatively slowly changing network conditions to function effectively. Many manually specified heuristic methods exist. However, these may lead to suboptimal decisions as they may be constructed upon incorrect assumptions about the target network environment.

Message routing learning methods provide an attractive alternative and have been explored in some depth in routing generally. An interesting example in the context of FANETs can be found in Zheng et al. (see [1]), who propose self-learning routing protocol based on reinforcement learning (RLSRP), which applies an online reinforcement learning (RL) method to the routing decision problem and shows improved performance across several metrics, including delivery latency. Channel modeling and resource allocation are also key networking problems that are challenging for first principles methods to solve that can be improved with learning. Unsupervised learning has been applied to channel modeling, which allows for the optimization of transmission power by accurately estimating the quality of links to other network participants.

AI/ML/DL-BASED COMMUNICATION BEHAVIORS

AI/ML/DL-based methods have proven effective at designing robot control policies for an increasing number of tasks. Recent work utilizes a data-driven approach to solve multi-robot problems, for example for multi-robot motion planning in the continuous domain or path finding in the discrete domain. Yet, research on learning how to synthesize robot-to-robot communication policies is nascent. From the point of view of an individual robot, its local decision-making system is incomplete since other agents' unobservable states affect future values. While the way information is shared is crucial to the system's performance, the problem is not well addressed by handcrafted (bespoke) approaches.

AI/ML/DL-based methods, instead, promise to find solutions that balance optimality and real-world efficiency, by bridging the gap between the qualities of full-information centralized approaches and partial information decentralized approaches. Key to the decentralization of centralized (optimal) policies is the property of permutation equivariance. Permutation equivariance ensures that at the robot network level, the set of actions automatically rearranges itself as the agents swap order. One of the earliest works that satisfy this property. This was concurrently developed by a line of work that builds on graph neural networks (GNNs), which are permutation equivariant by design. GNNs have since then shown promising results in learning explicit communication strategies that enable complex multi-agent coordination.

When deploying GNNs in the context of multi-robot systems, individual robots are modeled as nodes, the communication links between them as edges, and the internal state of each robot as graph signals. By sending messages over the communication links, each robot in the graph indirectly receives access to the global state. A key attribute of GNNs is that they compress data as it flows through the communication graph. In effect, this compresses the global state, affording agents access to relevant encodings of global data. Since encodings are performed locally (with parameters that can be shared across the entire graph), the policies are intrinsically decentralized. In cases where the downstream task is tightly coupled with the communication requirements, it is beneficial to optimize the communication strategy jointly with perception and action policies. This was done for multi-robot flocking and for multi-agent path planning. These frameworks implement a cascade of a convolutional neural network (CNN) and a GNN, which they jointly train so that image features and communication messages are learned in conjunction to better address the specific task. Recent work also shows how GNNs can be augmented by attention modules to produce message-aware communication strategies that allow robots to discern between important and less important message elements.

Approaches from within the multi-agent reinforcement learning (MARL) community tackle the learning of continuous communication protocols by formulating the problem as a decentralized partially observable Markov decision process (Dec-POMDP). The work learns a targeted multi-agent communication strategy by exploiting a signature-based soft attention mechanism (whereby message relevance is learned). Similarly, each robot learns to reason about other robots' states and to communicate trajectory information more efficiently (i.e., when and to whom), and applies the solution to the problem of collision avoidance. While efficient cooperative communication strategies are desirable, it is also shown how separate robot teams can learn to communicate with adversarial strategies that contribute to manipulative (noncooperative) behaviors. Clearly, underlying training paradigms need to be carefully designed to avoid such outcomes.

CHALLENGES AND OPEN PROBLEMS

We finally present some avenues of research and engineering that are worth exploring to address our critiques discussed so far. We categorize them into four broad Open Problems.

1. **Codesign**: An emergent theme throughout this survey is the lack of approaches that codesign the robot and its communication capabilities. A variant of this concept considers a basic parallel reconfiguration of a network as well as the robot's controller that can be beneficial when the robot moves across network stations. However, a true codesign scheme will jointly evolve all layers of the networking stack to favor the robotic task at hand. The design of a meta-system that can compute the limitations of robotic requirements as well as network capabilities and dynamically throttle both may be essential to safe deployment of robots into the real world. Any robotic control algorithm that uses explicit communications is vulnerable to failure if the network unexpectedly under-delivers and performs suboptimally if the network over-delivers – managing this resource allocation problem in a real-world multi-robot setting is a subject we will tackle in our future work.

2. **Data-driven optimization**: AI/ML/DL, and specifically RL, can drive the development of multi-robot communications into new and interesting paradigms. Existing approaches that already learn what/when to send (and whom to send to) still often depend on hand-designed architectures and specific task groups. With sufficiently large datasets, novel ML architectures also have the potential to learn to optimize multiple aspects of multi-robot systems at once (e.g., perception, action, and communication).

3. **Simulation-to-real for robot networks**: The problems in simulation-to-real transfer of robot coordination strategies are generally exacerbated by the "reality gap" found in communications [1]. Practical communication links suffer from message dropouts, asynchronous and out-of-order reception, and decentralized mesh topologies that may not offer reliability guarantees. Since multi-robot policies are typically trained in a synchronous fashion, these factors are hard to capture and simulate. Furthermore, very few studies have captured any of these network effects in a large-scale setting. Consequently, we find that embedding the reality gap of robot networking into data-driven approaches to multi-robot planning is an open research domain.

4. **New technologies/schemes**: There is a need for wireless data standards that specifically target the communication requirements of connected robots. The IEEE 1920 working group is a significant step in this direction, which was formed to propose a protocol that is intended for autonomous robotic networks. Such a protocol is likely to be founded on 802.11bd since it is already a significant leap forward from the legacy 802.11p standard used in vehicle-to-vehicle (V2V) standards today. Additionally, future 5G updates and 6G cellular communications promise dramatic improvements that hold the potential to bring cloud- and edge-computing at the forefront of many data-intensive multi-robot collaborations. Finally, we also note that

geographic routing in FANETs may be an enabling technology for practically dealing with highly dynamic routing topologies. This will, however, require holistic developments in robot control algorithms that work in tandem to avoid an additional information distribution problem.

TRUSTWORTHY ROBOTS AND AUTONOMOUS SYSTEMS

Trust is essential in designing autonomous and semiautonomous RASs [2], because of the "No trust, no use" concept. RASs should provide high-quality services, with four key properties that make them trustworthy: they must be (i) robust with regards to any system health-related issues, (ii) safe for any matters in their surrounding environments, (iii) secure against any threats from cyber spaces, and (iv) trusted for human-machine interaction. This chapter thoroughly analyzes the challenges in implementing the trustworthy RASs in respect of the four properties and addresses the power of AI in improving the trustworthiness of RASs. While we focus on the benefits that AI brings to human, we should realize the potential risks that could be caused by AI. This chapter introduces for the first time the set of key aspects of human-centered AI for RAS, which can serve as a cornerstone for implementing trustworthy RASs by design in the future.

SUMMARY

We have presented a survey of communication technologies and their role in enabling multi-robot applications [1]. We have broadly covered the various technologies that have played key roles in networked robotics and have also discussed how state-of-the-art robot applications typically deal with network constraints. Our approach to this has been mostly critical and, thus, has identified several deficiencies in the way robotics and networks have evolved. Toward the end, we also cover ML approaches and their role in developing data-driven communication strategies. We conclude the chapter with a list of challenges and open problems that the community currently faces and provide an outlook for how learning-based approaches can tackle several of them. We have also discussed briefly how trust is essential in designing autonomous and semiautonomous RASs.

REFERENCES

1. Gielis, H. Shanker, A., and Prorok, A., "A Critical Review of Communications in Multi-robot Systems," *Current Robotics Reports* 2022, 3, 213–225. https://doi.org/10.1007/s43154-022-00090-9
2. He, H. et al., "The Challenges and Opportunities of Artificial Intelligence for Trustworthy Robots and Autonomous Systems," *2020 the 3rd International Conference on Intelligent Robotics and Control Engineering*, 2020.

21 Virtual, Augmented, Mixed Reality, and Extended Reality

OVERVIEW

"Extended reality" (XR), the umbrella term that includes augmented reality (AR), mixed reality (MR), and virtual reality (VR). It is an immersive, completely artificial computer-simulated image and environment with real-time interaction. XR has a tremendous market size and will profoundly transform our lives by changing the way we interact with the physical world [1, 2]. However, existing XR devices are mainly tethered by cables which limit users' mobility and quality-of-experience (QoE). Wireless XR leverages existing and future wireless technologies, such as 5G, 6G, and Wi-Fi, to remove cables that are tethered to the head-mounted devices (HMDs). Such changes can free users and enable a plethora of applications. High-quality ultimate XR requires an uncompressed data rate up to 2.3 Tbps with a latency lower than 1 ms. AR applications must continually gather information about the user's surrounding environment via multiple sensors. Complicated algorithms must be used to make sense of sensor data of the environment. Artificial intelligence (AI) can simplify that process and make it more accurate than a model made exclusively by a human. Consequently, AR and AI will become more closely entwined. Although 5G has significantly improved data rates and reduced latency, it still cannot meet such high requirements. However, we are providing a roadmap toward wireless ultimate XR. The basics, existing products, and use cases of AR, MR, and VR are considered, upon which technical requirements and bottleneck of realizing ultimate XR using wireless technologies are identified. Challenges of utilizing 6G wireless systems and the next-generation Wi-Fi systems and future research directions are discussed.

BRIEF DESCRIPTION OF XR, AR, MR, AND VR

AR and VR have been evolving with the development of sensors, displays, and computers since the 1960s. Recently, MR is emerging as we have the capability to interact with virtual/digital objects in real environments. AR, MR, and VR are all spatial computing technologies which are encompassed by XR. Their differences mainly reside in the rendering format and percentage of virtual content, as shown in Figure 21.1. Today's XR technologies are mainly used for immersive gaming, remote assistance, and professional training. Customers have a wide variety of options.

Existing XR devices use HMDs, which have strict constraints on power consumption and weight. HMDs must be made thin and light to meet the requirements of QoE. Thus, most computing and storage tasks are offloaded to a computer or a server

DOI: 10.1201/9781003499480-21

		Extended Reality (XR)		
	Reality	Augmented Reality (AR)	Mixed Reality (MR)	Virtual Reality (VR)
Display	Naked Eye/ Optical Glasses	Translucent Display	Translucent Display	Occlusion Display
Display Example	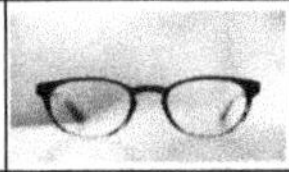			
Example	Real View of the World	Augmented Virtual Map	Interactive Virtual Contents	Virtual Games

FIGURE 21.1   Reality-virtual continuum – AR, MR, VR, and XR.

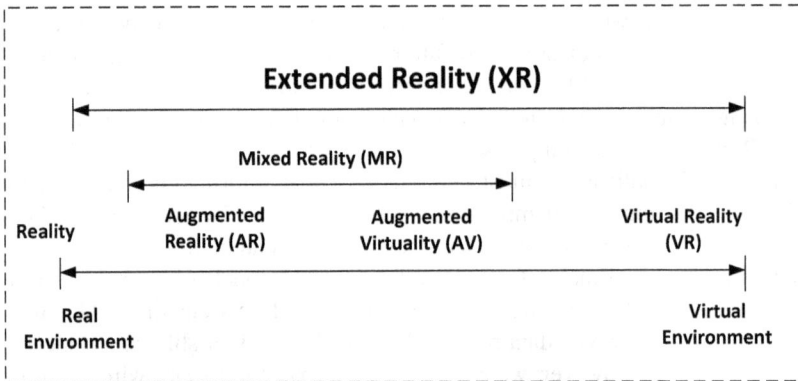

FIGURE 21.2   Reality and virtuality continuum.

to reduce the overall power consumption and weight of HMDs. Most existing XR devices use cables to connect HMDs with computers. Another way of representing reality-virtual continuum is depicted in Figure 21.2. A vann diagram of XR, AR, MR, and VR is shown in Figure 21.3.

We define AR, MR, VR, and XR below (referring to Figures 21.1–21.3):

- **Reality**: The surrounding environments and objects are real.
- **Real environment**: The circumstances, objects, or conditions by which one is surrounded. The realm of human experience comprising physical objects and excluding theoretical constructs, hypotheses, artificial environments, and virtual worlds.
- **Virtual environment**: A virtual environment is a computer-generated representation of a physical environment and may recreate stimuli that engage any number of the senses.

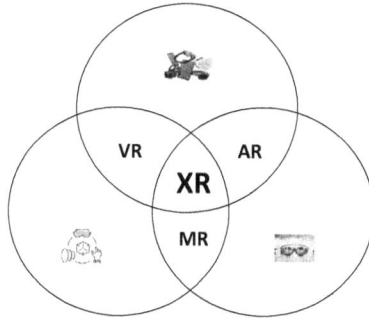

FIGURE 21.3   XR, AR, MR, and VR vann diagram.

- **Augmented reality**: The surrounding environments are real but enriched with virtual augmentations. The reality is the physical world that we observe without any virtual content, and AR overlays a virtual layer on top of reality.
- **Augmented virtuality**: The surrounding environments are virtual but enriched with real augmentations.
- **Mixed Reality**: MR is a broad concept containing all the technologies that mix real and virtual environments, including AR. AR and MR are used interchangeably in the literature due to the lack of a clear boundary. In this chapter, we consider MR as an advanced version of AR.
- **Virtual Reality**: All the contents presented by VR are virtual, which are not related to the user's real surrounding environment. VR synthesizes an immersive environment that can be isolated from the real world.
- **Extended Reality**: "XR" is a general term encompassing VR, AR, and MR. XR refers to all real and virtual combined environments between human and computer-generated input processed to create an interactive environment. XR, also referred to as cross-reality (CR), includes several immersive technologies (e.g., AR, VR, MR) and any other future realities that may emerge. AI, machine learning (ML), and deep learning (DL) (AI/ML/DL) are increasingly used for XR.

Note that the occlusion capability (Figure 21.1) of a see-through display is important in enhancing user's perception, visibility, and realism of the synthetic scene presented. Unlike video see-through displays, occlusion of a real scene in an optical see-through fashion is quite difficult to achieve, as the real scene is always seen through the partially transmissive optical combiner. In this chapter, four portions in ray paths of an optical see-through display are first identified between the light source and the eye. Corresponding to them, a few existing approaches for an occlusion display are then introduced that cut off the light in a different manner. On the other hand, translucent capability (Figure 21.1) means the material will allow light to pass through it but objects on the other side will not be clearly seen. Frosted glass is an example of a translucent material. The translucent display is provided by an opaque object that does not allow light to pass through.

EXTENDED REALITY

An XR system consists of the application, interaction, processing and rendering, and feedback. The XR system can include many different applications, such as social, medical, educational, and cultural, and can also involve themes around advertising, entertainment, the military, tourism, and heritage. Major components of an XR system are shown in Figure 21.4.

An XR system consists of the application, interaction, processing and rendering, and feedback, as seen in Figure 21.4. The XR system can include many different applications, such as social, medical, educational, and cultural, and can also involve themes around advertising, entertainment, the military, tourism, and heritage. When using this application, the user will be presented with a scene and will then interact with it. His or her interaction is captured via a range of visual, audio, or motion input devices and sensors.

The data from this interaction will be passed as input (in addition to other data from the environment if necessary) to the XR hardware devices, where further processing and rendering will be applied to create the desired output. The rendered scene(s) will be delivered as feedback to the user. When using XR applications, the user will be presented with a scene and will then interact with it. His or her interaction is captured via a range of visual, audio, or motion input devices and sensors. The data from this interaction will be passed as input (in addition to other data from the environment if necessary) to the XR hardware devices, where further processing and rendering will be applied to create the desired output. The rendered scene(s) will be delivered as feedback to the user.

Designing for XR applications is a complex task that requires knowledge from multiple disciplines in terms of interaction design, user experience, programming, and content creation including AI/ML/DL technologies. For example, this complexity increases for educational XR, as it requires additional competencies in learning objectives because the activities designed need to fit the objectives that are to be achieved. Even AI/ML/DL technology is used to enhance the detection of input like gestures or speech, performance improvements, automation, and other capabilities in XR having a powerful tool at the disposal of XR application designers and users

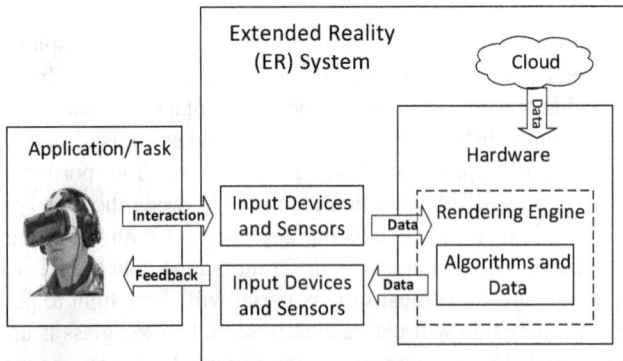

FIGURE 21.4 Major components of an extended reality system.

alike. XR can support a plethora of other applications, such as holographic teleportation, retail, tourism, fitness, and many more. In this way, XR technology goes much beyond AR, MR, and VR.

In the long run, XR has the potential to replace computers and portable devices and become general computing platforms. As mentioned above, wireless communication based on 5G/6G, and latest Wi-Fi systems, will play an important role in realizing mobile XR. Here we point out that XR not only will be used for entertainment, as it is the case currently, but also can transform the way we interact with the physical world in the future.

This significantly limits users' mobility and QoE. Wi-Fi (802.11b/g/n/ac) and Bluetooth are adopted by mainstream XR devices to provide wireless services. Due to limited data rates of Wi-Fi and Bluetooth, they can only support entry-level low-quality XR. 5G cellular networks and Wi-Fi wireless systems have demonstrated to achieve peak data rates of several Gbps. Using these networks, better wireless connections for XR can be realized. However, recent studies have shown that the ultimate XR requires uncompressed data rates of 2.3 Tbps with a latency lower than 1 millisecond which cannot be supported by existing 5G cellular networks and Wi-Fi.

We define that the uplink (UL) channel is from the HMD/optical HMD (OHMD) to the server and the downlink (DL) channel is in the opposite direction. For VR, the UL is used to send sensing information such as head moving and eye tracking. The DL is used to send rendered videos. Thus, the DL requires larger bandwidth compared to the UL. AR and MR OHMDs also send sensing information through the UL. In addition, they stream real-time videos of the real surrounding environment, which require a large bandwidth. Generally, AR and MR ULs require higher data rates than VR.

From Table 21.1 we notice that the current products need to be improved in the following aspects to realize wireless XR in the future. Existing XR devices' connectivity relies on Wi-Fi systems and cables. Most devices have the option of USB cables, which provide data communication and power. This significantly affects the user's mobility and user QoE.

Moreover, the wireless options are available using Wi-Fi-5 but its peak data rate is not sufficient to support future high-quality ultimate XR applications. Intel WiGig (60 GHz Wi-Fi) HMD will be used for Vive Cosmos Elite, which is based on millimeter-wave radios at 60 GHz. It can support three players with a range of 7 meters. Although this is a significant step toward wireless XR, this technology is not widely used for other XR products, and the number of users and operation range are limited. Existing XR devices have limited computing and storage capabilities which make it challenging to perform complex computing tasks, such as ML-based motion prediction and content caching. A computer or a server is necessary to run XR applications. This requires wireless communication between XR devices and servers.

The QoE is limited by power consumption and headset weight. High power consumptions of wireless communication and computation not only drain the battery fast but also generate heat problems which affect the user experience. Also, different from computers that are placed on desks, XR devices are wearable, and their weight should be minimized. Today's can only support two to three hours of operation, which is not sufficient for persistent applications. Their weight is around 500 grams,

TABLE 21.1

AR, MR, VR, and XR Key Performance Indicator

Specification	AR	MR	VR	XR	Remarks
Screen	Translucent	Translucent	Occlusion	Occlusion	XR KPIs are still emerging
Display	Optical head-mounted device	Optical head-mounted device	Head-mounted device	Optical head-mounted device	
Environment	Passive virtual and real	Passive virtual, active virtual, and real	Virtual	Passive virtual, active virtual, and real	
Uplink data rate	0.02–1.0 Gbps	0.02–1.0 Gbps	150 Kbps	1–2.3 Tbps (uncompressed)	
Downlink data rate	0.02–1.0 Gbps	0.02–1.0 Gbps	0.02–1.0 Gbps	1–2.3 Tbps (uncompressed)	
Latency	20 milliseconds	20 milliseconds	20–1000 milliseconds	1–8.3 milliseconds	
Refresh rate	~90 Hz	~90 Hz	~90 Hz	~90 Hz	
Pixel-per-degree	30–60	30–60	10–15	10–15	
Field-of-view	$20° - 50°$	$20° - 50°$	$100° - 150°$	$360°$	

which is much higher than wearable optical glasses – the weight of standard optical glasses is around 20 grams.

XR AND AI/ML/DL

Research on XR and AI is booming. It is out of scope for this book to describe even a part of it because it requires full treatment of this topic. Interestingly, AI researchers are also using XR for virtualize their tools and vice versa. However, we are summarizing where AI/ML/DL technologies are used for XR and vice versa:

- **AI Method applied for XR**: AI methods are applied for an XR problem (e.g., for redirected walking, viewport generation, and sickness prediction).
- **XR applied for AI**: XR technology is applied for AI problems (e.g., to visualize neural networks in VR).
- **Interaction with embodied AI**: Aiming to enhance interaction with intelligent virtual agents (VAs).
- **Application focus**: XR and AI are applied to problems but are not the focus of the presented research (e.g., an AR-based system that helps with tumor recognition and applies DL to estimate positions).

In another way of looking into it is the AI-XR combination, as AI and XR differ in their origin and primary objectives. However, their combination is emerging as a powerful tool for addressing prominent AI and XR challenges and opportunities for cross-development [3]. We have identified the main applications of the AI-XR

TABLE 21.2
AI-XR Combination Applications

AI-XR combination	Interpretation of XR generated data	Affective computing
		Medical training
		Gaming
	Conferring intelligence on XR	Cancer diagnosis
		Virtual patient
		Advanced visualization
		Smart home
	Training AI	Armed force training
		Autonomous cars and robotics
		Gaming
		Gaming

combination, including autonomous cars, robotics, military, medical training, cancer diagnosis, entertainment, and gaming applications, advanced visualization methods, smart homes, affective computing, and driver education and training. In addition, we found that the primary motivation for developing the AI-XR applications include (1) interpreting XR-generated data, (2) conferring intelligence on XR, and (3) training AI, as depicted in Table 21.2.

AUGMENTED REALITY

As shown in Figure 21.1, the reality is the physical world that we observe without any virtual content. AR overlays a virtual layer on top of reality. Next, we introduce two aspects of AR, namely, the content that AR provides and the device that can realize AR. Virtual contents in AR are presented in two formats. Virtual objects are placed in real environments, and users cannot easily distinguish them from real objects. The widely used Pokémon GO is an example. Virtual information, such as real-time maps, notations, and sensory data, is provided to help users understand the real environment and provide the desired assistance. For example, AR navigation information can be displayed in real-time to assist drivers. AR users can observe the real environment and virtual contents simultaneously. Currently, there are mainly two approaches to view AR contents:

- Nonimmersive AR using phones, tablets, or any other handheld smart devices with cameras.
- Immersive AR using smart glasses or other optical HMDs (OHMDs).

The nonimmersive AR allows users to watch virtual contents through cameras on smart devices. For example, consider smartphone-based applications, cameras capture real-time real environments, and smartphones augment virtual contents and display mixed environments to users. The immersive AR presents mixed environments directly in users' sight. Even users turn their heads around, they can still observe

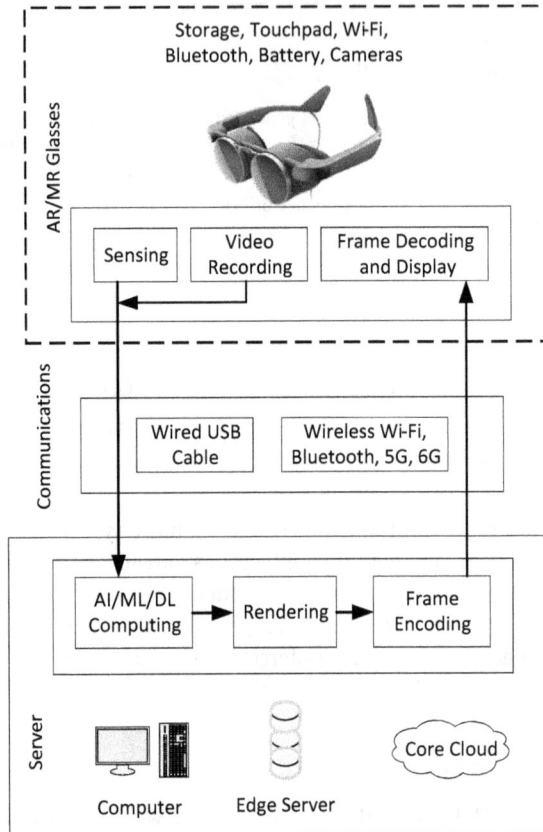

FIGURE 21.5 AR/MR system architecture.

virtual content. An example of AR OHMD is given in Figure 21.5 with VUZIX M4000 AR glasses.

MR has similar glasses and system architectures. The glasses send sensing information, such as head and eye tracking, and real-time videos of real environments to a server via wired or wireless communication. The server can be a computer, an edge server, or the core cloud, which performs environment and human understanding using AI, particularly ML algorithms, and renders virtual contents. The rendered video is sent back to the glasses for display. Note that, depending on the display technique, the AR can be divided into two:

- Optical see-through (OST)
- Video see-through (VST)

The OST glasses are translucent, as the one shown in
Figure 21.5. The VST glasses or headsets use a camera to capture real-time videos and then augment virtual contents onto them. A simple example of VST AR is

the smartphone-based application. More complicated VST AR can use HMDs. This chapter mainly focuses on the OST AR, which is extensively used in high-end AR devices.

MIXED REALITY

As shown in Figure 21.5, MR is a broad concept containing all the technologies that mix real and virtual environments, including AR. AR and MR are used interchangeably in the literature due to the lack of a clear boundary. In this chapter, we consider MR as an advanced version of AR. MR presents richer and more capable virtual contents than AR. MR allows users to interact with active virtual contents, while AR only displays passive virtual contents. Like AR, MR can also use smartphones and tablets. However, the QoE may not be acceptable. OHMDs are used extensively for MR because they can provide more immersive experiences compared with smartphones and tablets. MR can share the same system architecture as AR, as shown in Figure 21.5. However, the AI computing of MR for the environment and human understanding is more complex than that of AR since it provides more interactions with virtual contents. Computer virtualization architecture with MR virtual inputs and virtual display is depicted in Figure 21.6.

Today's computers and portable devices have integrated inputs, computing units, storage, and outputs. Laptops and smart devices have significantly changed our lives thanks to their small size and lightweight. MR will provide a completely different computing platform by offloading computing, storage, and many other functions to the edge and cloud. The inputs, such as keyboard and trackpad, and outputs will become virtual using MR, as shown in Figures 21.5 and 21.6. The computer manufacturers will not physically produce computers; instead, they will provide virtual computers in the edge and cloud. Users can utilize MR glasses with network access to create virtual inputs and displays. As 5G and 6G will provide ubiquitous wireless services, users can have access to their computers anytime anywhere. Besides computers, smartwatches and cell phones can also be accessed using MR devices.

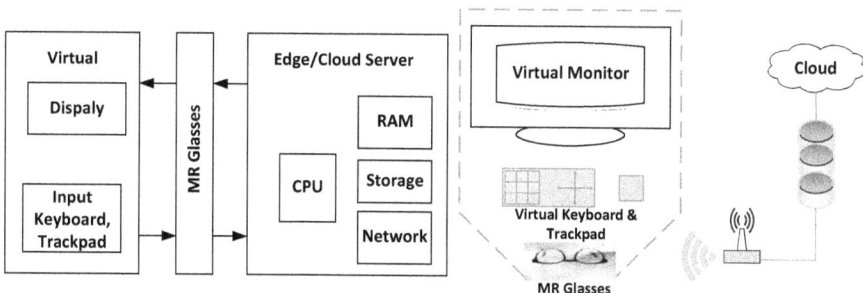

FIGURE 21.6 Computer virtualization architecture with MR virtual inputs and virtual display.

VIRTUAL REALITY

All the contents presented by VR are virtual, which are not related to the user's real surrounding environment. VR synthesizes an immersive environment that can be isolated from the real world. Occluded HMDs are used to block the real surrounding environment and provide the user with immersive experiences. Although VR can also be presented in smartphones and tablets, due to the nonimmersive low-quality experience, we do not consider them in this chapter. An illustration of VR system architecture is shown in Figure 21.7.

VR HMD sends sensing information, such as head and eye tracking, to the server which performs AI computing to understand users' behaviors. The connection between the HMD and the server can also be wired or wireless. VR does not require videos of the real environment since all the presented contents are virtual. Like AR and MR, the server can be a computer, an edge server, or the core cloud. Pre-created

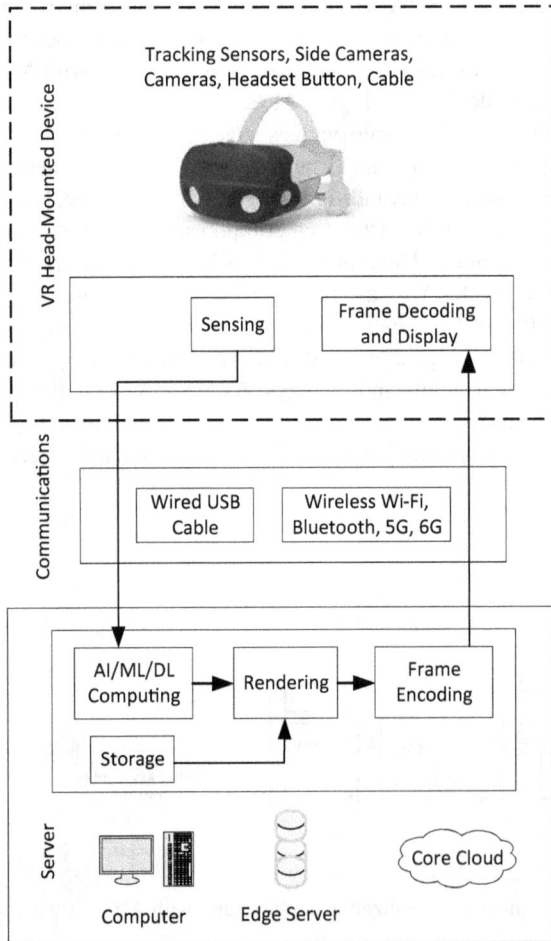

FIGURE 21.7   VR system architecture.

videos are saved in the storage which is rendered based on sensing information. The rendered video is sent back to the HMD for decoding and display.

TECHNICAL DIFFERENCES

Although AR, MR, and VR have common features, we highlight the following major technical differences which affect the wireless system design.

- **Display**: AR and MR use OST-based translucent display installed on OHMDs. Users can observe the real environment. VR uses an occlusion display installed on HMDs; the real environment is not visible.
- **Human understanding**: Interactions with virtual contents highly rely on external inputs. Voice control, touchpad, head tracking, eye tracking, and hand tracking are widely used in AR, MR, and VR. Currently, AR and MR are mainly used for assistance and training, while VR is mainly used for gaming. Therefore, AR and MR prefer an all-in-one format, which can be worn conveniently. Inputs, such as touchpad and buttons, are integrated into the glasses. On the contrary, VR uses external controllers to provide more immersive gaming experiences.
- **Environment Understanding**: VR does not need to understand the user's real surrounding environment. Differently, AR and MR mix virtual contents with the real surrounding environment. Especially, when a virtual object is placed in a real setting, it must be put at a suitable location, e.g., a virtual cup should be placed on a table rather than in the air. Thus, AR and MR must understand the environment. This is achieved by sending real-time videos by HMD cameras to the server, where AI algorithms are used. The computation in a VR server is to render videos based on users' field-of-view (FoV) and inputs, whereas AR and MR servers recognize objects and create extra useful information using AI.
- **Uplink vs. downlink**: We define that the UL channel is from the HMD/OHMD to the server and the DL channel is in the opposite direction. For VR, the UL is used to send sensing information such as head moving and eye tracking. The DL is used to send rendered videos. Thus, the DL requires larger bandwidth compared to the UL. AR and MR OHMDs also send sensing information through the UL. In addition, they stream real-time videos of the real surrounding environment, which require a large bandwidth. Generally, AR and MR ULs require higher data rates than VR.
- **Latency tolerance**: AR and MR mix virtual contents with the real environment. The real environment is highly dynamic, e.g. the user is looking at a moving vehicle, AR and MR have to respond to these dynamics. Thus, they require ultra-low latency of video rendering. Moreover, MR is more challenging than AR since it interacts with virtual contents; it has more strict requirements on latency. The latency tolerance of VR depends on the specific application. For high-interactive VR, such as gaming, the latency tolerance is low, and it requires ultralow latency. For low-interactive VR, such as virtual movie theaters, the latency tolerance is high since the user barely interacts with the movie.

EXISTING XR DEVICES

There are various XR products on the market. We cannot enumerate all of them due to the limited space here.

Augmented reality devices: These are used for controlling UAVs (unmanned aerial vehicles), remote support for field technicians, operations, and tele-medicine. By wearing AR glasses, operators and technicians do not need to hold or look at smart devices, which improve their efficiency. Various AR products are available or under development, such as Epson Moverio BT300, VUZIX M4000, Apple Glasses, and Google Glass.

Mixed reality devices: Microsoft HoloLens is a representative product. Like AR, MR devices also aim to boost productivity for manufacturing, engineering and construction, healthcare, and education. The vendor also provides software support, and users can develop their own applications. Generally, MR glasses are more expensive than AR glasses due to their complicated functions.

Virtual reality devices: These are mainly used for gaming and movies. The Sony PlayStation, Oculus Quest series, and HTC Vive series are popular products. Users can purchase VR games and movies in online stores. We notice that the current products need to be improved in the following aspects to realize wireless XR in the future.

Extended reality devices: The existing XR devices' connectivity relies on Wi-Fis and cables. Most devices have the option of USB cables, which provide data communication and power. This significantly affects the user's mobility and user QoE. Moreover, the wireless options are available using Wi-Fi-5, but its peak data rate is not sufficient to support high-quality ultimate XR applications. Intel WiGig will be used for Vive Cosmos Elite which is based on millimeter-wave (mm-Wave) radios at 60 GHz. It can support three players with a range of 7 meters. This is not widely used for other XR products.

The existing XR devices have limited computing and storage capabilities, which makes it challenging to perform complex computing tasks, such as ML-based motion prediction and content caching. A computer or a server is necessary to run XR applications.

The QoE is limited by power consumption and headset weight. High power consumption of wireless communication and computation not only drains the battery fast but also generates heating which affects the user experience. Also, different from computers that are placed on desks, XR devices are wearable, and their weight should be minimized. Today's XR devices can only support two to three hours of operation, which is not sufficient for persistent applications. Their weight is around 500 grams which is much higher than wearable optical glasses – the weight of standard optical glasses is around 20 grams.

USE CASES

In this section, we introduce representative use cases of XR, including AR, MR, and VR. With the 6G wireless systems and the next-generation Wi-Fis, we anticipate that these applications can be fully supported by wireless technologies.

Augmented Reality

Sports: AR can improve the performance of athletes and the QoE of the audience. With AR glasses, athletes can receive real-time AI support. For example, the optimal pass route can be displayed in soccer players' glasses, so that the success pass rate can be improved. The audience can observe the information that are not available before, such as players' names, ratings, and background information. For instance, sports fans sitting in the back of a big stadium cannot clearly see players. Their view can be augmented with players' names and performances close to the corresponding player.

Automotive industry: AR plays an import role in the life cycle of motor vehicles, including design, manufacturing, sales, driving, and maintenance. AR can help designers choose different options by virtually manufacturing the car. Also, it can virtually change the car model for customers to select. AR has been used to provide collision warning, driver assistance, navigation, and lane departure warning for drivers with OHMDs. Moreover, it is usually challenging for a driver to identify problems of a car due to the lack of professional training. With AR devices, AI can identify the problems and display them. Additional information such as nearby car repair shops or contact information can also be shown.

Mixed Reality

Computing platform: Today's computers and portable devices have integrated inputs, computing units, storage, and outputs. Laptops and smart devices have significantly changed our lives thanks to their small size and lightweight. MR will provide a completely different computing platform by offloading computing, storage, and many other functions to the edge and cloud. The inputs, such as keyboard and trackpad, and outputs will become virtual using MR, as shown in Figures 21.5 and 21.6. The computer manufacturers will not physically produce computers; instead, they will provide virtual computers in the edge and cloud. Users can utilize MR glasses with network access to create virtual inputs and displays. As 5G and 6G will provide ubiquitous wireless services, users can have access to their computers anytime anywhere. Besides computers, smartwatches and cell phones can also be accessed using MR devices.

Healthcare: MR will provide better medical services to the board community, especially for those without convenient access to medical providers. Doctors will leverage different medical examination results which can be projected in their MR glasses. AI-enabled identification and classification can also help doctors evaluate patients' health status and perform surgeries in real-time.

3D design: Existing 3D computer-aided design (CAD) is conducted in a computer with 2D displays, which physically limits productivity. With MR, 3D CAD can be performed in a 3D space. 3D printing is a useful tool in various contexts, but the 3D model design is challenging for ordinary people without training. MR will reduce the complexity of 3D CAD design and make 3D printing more accessible to general users.

VIRTUAL REALITY

Personal movie theater: VR allows users to virtually sit in a movie theater with a wide virtual screen. Multiple users can watch a movie together just like they were in a real movie theater. This is a low-interactive VR application. Users do not interact with the virtual world. Therefore, it is relatively simple to implement. Most existing VR devices support this application.

Gaming: VR can place users into an immersive virtual world. Different from existing computer games using a keyboard and a mouse to control a character in the game, users play VR games as if they were in the real world. This is a high-interactive VR application, which has strict requirements on latency. Otherwise, users may feel nausea while they are playing.

TECHNICAL REQUIREMENTS

In this section, we first review the key performance indicators (KPIs) of XR. Then, we study the KPIs of the existing XR and the ultimate XR. We focus on the parameters that are related to wireless communication and networking. Although not listed here, XR can support a plethora of other applications, such as remote education, holographic teleportation, retail, tourism, fitness, and many more.

Basics of XR Parameters

As discussed earlier, the human perception is based on five senses: sight, hearing, touch, smell, and taste. To create a fully or partially virtual environment, we need to synthesize all these senses. However, the sight is the most challenging sense because it requires a large amount of multimedia data. Thus, the existing XR KPIs are mainly related to videos. The FoV is the angle of the maximum area that we can observe. Each human eye can cover nearly 130°.

With two eyes, we can observe nearly 180°. In VR, a widely used term is '360°video,' which records every direction at the same time. This can be achieved by using special cameras or multiple regular cameras. Human eyes cannot observe such a wide angle without turning heads.

The pixels-per-degree (PPD) is the number of pixels that are in the view for each degree. A large PPD means there are more pixels, and the video is sharp and clear, vice versa. The resolution is the measurement of a video frame's width and height in pixels. Note that resolution, FoV, and PPD are related as follows:

$$\text{Resolution} = \text{horizontal degree} \times \text{PPD} \times \text{vertical degree} \times \text{PPD}$$

Thus, for a given resolution, a large FoV results in a small PPD. Existing AR and MR glasses have a small FoV, i.e., 20°–50°, but their PPD is above 30. This provides the user with clear virtual contents in the presence of real environments. On the contrary, VR displays have a large FoV, i.e., 100°–150°, to provide users with immersive experiences. As a result, the PPD of VR displays is relatively small, which is around 10–15.

The refresh rate is the number of video frames that can be displayed in one second. A low refresh rate for XR may result in headaches or nausea. Usually, a 90 Hz refresh rate is suggested for XR devices. The data rate can be obtained based on the KPIs of XR videos, including the refresh rate, the resolution, and the number of bits of color.

The latency is the response time of XR devices when there is a change caused by the real environment or user. It is determined by the specific XR application. For example, the display currently shows frame A, and, meanwhile, the HMD/OHMD camera captures a new frame B or sensors receive new inputs from the real environment or user. Now, frame B needs to be rendered with virtual content, and this should be reflected in the next frame that is displayed. The latency that can be tolerated is the time from displaying frame A to displaying rendered frame B. Depending on the refresh rate, the latency tolerance can be as low as several milliseconds.

Next, we mainly focus on the data rate and latency, which are the two key parameters that affect wireless system design. Wireless XR will evolve with the development of wireless technologies. Currently, we are at the stage 1, where XR is moving from wired connections to wireless connections. The ultimate XR at the stage 3 will require several Tbps throughputs and ultra-low latency, which can be lower than 1 milliseconds. The stage 2 is a transition stage between stage 1 and stage 3. A summary of existing typical XR KPIs is given in Table 19.41. Note that AR can be considered as a simple version of MR. Finally, the advanced AR technologies may be merged into MR. Thus, the ultimate XR only consists of MR and VR.

Data Rates

Due to the environment and human understanding, AR and MR require similar UL and DL data rates. The OHMD must send real-time videos to the server, and the server sends back rendered videos. The UL of VR requires very low data rates, e.g. less than 150 kbps, since it only transmits sensing information. The DL of VR requires similar data rates as AR and MR.

> **Existing XR**: Although current AR, MR, and VR displays have different FoV, their resolution and refresh rates are comparable, which require similar data rates. The resolution of the XR display is determined by the FoV and PPD. We use HTC Vive Cosmos Elite as an example to evaluate current requirements for wireless XR. Consider the $1,440 \times 1,700$ resolution (per eye) with a refresh rate of 90 Hz and 8 bits of color. The required data rate without compression is 10.6 Gbps. The data rate is obtained using:
>
> $$1440 \times 1700 \times 3 \times 8 \left(\text{bits of color}\right) \times 2 \left(2\,\text{eyes}\right) \times 90 \left(\text{refresh rate}\right) = 10.6\,\text{Gbps}$$
>
> Using standard video lossy compression techniques with a 300 : 1 rate, the required data rate can be reduced to 35.3 Mbps. Intel WiGig wireless adapter can support 8 Gbps data rates, which are sufficient to provide reliable wireless connections. However, this is only an entry-level VR that has relatively low PPD, refresh rate, and bit of color.
>
> **Ultimate XR**: The ultimate (or extreme) XR, which is the stage 3, requires $360° \times 180°$ full view with 120 Hz refresh rate, 64 PPD, and 12 bits of color. Although a refresh rate higher than 120 Hz can improve the video quality, most users may not be able to distinguish the difference. Thus, the required

data rate without compression is 2.3 Tbps. Using video lossy compression at the rate of 300 : 1, the reduced data rate is 7.7 Gbps. To reduce the required data rate, an FoV of $110° \times 110°$ can be used. The updated data rates are 428.2 Gbps without compression and 1.4 Gbps with a compression rate of 300 : 1.

Latency

Existing XR: We can divide XR into two categories based on latencies, namely:
(a) AR, MR, and high-interactive VR, and
(b) low-interactive VR.

The former has strict requirements on latency, and the latter can tolerate a certain latency. The latency is affected by the refresh rate. Currently, a 90 Hz refresh rate is widely used, which requires a latency smaller than 11 milliseconds. As shown in Table 21.1, some low-interactive VR can tolerate around 1,000 milliseconds latency. This is because the HMD can use a buffer to save multiple frames and play them with a certain delay. This can effectively address network jitter. The buffer size and delay can be determined by the specific application's latency tolerance.

Ultimate XR: Based on the refresh rate of ultimate XR, the latency should be smaller than 8.3 milliseconds. Note that the latency consists of wireless communication, sensing data fusion, computing, access to edge or cloud servers, and display response time. Since the latency caused by each party is highly stochastic, the communication and networking latency should be much smaller than 8.3 milliseconds to provide a high QoE.

CHALLENGES AND FUTURE RESEARCH DIRECTIONS

Depending on the environment, we divide XR applications into two categories: local area VR, AR, and MR (LAVAR) and wide area VR, AR, and MR (WAVAR). LAVAR supports applications in small areas, such as apartments, offices, and retail stores, whereas WAVAR supports applications in large areas, such as sports stadiums and autonomous vehicles. Note that WAVAR is a broader concept than the mobile AR (MAR), and it aims to provide ubiquitous wireless services for XR. Next, we study the grand challenges of realizing the wireless ultimate XR. Potential solutions and future research directions are also provided.

Data Rates

Our vision is that WAVAR will use 6G wireless systems, whereas LAVAR will use the next-generation Wi-Fis. In this way, we can achieve seamless ubiquitous connectivity. Note that using only 6G wireless systems may not be practical because the mm-Wave, Terahertz, and visible light signals experience significant propagation losses due to building blockages. Deploying more 6G base stations cannot effectively solve this problem because the base stations are much more expensive than the Wi-Fi access points. It is more economical to use Wi-Fi for LAVAR instead of 6G wireless systems.

Though 5G promises a peak data rate of 20 Gbps, recent network measurements show that the achievable data rates are around 0.1–2.0 Gbps. Since the requirement of existing entry-level XR is lower than 1.0 Gbps, 5G can provide sufficient data rates. However, the ultimate XR requires much higher data rates than that provided by 5G. The envisioned 6G wireless system has a peak data rate of 1 Tbps and an experience data rate of 1.0 Gbps. Such high data rates will enable the use of high-quality ultimate WAVAR. Most existing XR devices support Wi-Fi-5 which cannot provide sufficient data rates for ultimate XR applications.

LAVAR will employ the next-generation Wi-Fi systems, such as 802.11be (around 46 Gbps) and 802.11ay (around 100 Gbps). Such high data rates together with the data compression techniques, such as H.266 (versatile video coding), and Wi-Fi systems can support ultimate LAVAR. WAVAR and LAVAR have the following specific challenges to achieve and maintain high data rates.

Novel wireless system design: The high data rates in 6G wireless systems rely on novel wireless communication systems, such as Terahertz, mm-Wave, and visible light communication (VLC). Mm-Wave bands have received significant attention in 5G systems; they may still play an important role in 6G. Terahertz wireless communication systems have been developed for more than a decade, but there are still open research problems, such as optimal resource allocation in the Terahertz band, co-design of sensing, communication, and intelligence, and beamforming among others.

Unreliable/blocked wireless environment: A novel design is required to avoid blockages in indoor and outdoor environments. For example, in VR gaming, the human body may block mm-Wave or Terahertz signals intermittently. Reconfigurable intelligent surfaces (RIS) have been used to create extra propagation paths. This increases the system reliability by providing redundant propagation paths in case of blockages. It is challenging to provide reliable high data rates considering the stochastic nature of the wireless channel. Adaptive protocols are needed to optimally control communication systems to meet required QoE.

Multiusers: WAVAR must support many devices (tens of thousands) simultaneously. Consider the AR sports where fans can see players' names and performances that are displayed close to players in real-time. A potential solution can divide the large stadium into small LAVAR and use multiple Wi-Fi access points to provide AR service. Interference should be considered when planning router locations.

Latency

The major sources that generate latency in wireless XR systems: For AR, MR, and high-interactive VR, video frames are displayed immediately, which can tolerate extremely low latency, while for low-interactive VR video frames can be buffered, and the latency tolerance is high.

Wireless and wireline latency minimization: Today's 5G end-to-end delay is much higher than 10 milliseconds which needs significant improvements. For example, recent network measurements show that the end-to-end delay in 5G networks is around 21.8 milliseconds and 27.4 milliseconds. 6G proposes to reduce the delay to around 1 millisecond. The radio access networks using mm-Wave and Terahertz can achieve very low latency, but it is usually neglected that the wireline communication also needs to be updated to support 6G networks. Also, some cellular core network functions can be moved to base stations to reduce the access delay. The device-to-device (D2D) communication at mm-Wave and Terahertz bands can further reduce the delay by allowing the user to directly communicate with the local server.

Trade-off between video encoding and wireless communication: High-quality video encoding and decoding may take longer than 10 milliseconds, which is even larger than the overall latency requirements. Usually, on the one hand, video is encoded/compressed before wireless transmission to reduce the communication latency. On the other hand, video encoding and decoding increase the latency, which is usually much smaller than the communication latency in wireless networks. Since 5G and 6G have high communication data rates, the communication latency can be significantly reduced. It is not clear whether we still need high compression ratios if the communication channel can allow high volume data transmission. Adaptive encoding algorithms considering wireless communication channels can be more efficient.

Edge computing and caching: Wireless XR will leverage edge computing and caching due to the following reasons.

- High-bandwidth cloud computing services are expensive, which may cost several thousand dollars per month. Compared to existing XR devices, which cost around $300 to $5,000, cloud computing services with GPU servers may not be practical.
- The latency is also affected by the path length. Using cloud services may create significant traffic in the network and increase latency.

AR, MR, and high-interactive VR rely on edge servers because the communication with the core cloud may generate significant latency, which cannot meet the latency requirements. Edge servers are close to users which incur short delays. Also, local information, such as indoor environment and street information, can be cached in edge servers, which will significantly reduce the latency and improve computing efficiency. For example, the pictures or videos created in the same apartment have significant identical contents which can be cached and reused.

Software-defined networking (SDN), network function virtualization (NFV), and automatic network slicing: 6G wireless systems are envisioned to support a wide variety of applications, and XR is only one of them. Considering the fast-growing network traffic, the automatic network slicing with the support of SDN and NFV is necessary to prioritize XR applications to reduce the latency. Optimal network slicing algorithms for XR applications are desirable in the era of 6G to efficiently manage and use networking resources.

Mobility

Mobile WAVAR can be used for navigation for automobiles and pedestrians. Due to the short range of mm-Wave and Terahertz wireless systems, the mobility incurs frequent handoffs which cause long latencies. The soft handoff that allows multiple connections is necessary for seamless connections. Also, DL-based motion prediction in conjunction with network scheduling can be used to plan resource allocation for the user. UAVs provide a large coverage area which can reduce the number of handoffs. The UAV trajectory and location can be jointly designed with XR users' mobility. Motion prediction also allows LAVAR to pre-render content and reduce the latency. Since mm-Wave and Terahertz wireless systems are highly directional, beam steering considering the user's mobility is challenging. Motion prediction is necessary for accurate and efficient beamforming.

Weight and Power Consumption

Different from laptops and smartphones, XR HMDs are worn on the head. The weight should be small to improve the QoE. A heavy HMD may not be accessible to everyone. However, advanced computation, communication, sensing, and display require bulky devices. Moreover, high-power consumption also requires large batteries to prolong the operation time. All these factors can increase the weight of HMD. Also, the high-power consumption may generate heating, which makes the HMD not wearable. To make wireless XR practical, low-power communication, computation, and networking protocols must be employed. Simultaneous wireless information and power transfer at the mm-Wave and Terahertz bands have the potential to partially address this issue.

Collaborative XR

Collaborative XR will enable multiple users to work on the same task simultaneously. For example, LAVAR can support several doctors to work collaboratively on a surgery. This is a challenging problem because it requires higher network throughput and ultra-low latency. Serving multiple users simultaneously includes various wireless communication and networking problems, such as synchronization and end-to-end latency minimization. Using one mobile edge server may not be sufficient, and multiple RISs are necessary to control wireless signal propagation. The intelligent communication environments can be an efficient solution to meet such high requirements.

Other Challenges

XR is a complex system, and there are many other challenges that are directly or indirectly related to wireless communication and networking. Security and privacy of XR are of paramount importance, especially AR and MR that combines real and virtual environments. For example, if an attacker modifies the traffic light, speed limit, and road symbol signs, users or autonomous vehicles may be misled to make life-threatening decisions. The data storage and communication must be protected, and intelligent applications can be installed in XR devices to detect and correct malicious information.

Wireless sensing can be integrated into XR devices to reduce the use of peripheral sensors. In this way, XR devices can be made more compact. The use of Terahertz wireless communication for high-data-rate communication can also provide unprecedented wireless sensing accuracies due to its short wavelength. Although the human body can block Terahertz signals, this also provides information about the motion of the human body. In conjunction with optical cameras, this can provide accurate motion sensing.

Operating systems dedicated for XR are also desirable to efficiently manage applications, hardware, energy, data communication, security and privacy, and display among others. XR devices will support a plethora of applications simultaneously and integrate a wide variety of intelligent things. The future networked XR will connect many XR devices. Such complicated systems require operating systems to manage resources and tasks accordingly.

SUMMARY

XR consisting of AR, MR, and VR will soon become the next-generation mobile computing platform that can make rapid and profound changes in our lives, just as the changes that laptops and smartphones have brought to us. However, today's XR devices are mainly tethered using cables which limit their mobility and potentials. In this chapter, we introduce wireless XR systems and discuss their requirements of wireless data rates and latency, as well as their use cases. Research challenges and potential solutions to realize the environed indoor and outdoor applications are provided. 6G wireless systems and the next-generation Wi-Fis will allow XR users to move without hindrance, and they can also support multiple users simultaneously, which are the enablers of high-quality ultimate XR.

REFERENCES

1. Akyildiz, F. I. and Guo, H. "Wireless Extended Reality (XR): Challenges and New Research Directions,"1921, Truva Inc., Alpharetta, GA 30022, USA and Engineering Department, Norfolk State University, VA, 23504, USA Emails: ian@truvainc.com, hguo@nsu.edu
2. Akyildiz, F. I. and Guo, H. "Wireless Communication Research Challenges for Extended Reality (XR)," *ITU Journal on Future and Evolving Technologies* 2022, 3(1). https://www.itu.int/en/journal/j-fet
3. Reiners, D. et al., "The Combination of Artificial Intelligence and Extended Reality: A Systematic Review," September 2021. https://doi.org/10.3389/frvir.2021.721933

22 Unmanned Aerial Vehicles Communication Systems

OVERVIEW

Unmanned aerial vehicle (UAVs) communication systems consist of audio, video, and data payload and help drones and their operators to achieve their desired results. Note that UAVs are composed of a variety of essential components. While all components harmoniously perform a specific purpose that contributes to operational flight, the most important component is the communications systems. Without these systems, not only would unmanned flight be considered unobtainable, but would also make collecting and transmitting aerial visuals and communications data impossible.

As UAVs continue to position themselves as the preeminent aerial data collection platform across a wide variety of industries, their communications systems grow proportionally in importance. Without highly adaptable and reliable communications systems, operators are left at a stark disadvantage for obtaining aerial visuals and data. As it stands, radio frequency (RF) communications are the most optimized solution for reliable drone communications systems. A combination of their small size, weight, minimized power consumption, and robust communications link makes them the most suitable solution for most civilian UAVs.

Civilian UAV communications systems typically operate on frequencies of 2.4 GHz and 5.8 GHz. UAV communications systems work by using one frequency to control the aerial vehicle from the ground via a remote pilot while the other frequency is used to beam data or relay first-person view (FPV) video. By utilizing high-quality, reliable communications links, civilian UAVs can relay aerial visuals and data to those on the ground with ease, while remaining in flight.

However, in defense applications, different types of drones are implemented. Defense drones vary greatly from civilian UAVs, as their missions are frequently longer duration or require striking capabilities in addition to providing aerial visuals of the battlefield below. One concern in the use of drones in defense applications is the occurrence of signal jamming. When signal jamming is used, this cuts the remote pilot and operations base off from visuals of what the drone is seeing. While it seems like a disastrous scenario, most defense drones are engineered to return to base after a loss of communications contact. Defense operators have found another solution to potential signal jamming – redundant on-board navigation systems that don't rely on global positioning system (GPS). By eliminating available GPS data, jamming becomes an infrequent occurrence, allowing defense drones to complete their missions and return to base safely.

DOI: 10.1201/9781003499480-22

The quadrotor's (drone's) attitude, altitude, and horizontal linearized dynamics result in a set of piecewise affine models, enabling the controller to account for a larger part of the quadrotor's flight envelope while modeling the effects of atmospheric disturbances as additive-affine terms in the system. A drone's attitude is the three-dimensional orientation of a vehicle with respect to a specified reference frame. Attitude systems include the sensors, actuators, avionics, algorithms, software, and ground support equipment used to determine and control the attitude of a vehicle. Although drones are available in many different forms, we have only focused on the quadrature drone. Drone flights are bound by specific regulations that dictate their maximum altitude. In the United States, as well as many other countries, the law sets a strict limit of 400 feet above ground level. Despite this legal ceiling, drones can technically reach altitudes up to 10 kilometers (33,000 feet).

Quadrotor Drone

Quadrotors, also known as quadrotors, are rotary wing UAVs capable of vertical take-off and landing. It belongs to a more general class of aerial vehicles called multicopter or multirotor. Small quadcopters are easy to build because of low cost, low inertial force, and simple flight control system. Quadcopters provide stable flight performance, making them ideal for surveillance and aerial photography. As their use becomes widespread worldwide, the number of studies conducted to enable autonomous tasks is growing.

Reference Frames

We have discussed specifically the quadrotor robot aircraft. The various reference frames and coordinate systems are used to describe the position of orientation of aircraft, and the transformation between these coordinate systems. Note that, in classical physics and special relativity, an inertial frame of reference is defined as a basic frame of reference that is not undergoing acceleration where Newton's law holds true. However, it is necessary to use several different coordinate systems for the following reasons [1]:

- Newton's equations of motion are given the coordinate frame attached to the quadrotor.
- Aerodynamics forces and torques are applied in the body frame.
- On-board sensors like accelerometers and rate gyros measure information with respect to the body frame.
- Alternatively, GPS measures position, ground speed, and course angle with respect to the inertial frame.

Most mission requirements, like loiter points and flight trajectories, are specified in the inertial frame. In addition, map information is also given in an inertial frame. One coordinate frame is transformed into another through two basic operations: rotations and translations. Translations are usually expressed as matrices providing relationships between the two frames in terms of rotation angles. We also derive the Coriolis formula, which is the basis for transformations between translating and rotating frames.

Rotational Matrices

First, we will consider a general rotation matrix. Figure 22.1 shows a rotation frame $\mathcal{F}^1$ (specified by $\hat{i}^0$, $\hat{j}^0$, and $\hat{k}^0$) that rotates with respect to the inertial plane $\mathcal{F}^0$ specified by $\left(\hat{i}^1, \hat{j}^1, \text{and } \hat{k}^1\right)$ through an arbitrary angle ψ toward the right-hand side (anticlockwise) centering O such that brings $\hat{k}^1$ to $\hat{k}^0$ $\left(i.e., \hat{k}^1 = \hat{k}^0\right)$. It is seen that there is a relationship in different axis: $(x, y, \text{and } z) \rightarrow \left(\left(\hat{i}^0, \hat{j}^0, \text{and } \hat{k}^0\right) \rightarrow \left(\hat{i}^1, \hat{j}^1, \text{and } \hat{k}^1\right)\right)$. We observe that the same vector p (Figure 22.1) is expressed in both the $\mathcal{F}^0$ and $\mathcal{F}^1$, respectively. Since both frames express the same vector, we can write as follows:

$$p_x^1 \hat{i}^1 + p_y^1 \hat{j}^1 + p_z^1 \hat{k}^1 = p = p_x^0 \hat{i}^0 + p_y^0 \hat{j}^0 + p_z^0 \hat{k}^0$$

Taking the dot product of both sides with $\hat{i}^1$, $\hat{j}^1$, and $\hat{k}^1$, respectively, the following is obtained:

$$\hat{i}^1 \cdot p_x^1 \hat{i}^1 + \hat{i}^1 \cdot p_y^1 \hat{j}^1 + \hat{i}^1 \cdot p_z^1 \hat{k}^1 = \hat{i}^1 \cdot p_x^0 \hat{i}^0 + \hat{i}^1 \cdot p_y^0 \hat{j}^0 + \hat{i}^1 \cdot p_z^0 \hat{k}^0$$
$$\hat{j}^1 \cdot p_x^1 \hat{i}^1 + \hat{j}^1 \cdot p_y^1 \hat{j}^1 + \hat{j}^1 \cdot p_z^1 \hat{k}^1 = \hat{j}^1 \cdot p_x^0 \hat{i}^0 + \hat{j}^1 \cdot p_y^0 \hat{j}^0 + \hat{j}^1 \cdot p_z^0 \hat{k}^0$$
$$\hat{k}^1 \cdot p_x^1 \hat{i}^1 + \hat{k}^1 \cdot p_y^1 \hat{j}^1 + \hat{k}^1 \cdot p_z^1 \hat{k}^1 = \hat{k}^1 \cdot p_x^0 \hat{i}^0 + \hat{k}^1 \cdot p_y^0 \hat{j}^0 + \hat{k}^1 \cdot p_z^0 \hat{k}^0$$

We know that $\hat{i}^1 \cdot \hat{i}^1 = \hat{j}^1 \cdot \hat{j}^1 = \hat{k}^1 \cdot \hat{k}^1 = 1$, and $\hat{i}^1 \cdot \hat{j}^1 = \hat{j}^1 \cdot \hat{k}^1 = \hat{k}^1 \cdot \hat{i}^1 = 0$. Stacking the result into matrix form gives:

$$p^1 \triangleq \begin{bmatrix} p_x^1 \\ p_y^1 \\ p_z^1 \end{bmatrix} = \begin{bmatrix} \hat{i}^1 \cdot \hat{i}^0 & \hat{i}^1 \cdot \hat{j}^0 & \hat{i}^1 \cdot \hat{k}^0 \\ \hat{j}^1 \cdot \hat{i}^0 & \hat{j}^1 \cdot \hat{j}^0 & \hat{j}^1 \cdot \hat{k}^0 \\ \hat{k}^1 \cdot \hat{k}^0 & \hat{k}^1 \cdot \hat{j}^0 & \hat{k}^1 \cdot \hat{k}^0 \end{bmatrix} \begin{bmatrix} p_x^0 \\ p_y^0 \\ p_z^0 \end{bmatrix} = R_0^1 \begin{bmatrix} p_x^0 \\ p_y^0 \\ p_z^0 \end{bmatrix} = R_0^1 p^0 \qquad (22.1)$$

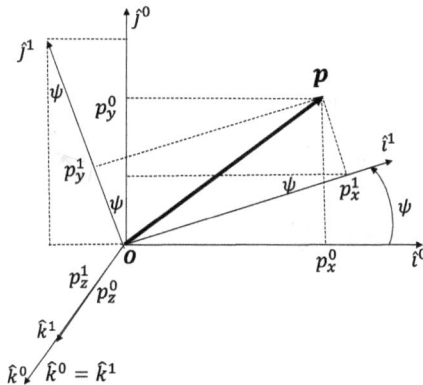

FIGURE 22.1 Rotation of frames with respect to z-axis.

We define the rotation geometry of Figure 22.1:

$$R_0^1 \triangleq \begin{bmatrix} \hat{i}^1 \cdot \hat{i}^0 & \hat{i}^1 \cdot \hat{j}^0 & \hat{i}^1 \cdot \hat{k}^0 \\ \hat{j}^1 \cdot \hat{i}^0 & \hat{j}^1 \cdot \hat{j}^0 & \hat{j}^1 \cdot \hat{k}^0 \\ \hat{k}^1 \cdot \hat{k}^0 & \hat{k}^1 \cdot \hat{j}^0 & \hat{k}^1 \cdot \hat{k}^0 \end{bmatrix} = \begin{bmatrix} \cos\psi & \sin\psi & 0 \\ -\sin\psi & \cos\psi & 0 \\ 0 & 0 & 1 \end{bmatrix}$$

The notation R_0^1 denotes a rotation matrix from coordinate frame $\mathcal{F}^0$ to coordinate frame $\mathcal{F}^1$ or right-handed rotation about z-axis. More appropriately, we can rewrite the notation as follows:

$$^z R_0^1(\psi) \triangleq \begin{bmatrix} \cos\psi & \sin\psi & 0 \\ -\sin\psi & \cos\psi & 0 \\ 0 & 0 & 1 \end{bmatrix}$$

Similarly, proceeding with a right-handed rotation of the coordinate system about the y-axis, Figure 22.2 depicts the rotation. Proceeding as we have done before, we have:

$$^y R_0^1(\theta) \triangleq \begin{bmatrix} \cos\theta & 0 & -\sin\theta \\ 0 & 1 & 0 \\ \sin\theta & 0 & \cos\theta \end{bmatrix}$$

Figure 22.3 shows the right-handed rotation of the coordinate system about the x-axis. The corresponding rotation matrix is provided by:

$$^x R_0^1(\phi) \triangleq \begin{bmatrix} 1 & 0 & 0 \\ 0 & \cos\phi & \sin\phi \\ 0 & -\sin\phi & \cos\phi \end{bmatrix}$$

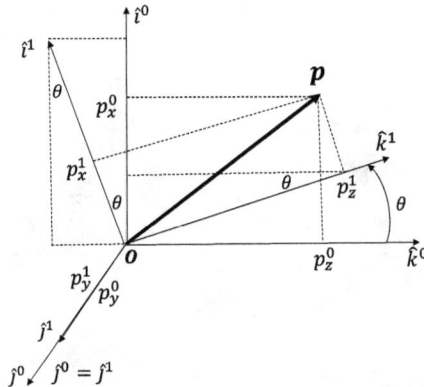

FIGURE 22.2   Rotation of frame with respect to y-axis.

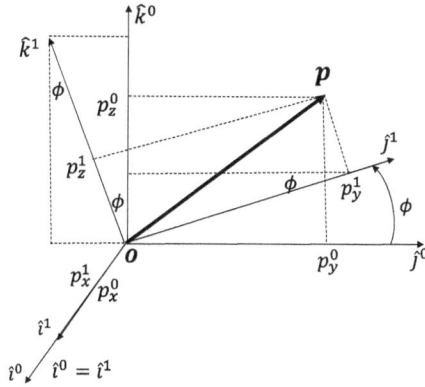

FIGURE 22.3 Rotation of frame with respect to *x*-axis.

The matrix R_0^1 shown in the above equations are the examples of a more general class of rotation matrices. These rotation matrices are orthogonal matrices and have the following properties:

$$\left(R_a^b\right)^{-1} = \left(R_a^b\right)^T = R_b^a$$

$$R_b^c R_a^b = R_a^c$$

$$\left(R_a^b\right)\left(R_a^b\right)^T = \left(R_a^b\right)^T \left(R_a^b\right) = \delta$$

$$\det R_a^b = 1$$

Rotation Formula

The rotation formula transforms a given coordinate system by rotating it through a counterclockwise angle Ψ about an axis $\hat{n}$. Referring to the above figure, the equation for the "fixed" vector in the transformed coordinate system (i.e., Figure 22.4)

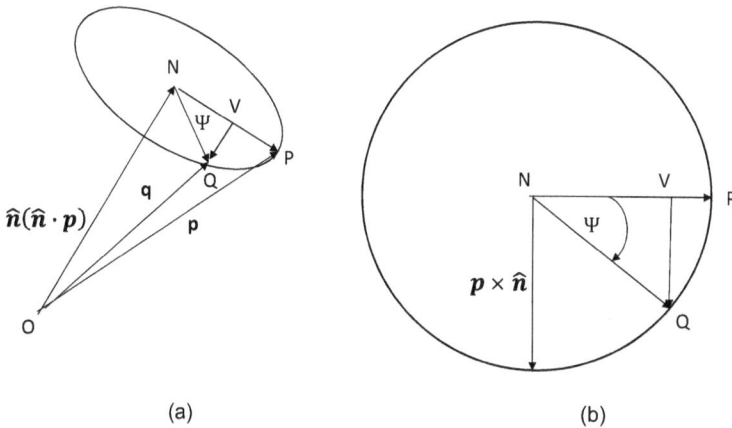

(a)

(b)

FIGURE 22.4 Coordinate systems for rotating system.

corresponds to an alias transformation. A transformation in which the coordinate system is changed, leaving vectors in the original coordinate system "fixed" while changing their representation in the new coordinate system. In contrast, a transformation in which vectors are transformed in a fixed coordinate system is called an alibi transformation.

From Figures 22.4(a) and (b), we can write:

$$
\begin{aligned}
q &= \overrightarrow{ON} + \overrightarrow{NV} + \overrightarrow{VQ} \\
&= \hat{n}(\hat{n}\cdot p) + \left[p - \hat{n}(\hat{n}\cdot p) \right]\cos\Psi + p\times\hat{n}\sin\Psi \qquad (22.2)\\
&= p\cos\Psi + \hat{n}(\hat{n}\cdot p)(1 - \cos\Psi) + \cos\Psi + p\times\hat{n}\sin\Psi
\end{aligned}
$$

Rearranging Equation (22.2) becomes:

$$
q = (1 - \cos\Psi)(p\cdot\hat{n})\hat{n} + \cos\Psi p - \sin\Psi(\hat{n}\times p) \qquad (22.3)
$$

Equation (22.3) is known as the rotation formula. An example will make it clearer. Let us consider the left-handed of p^0 in frame $\mathcal{F}^0$ about the z-axis, as shown in Figure 22.5.

Using rotation formula of Equation (22.3), we get:

$$
\begin{aligned}
q^0 &= (1 - \cos\psi)(\mathbf{p}\cdot\hat{\mathbf{n}})\hat{\mathbf{n}} + \cos\psi\mathbf{p} - \sin\psi(\hat{\mathbf{n}}\times\mathbf{p}) \\
&= (1 - \cos\psi)\, p_z^0 \begin{bmatrix} 0 \\ 0 \\ 1 \end{bmatrix} + \cos\psi \begin{bmatrix} p_x^0 \\ p_y^0 \\ p_z^0 \end{bmatrix} - \sin\psi \left[n\hat{k}^0 \right] \times \begin{bmatrix} p_x^0 \hat{i}^0 \\ p_y^0 \hat{j}^0 \\ p_z^0 \hat{k}^0 \end{bmatrix}
\end{aligned}
$$

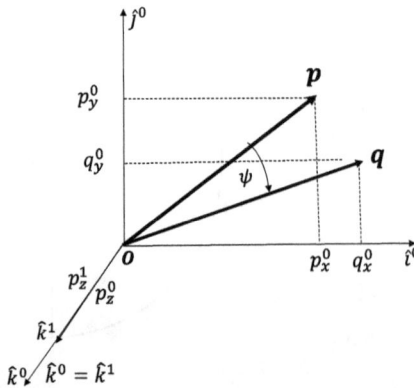

FIGURE 22.5   Rotation of **p** about the z-axis.

Actually,

$$\left[n\hat{k}^0\right] \times \begin{bmatrix} p_x^0 \hat{i}^0 \\ p_y^0 \hat{j}^0 \\ p_z^0 \hat{k}^0 \end{bmatrix} = \begin{bmatrix} np_x^0\left(\hat{k}^0 \times \hat{i}^0\right) \\ np_y^0\left(\hat{k}^0 \times \hat{j}^0\right) \\ np_z^0\left(\hat{k}^0 \times \hat{k}^0\right) \end{bmatrix} => n = 1 => \begin{bmatrix} p_x^0 \hat{j}^0 \\ -p_y^0 \hat{i}^0 \\ 0 \end{bmatrix} => \text{clockwise}; => p_x^0 \hat{j}^0 \equiv p_y^0,$$

$$p_y^0 \hat{i}^0 \equiv p_x^0 => \begin{bmatrix} p_y^0 \\ -p_x^0 \\ 0 \end{bmatrix} => \text{Sign change} => \begin{bmatrix} -p_y^0 \\ p_x^0 \\ 0 \end{bmatrix}$$

$$q^0 = \left(1 - \cos\psi\right)p_z^0 \begin{bmatrix} 0 \\ 0 \\ 1 \end{bmatrix} + \cos\psi \begin{bmatrix} p_x^0 \\ p_y^0 \\ p_z^0 \end{bmatrix} - \sin\psi \begin{bmatrix} -p_y^0 \\ p_x^0 \\ 0 \end{bmatrix} = \begin{bmatrix} \cos\psi & \sin\psi & 0 \\ -\sin\psi & \cos\psi & 0 \\ 0 & 0 & 1 \end{bmatrix}$$

$$p^0 = {}^zR_0^1(\psi)p^0$$

Note that the rotation matrix R_0^1 can be interpreted in two different ways. The first interpretation is that it transforms the fixed vector p from an expression in frame $\mathcal{F}^0$ to an expression in frame $\mathcal{F}^1$, where $\mathcal{F}^1$ has been obtained from $\mathcal{F}^0$ by a right-handed rotation. The second interpretation is that it rotates a vector p though a left-handed rotation to a new vector q in the same reference frame. Right-handed rotations of vectors are obtained by using $\left(R_0^1\right)^T$.

Euler Angles

The orientation of a triad is defined by three Euler angles, measured in degrees. We begin by the body axes $Oxyz$ coinciding with the space-fixed axes $OXYZ$ in Figure 22.6, and from that position, we give the body system a rotation by an angle ψ (the *precession* angle) about the Z-axis, so that the body system takes the position OKY_1Z.

The last system is then given a rotation by an angle θ (the angle of *nutation*) about OK. This brings the body system to the position OKY_2z. We now fix the z-axis in the

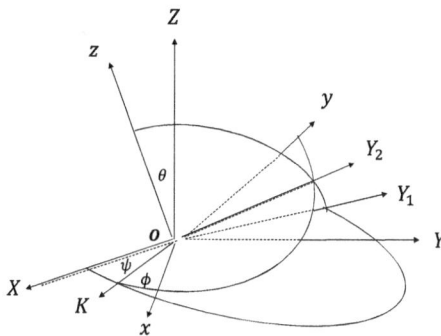

FIGURE 22.6 Euler's angles.

body and give the body system a rotation by an angle ϕ (the angle of *proper rotation*) about the z-axis to reach its final position $Oxyz$, fixed in the body. In this way, ψ is the angle of rotation of the body about the space axis Z, θ is the angle between z and Z, and ϕ is the angle of rotation about z. The line OK is the intersection of the two planes Oxy and OXY. It is called the *line of nodes*.

Euler Angles Rotation

An arbitrary rotation from $I = \left(\bar{i}_1, \bar{i}_2, \bar{i}_3\right)$ to $\mathcal{E} = \left(\bar{e}_1, \bar{e}_2, \bar{e}_3\right)$ can be viewed as a succession of three planar rotations about three different axes.

Euler 3-2-1 Rotation Sequence

First, we are using the 3-2-1 sequence, which is popular for airplane flight mechanics. Euler angles, both at input and at output, are measured in degrees. This defines three consecutive planar rotations:

- A planar rotation of magnitude ψ, called heading, about axis $\bar{i}_3$ brings basis $I = \left(\bar{i}_1, \bar{i}_2, \bar{i}_3\right)$ to basis $A\left(\bar{a}_1, \bar{a}_2, \bar{a}_3\right)$.
- A planar rotation of magnitude θ, called attitude, about axis $\bar{a}_2$ brings basis $A\left(\bar{a}_1, \bar{a}_2, \bar{a}_3\right)$ to basis $B\left(\bar{b}_1, \bar{b}_2, \bar{b}_3\right)$.
- A planar rotation of magnitude ϕ, called bank, about axis $\bar{b}_1$ brings basis $B\left(\bar{b}_1, \bar{b}_2, \bar{b}_3\right)$ to basis $\mathcal{E} = \left(\bar{e}_1, \bar{e}_2, \bar{e}_3\right)$.

The sequence of mode is defined as $(3 - 2 - 1) : \left(\bar{i}_3 - \bar{a}_2 - \bar{b}_1\right)$. It implies that in Euler angles rotation sequence: first, heading = rotation about y axis; second, attitude = rotation about z axis; and last bank = rotation about x axis. Figure 22.7 enumerates the sequence of 3-2-1 Euler angles rotation.

1. A planar rotation of magnitude ψ, called heading, about axis $\bar{i}_3$ brings basis $I = \left(\bar{i}_1, \bar{i}_2, \bar{i}_3\right)$ to basis $A\left(\bar{a}_1, \bar{a}_2, \bar{a}_3\right)$. The corresponding rotation is shown in Figure 22.8.

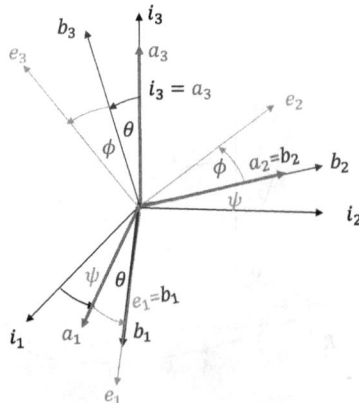

FIGURE 22.7 Euler angles rotation in 3-2-1 sequence.

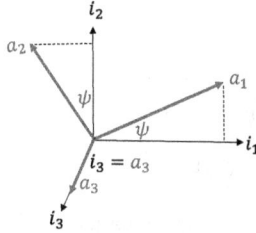

FIGURE 22.8   Planar rotation of heading about $\bar{i}_3$-axis.

From Figure 22.8, we have:

$$a_1 = i_1 \cos\psi + i_2 \sin\psi$$
$$a_2 = -i_1 \sin\psi + i_2 \cos\psi$$
$$a_3 = i_3$$

$$\begin{bmatrix} a_1 \\ a_2 \\ a_3 \end{bmatrix} = \begin{bmatrix} \cos\psi & \sin\psi & 0 \\ -\sin\psi & \cos\psi & 0 \\ 0 & 0 & 1 \end{bmatrix} \begin{bmatrix} i_1 \\ i_2 \\ i_3 \end{bmatrix}$$

The corresponding direction cosine matrix or rotation matrix is:

$$R_i^a(\psi) \triangleq \begin{bmatrix} \cos\psi & \sin\psi & 0 \\ -\sin\psi & \cos\psi & 0 \\ 0 & 0 & 1 \end{bmatrix}$$

2. A planar rotation of magnitude θ, called attitude, about axis $\bar{a}_2$ brings basis $A(\bar{a}_1, \bar{a}_2, \bar{a}_3)$ to basis $B(\bar{b}_1, \bar{b}_2, \bar{b}_3)$. The corresponding rotation is shown in Figure 22.9.

From Figure 22.9, we have:

$$b_1 = a_1 \cos\theta - a_3 \sin\theta$$
$$b_2 = a_2$$
$$b_3 = a_1 \sin\theta + a_3 \cos\theta$$

$$\begin{bmatrix} b_1 \\ b_2 \\ b_3 \end{bmatrix} = \begin{bmatrix} \cos\theta & 0 & -\sin\theta \\ 0 & 1 & 0 \\ \sin\theta & 0 & \cos\theta \end{bmatrix} \begin{bmatrix} a_1 \\ a_2 \\ a_3 \end{bmatrix}$$

$$-\pi/2 \le \theta \le \pi/2$$

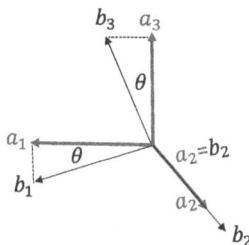

FIGURE 22.9   Planar rotation of heading about $\bar{a}_2$-axis.

The corresponding direction cosine matrix or rotation matrix is:

$$R_a^b(\psi) \triangleq \begin{bmatrix} \cos\theta & 0 & -\sin\theta \\ 0 & 1 & 0 \\ \sin\theta & 0 & \cos\theta \end{bmatrix}$$

3. A planar rotation of magnitude ϕ, called bank, about axis $\bar{b}_1$ brings basis $B(\bar{b}_1, \bar{b}_2, \bar{b}_3)$ to basis $\mathcal{E} = (\bar{e}_1, \bar{e}_2, \bar{e}_3)$. The corresponding rotation is shown in Figure 22.10.

From Figure 22.10, we have:

$$e_1 = b_1$$
$$e_2 = b_2 \cos\phi + b_3 \sin\phi$$
$$e_3 = b_2 \sin\phi + b_3 \cos\phi$$
$$0 \le \phi \le 2\pi$$

$$\begin{bmatrix} e_1 \\ e_2 \\ e_3 \end{bmatrix} = \begin{bmatrix} 1 & 0 & 0 \\ 0 & \cos\phi & -\sin\phi \\ 0 & -\sin\phi & \cos\phi \end{bmatrix} \begin{bmatrix} b_1 \\ b_2 \\ b_3 \end{bmatrix}$$

The corresponding direction cosine matrix or rotation matrix is:

$$R_b^e(\phi) \triangleq \begin{bmatrix} 1 & 0 & 0 \\ 0 & \cos\phi & \sin\phi \\ 0 & -\sin\phi & \cos\phi \end{bmatrix}$$

Combining all of them, we obtain:

$$\begin{bmatrix} e_1 \\ e_2 \\ e_3 \end{bmatrix} = \begin{bmatrix} 1 & 0 & 0 \\ 0 & \cos\phi & \sin\phi \\ 0 & -\sin\phi & \cos\phi \end{bmatrix} \begin{bmatrix} b_1 \\ b_2 \\ b_3 \end{bmatrix} = \begin{bmatrix} 1 & 0 & 0 \\ 0 & \cos\phi & \sin\phi \\ 0 & -\sin\phi & \cos\phi \end{bmatrix} \begin{bmatrix} \cos\theta & 0 & -\sin\theta \\ 0 & 1 & 0 \\ \sin\theta & 0 & \cos\theta \end{bmatrix} \begin{bmatrix} a_1 \\ a_2 \\ a_3 \end{bmatrix}$$

$$= \begin{bmatrix} 1 & 0 & 0 \\ 0 & \cos\phi & \sin\phi \\ 0 & -\sin\phi & \cos\phi \end{bmatrix} \begin{bmatrix} \cos\theta & 0 & -\sin\theta \\ 0 & 1 & 0 \\ \sin\theta & 0 & \cos\theta \end{bmatrix} \begin{bmatrix} \cos\psi & \sin\psi & 0 \\ -\sin\psi & \cos\psi & 0 \\ 0 & 0 & 1 \end{bmatrix} \begin{bmatrix} i_1 \\ i_2 \\ i_3 \end{bmatrix}$$

$$= \begin{bmatrix} \cos\theta & 0 & -\sin\theta \\ \sin\phi\sin\theta & \cos\phi & \sin\phi\cos\theta \\ \cos\phi\sin\theta & -\sin\phi & \cos\phi\cos\theta \end{bmatrix} \begin{bmatrix} \cos\psi & \sin\psi & 0 \\ -\sin\psi & \cos\psi & 0 \\ 0 & 0 & 1 \end{bmatrix} \begin{bmatrix} i_1 \\ i_2 \\ i_3 \end{bmatrix}$$

$$= \begin{bmatrix} \cos\theta\cos\psi & \cos\theta\sin\psi & -\sin\theta \\ \sin\phi\sin\theta\cos\psi - \cos\phi\sin\psi & \sin\phi\sin\theta\sin\psi + \cos\phi\cos\psi & \sin\phi\cos\theta \\ \cos\phi\sin\theta\cos\psi + \sin\phi\sin\psi & \cos\phi\sin\theta\sin\psi - \sin\phi\cos\psi & \cos\phi\cos\theta \end{bmatrix} \begin{bmatrix} i_1 \\ i_2 \\ i_3 \end{bmatrix}$$

$$\triangleq R_i^e(\psi,\theta,\phi) \begin{bmatrix} i_1 \\ i_2 \\ i_3 \end{bmatrix}$$

$$\tag{22.4}$$

$$\mathcal{E} = R_i^e(\psi,\theta,\phi) I \tag{22.5}$$

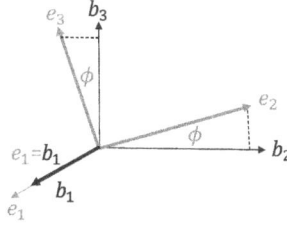

FIGURE 22.10 Planar rotation of heading about $\bar{b}_1$-axis.

where,

$$R_i^e\left(\psi,\theta,\phi\right) \triangleq \begin{bmatrix} \cos\theta\cos\psi & \cos\theta\sin\psi & -\sin\theta \\ \sin\phi\sin\theta\cos\psi - \cos\phi\sin\psi & \sin\phi\sin\theta\sin\psi + \cos\phi\cos\psi & \sin\phi\cos\theta \\ \cos\phi\sin\theta\cos\psi + \sin\phi\sin\psi & \cos\phi\sin\theta\sin\psi - \sin\phi\cos\psi & \cos\phi\cos\theta \end{bmatrix}$$

(22.6)

$$R_i^e\left(\psi,\theta,\phi\right) \triangleq \begin{bmatrix} c\theta c\psi & c\theta s\psi & -s\theta \\ s\phi s\theta c\psi - c\phi s\psi & s\phi s\theta s\psi + c\phi c\psi & s\phi c\theta \\ c\phi s\theta c\psi + s\phi s\psi & c\phi s\theta s\psi - s\phi c\psi & c\phi c\theta \end{bmatrix}$$ (22.7)

where $c\phi \triangleq \cos\phi$ and $s\phi \triangleq \sin\phi$.

From Equations (22.4), (22.6), and (22.7), we get as follows:

$$\tan\psi = \frac{R_{12}}{R_{11}}$$

$$\sin\theta = -R_{13}$$

$$\tan\phi = \frac{R_{23}}{R_{33}}$$

$$\psi = \tan^{-1}\left(\frac{R_{12}}{R_{11}}\right)$$

$$\theta = \sin^{-1}\left(-R_{13}\right)$$

$$\phi = \tan^{-1}\left(\frac{R_{23}}{R_{33}}\right)$$

$$\dot{\psi} \triangleq p = \frac{d}{dt}\left(\tan^{-1}\left(\frac{R_{12}}{R_{11}}\right)\right)$$

$$\dot{\theta} \triangleq q = \frac{d}{dt}\left(\sin^{-1}\left(-R_{13}\right)\right)$$

$$\dot{\phi} \triangleq r = \frac{d}{dt}\left(\tan^{-1}\left(\frac{R_{23}}{R_{33}}\right)\right)$$

Can we lead the following equation?

$$
\begin{bmatrix} p \\ q \\ r \end{bmatrix} = \begin{bmatrix} -\sin\phi & 0 & 1 \\ \sin\phi\cos\theta & \cos\phi & 0 \\ \cos\phi\cos\theta & -\sin\phi & 0 \end{bmatrix} \begin{bmatrix} \dot\psi \\ \dot\theta \\ \dot\phi \end{bmatrix}
$$

Of course, once the above equation is obtained, we will have:

$$
\begin{bmatrix} \dot\psi \\ \dot\theta \\ \dot\phi \end{bmatrix} = \frac{1}{\cos\theta} \begin{bmatrix} 0 & \sin\phi & \cos\phi \\ 0 & \cos\phi\cos\theta & -\sin\phi\cos\theta \\ \cos\theta & \sin\phi\sin\theta & \cos\phi\sin\theta \end{bmatrix} \begin{bmatrix} p \\ q \\ r \end{bmatrix}
$$

Euler 3-1-3 Rotation Sequence
Similar to the 3-2-1 sequence, we can define the 3-1-3 sequence as follows:

- A planar rotation of magnitude ϕ, called precession, about axis $\bar{i_3}$ brings basis $I = \left(\bar{i_1}, \bar{i_2}, \bar{i_3}\right)$ to basis $A\left(\bar{a_1}, \bar{a_2}, \bar{a_3}\right)$.
- A planar rotation of magnitude θ, called nutation, about axis $\bar{a_1}$ brings basis $A\left(\bar{a_1}, \bar{a_2}, \bar{a_3}\right)$ to basis $B\left(\bar{b_1}, \bar{b_2}, \bar{b_3}\right)$.
- A planar rotation of magnitude ψ, called spin, about axis $\bar{b_3}$ brings basis $B\left(\bar{b_1}, \bar{b_2}, \bar{b_3}\right)$ to basis $\mathcal{E}\left(\bar{e_1}, \bar{e_2}, \bar{e_3}\right)$.

Figure 22.11 shows the Euler 3-1-3 angular rotation.

1. A planar rotation of magnitude ϕ, called precession, about axis $\bar{i_3}$ brings basis $I = \left(\bar{i_1}, \bar{i_2}, \bar{i_3}\right)$ to basis $A\left(\bar{a_1}, \bar{a_2}, \bar{a_3}\right)$. The corresponding rotation is shown in Figure 22.11.

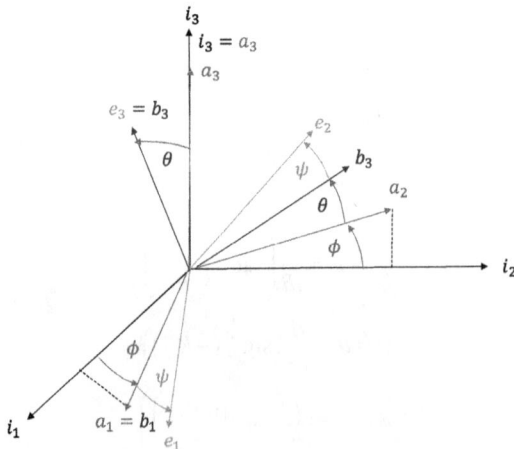

FIGURE 22.11   Euler angles rotation in 3-1-3 sequence.

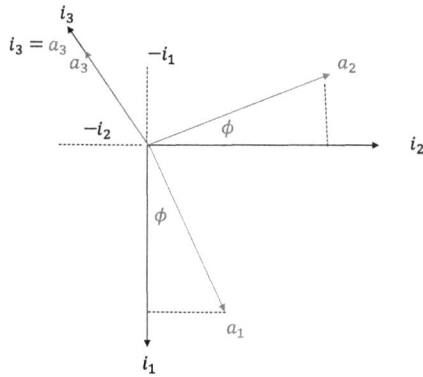

FIGURE 22.12   Planar rotation of precession about $\bar{i}_3$–axis.

From Figure 22.12, we have:

$$a_1 = i_1 \cos\phi + i_2 \sin\phi$$
$$a_2 = -i_1 \sin\phi + i_2 \cos\phi$$
$$a_3 = i_3$$

$$\begin{bmatrix} a_1 \\ a_2 \\ a_3 \end{bmatrix} = \begin{bmatrix} \cos\phi & \sin\phi & 0 \\ -\sin\phi & \cos\phi & 0 \\ 0 & 0 & 1 \end{bmatrix} \begin{bmatrix} i_1 \\ i_2 \\ i_3 \end{bmatrix}$$

The corresponding direction cosine matrix or rotation matrix is:

$$R_i^a(\phi) \triangleq \begin{bmatrix} \cos\phi & \sin\phi & 0 \\ -\sin\phi & \cos\phi & 0 \\ 0 & 0 & 1 \end{bmatrix}$$

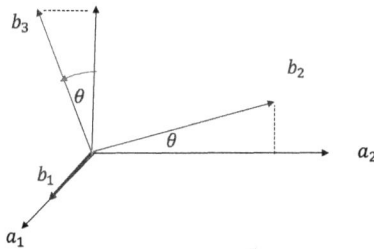

FIGURE 22.13   Planar rotation of precession about $\bar{a}_1$ – axis.

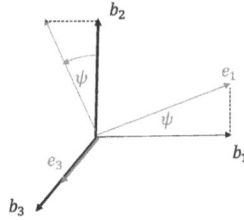

FIGURE 22.14 Planar rotation of precession about $\bar{b}_3$ – axis.

Figures 22.13 and 22.14 depict other two planner rotations.

$$b_2 = a_2 \cos\theta + a_3 \sin\theta$$
$$b_3 = -a_2 \sin\theta + a_3 \cos\theta$$
$$b_1 = a_1$$

$$\begin{bmatrix} b_1 \\ b_2 \\ b_3 \end{bmatrix} = \begin{bmatrix} 1 & 0 & 0 \\ 0 & \cos\theta & \sin\theta \\ 0 & -\sin\theta & \cos\theta \end{bmatrix} \begin{bmatrix} a_1 \\ a_2 \\ a_3 \end{bmatrix}$$

$$e_1 = b_1 \cos\psi + b_2 \sin\psi$$
$$e_2 = -b_1 \sin\psi + b_2 \cos\psi$$
$$e_3 = b_3$$

$$\begin{bmatrix} e_1 \\ e_2 \\ e_3 \end{bmatrix} = \begin{bmatrix} \cos\psi & \sin\psi & 0 \\ -\sin\psi & \cos\psi & 0 \\ 0 & 0 & 1 \end{bmatrix} \begin{bmatrix} b_1 \\ b_2 \\ b_3 \end{bmatrix}$$

The Euler angles introduced above correspond to the following sequence of planar rotations to indicate the sequence of body axes about which the three successive rotations are taking place. Clearly, Euler angles could be defined in several different manners: the first rotation could occur about either of the three axes, $I = \left(\bar{i}_1, \bar{i}_2, \bar{i}_3 \right)$, offering three choices. Because two consecutive rotations cannot take place about the same axis, two alternatives are possible for the second rotation. Two choices are again possible for the last rotation.

In all, $3 \times 2 \times 2 = 12$ possible choices exist, corresponding to sequences labeled 1-2-1, 1-2-3, 1-3-1, 1-3-2, 2-1-2, 2-1-3, 2-3-1, 2-3-2, 3-1-2, 3-1-3, 3-2-1, and 3-2-3. Two of these sequences, 3-2-1 and 3-1-2, are discussed above.

The representation of rotation in terms of three Euler angles shows that the direction cosine matrix can be expressed in terms of three parameters only. This representation, however, presents several drawbacks. First, Euler angles can be defined in several different manners, and the choice of the rotation sequence is entirely arbitrary. Furthermore, the expression for the direction cosine matrix, as seen for this example in Equation (22.6), is rather complicated and involves the evaluation of numerous trigonometric functions. Finally, singularities will occur in the evaluation of Euler angles from a direction cosine matrix for all 12 possible sequences.

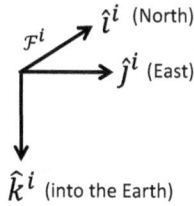

FIGURE 22.15 The inertial coordinate frame. The x-axis points North, the y-axis points East, and the z-axis points into the Earth.

Quadrotor Coordinate Frames

For quadrotors, there are several coordinate systems that are of interest. In this section, we will define and describe the following coordinate frames: the inertial frame, the vehicle frame, the vehicle-1 frame, the vehicle-2 frame, and the body frame. Throughout the quadrotor discussion, we assume flat, nonrotating Earth: a valid assumption for quadrotors.

Inertial Frame

The inertial coordinate system is an Earth-fixed coordinate system with origin at the defined home location. As shown in Figure 22.15, the unit vector $\hat{i}^i$ is directed toward the North, $\hat{j}^i$ is directed toward the East, and $\hat{k}^i$ is directed toward the center of the Earth.

Vehicle Frame

We have considered airplane and quadrotor vehicle. The origin of the vehicle frame is at the center of mass of the quadrotor. However, the axes of $\mathcal{F}^v$ are aligned with the axis of the inertial frame $\mathcal{F}^i$. In other words, the unit vector $\hat{i}^v$ is directed toward the North, $\hat{j}^v$ is directed toward the East, and $\hat{k}^v$ is directed toward the center of the Earth, as shown in Figure 22.16.

Quadrotor Vehicle-1 Frame

The origin of the vehicle-1 frame is identical to the vehicle frame, that is, the center of gravity. However, $\mathcal{F}^{v1}$ is positively rotated about $\hat{k}^v$ by the yaw angle ψ so that if the airframe is not rolling or pitching, then $\hat{i}^{v1}$ would point out the nose of the

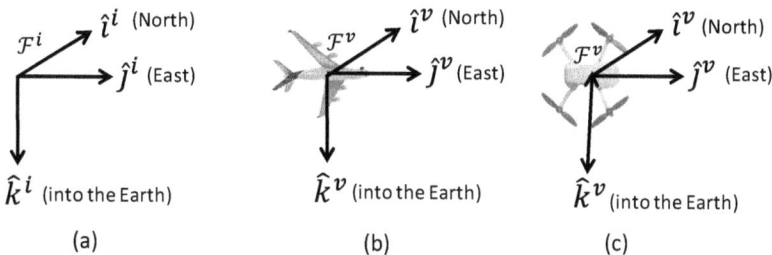

FIGURE 22.16 The vehicle coordinate frame. The x-axis points North, the y-axis points East, and the z-axis points into the Earth: (a) inertial frame, (b) airplane vehicle frame, and (c) quadrotor vehicle frame.

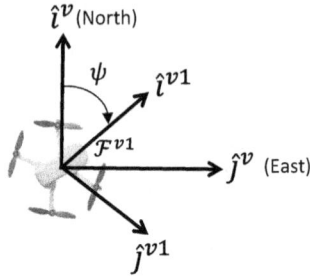

FIGURE 22.17 The vehicle-1 frame. If the roll and pitch angels are zero, then the x-axis points out the nose of the quadrotor, the y-axis points out the right wing, and the z-axis points into the Earth.

airframe, $\hat{j}^{v1}$ points out the right wing, and $\hat{k}^{v1}$ is aligned with $\hat{k}^{v}$ and points into the Earth. The vehicle-1 frame is shown in Figure 22.17.

The transformation from $\mathcal{F}^{v}$ to $\mathcal{F}^{v1}$ is given by:

$$p^{v1} = R_v^{v1}(\psi)p^v$$

where

$$R_v^{v1} = \begin{bmatrix} \cos\psi & \sin\psi & 0 \\ -\sin\psi & \cos\psi & 0 \\ 0 & 0 & 1 \end{bmatrix}$$

Quadrotor Vehicle-2 Frame

The origin of the vehicle-2 frame is again the center of gravity and is obtained by rotating the vehicle-1 frame in a right-handed rotation about the $\hat{j}^{v1}$ axis by the pitch angle θ. If the roll angle is zero, then $\hat{i}^{v2}$ points out the nose of the airframe, $\hat{j}^{v2}$ points out the right wing, and $\hat{k}^{v2}$ points out the belly, as shown in Figure 22.18.

The transformation from $\mathcal{F}^{v1}$ to $\mathcal{F}^{v2}$ is given by:

$$p^{v2} = R_{v1}^{v2}(\theta)p^{v1}$$

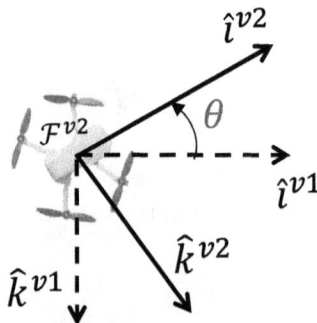

FIGURE 22.18 The vehicle-2 frame. If the roll angel is zero, then the x-axis points out the nose of the airframe, the y-axis points out the right wing, and the z-axis points out the belly.

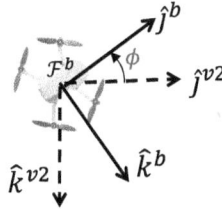

FIGURE 22.19 The body frame. The x-axis points out the nose of the airframe, the y-axis points out the right wing, and the z-axis points out the belly.

where

$$R_{v1}^{v2}(\theta) = \begin{bmatrix} \cos\theta & 0 & -\sin\theta \\ 0 & 1 & 0 \\ \sin\theta & 0 & \cos\theta \end{bmatrix}$$

Quadrotor Body Frame

The body frame is obtained by rotating the vehicle-2 frame in a right-handed rotation about $\hat{i}^{v2}$ by the roll angle ϕ. Therefore, the origin is the center-of-gravity, $\hat{i}^{b}$ points out the nose of the airframe, $\hat{j}^{b}$ points out the right wing, and $\hat{k}^{b}$ points out the belly. The body frame is shown in Figure 22.19.

The transformation from $\mathcal{F}^{v2}$ to $\mathcal{F}^{b}$ is given by:

$$p^{b} = R_{v2}^{b}(\phi) p^{v2}$$

where

$$R_{v2}^{b}(\phi) = \begin{bmatrix} 1 & 0 & 0 \\ 0 & \cos\phi & \sin\phi \\ 0 & -\sin\phi & \cos\phi \end{bmatrix}$$

The transformation from the vehicle frame to the body frame is given by:

$$R_{v}^{b}(\phi,\theta,\psi) = R_{v2}^{b}(\phi) R_{v1}^{v2}(\theta) R_{v}^{v1}$$

$$= \begin{bmatrix} 1 & 0 & 0 \\ 0 & \cos\phi & \sin\phi \\ 0 & -\sin\phi & \cos\phi \end{bmatrix} \begin{bmatrix} \cos\theta & 0 & -\sin\theta \\ 0 & 1 & 0 \\ \sin\theta & 0 & \cos\theta \end{bmatrix} \begin{bmatrix} \cos\psi & \sin\psi & 0 \\ -\sin\psi & \cos\psi & 0 \\ 0 & 0 & 1 \end{bmatrix}$$

$$= \begin{bmatrix} c\theta c\psi & c\theta s\psi & -s\theta \\ s\phi s\theta c\psi - c\phi s\psi & s\phi s\theta s\psi + c\phi c\psi & s\phi c\theta \\ c\phi s\theta c\psi + s\phi s\psi & c\phi s\theta s\psi - s\phi c\psi & c\phi c\theta \end{bmatrix}$$

$$(22.8)$$

where $c\phi \triangleq \cos\phi$ and $s\phi \triangleq \sin\phi$. Note that Equations (22.7) and (22.8) are the same.

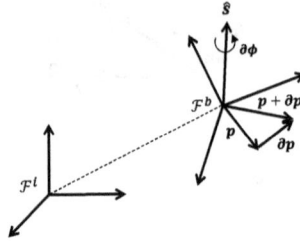

FIGURE 22.20   Derivation of the equation of Coriolis.

Equation of Coriolis

In this section, we provide a simple derivation of the famous equation of Coriolis. Suppose that we are given two coordinate frames $\mathcal{F}^i$ and $\mathcal{F}^b$, as shown in Figure 22.20.

For example, $\mathcal{F}^i$ might represent the inertial frame and $\mathcal{F}^b$ might represent the body frame of a quadrotor. Suppose that the vector p is moving in $\mathcal{F}^b$ and that $\mathcal{F}^b$ is rotating and translating with respect to $\mathcal{F}^i$. Our objective is to find the time derivative of p as seen from frame $\mathcal{F}^i$. We will derive the appropriate equation through two steps. Assume first that $\mathcal{F}^b$ is not rotating with respect to $\mathcal{F}^i$. Denote the time derivative of p in frame as seen from frame $\mathcal{F}^i$.

We will derive the appropriate equation through two steps. Assume first that $\mathcal{F}^b$ is not rotating with respect to $\mathcal{F}^i$. Denoting the time derivative of p in frame $\mathcal{F}^i$ as $\left(\dfrac{d}{dt_i}p\right)$, we get:

$$\frac{d}{dt_i}p = \frac{d}{dt_b}p \tag{22.9}$$

On the other hand, assume that p is fixed in $\mathcal{F}^b$ but that $\mathcal{F}^b$ is rotating with respect to $\mathcal{F}^i$, and let $\hat{s}$ be the instantaneous axis of rotation and $\partial\phi$ the (right-handed) rotation angle. Then the rotation formula in Equation (22.3) gives:

$$p + \partial p = \left(1 - \cos\left(-\partial\phi\right)\right)\left(p\cdot\hat{s}\right)\hat{s} + \cos\left(-\partial\phi\right)p - \sin\left(-\partial\phi\right)\left(\hat{s} \times p\right)$$

Using a small angle approximation $\partial\phi \to 0, \cos\left(-\partial\phi\right) \to 1, \sin\left(-\partial\phi\right) \to -\partial\phi$, (in radians),

$$p + \partial p \approx p + \partial\phi\left(\hat{s} \times p\right)$$
$$\partial p \approx \partial\phi\left(\hat{s} \times p\right)$$

Taking time derivative ∂t_i of both sides,

$$\frac{\partial p}{\partial t_i} \approx \frac{\partial \phi}{\partial t_i}(\hat{s} \times p) \approx \frac{\partial \phi}{\partial t_i}(\hat{s} \times p) \approx \dot{\phi}\hat{s} \times p \qquad (22.10)$$

Taking the limit as $\partial t_i \to 0$ and defining the angular velocity of $\mathcal{F}^b$ with respect to $\mathcal{F}^i$ as we know:

$$\omega_{b/i} \triangleq \hat{s}\dot{\phi}$$

We get,

$$\frac{\partial p}{\partial t_i} = \omega_{b/i} \times p \qquad (22.11)$$

Since differentiation is a linear operator, we can combine Equations (22.9) and (22.11) to obtain:

$$\frac{d}{dt_i} p = \frac{d}{dt_b} p + \omega_{b/i} \times p \qquad (22.12)$$

which is the equation of Coriolis.

AERIAL QUADROTOR ROBOT KINEMATICS AND DYNAMICS

We derive the expressions for the kinematics and the dynamics of a rigid body specifically for Quadrotor.

Quadrotor State Variable

The state variables of the quadrotor are the following 12 quantities:

p_n = the inertial (north) position of the quadrotor along $\hat{i}^i$ in $\mathcal{F}^i$,
p_e = the inertial (east) position of the quadrotor along $\hat{j}^i$ in $\mathcal{F}^i$,
h = the altitude of the aircraft measured along $\hat{k}^i$ in $\mathcal{F}^i$,
u = the body frame velocity measured along $\hat{i}^b$ in $\mathcal{F}^b$,
v = the body frame velocity measured along $\hat{j}^b$ in $\mathcal{F}^b$,
ω = the body frame velocity measured along $\hat{k}^b$ in $\mathcal{F}^b$,
ϕ = the roll angle defined with respect to $\mathcal{F}^{v2}$,
θ = the pitch angle defined with respect to $\mathcal{F}^{v1}$,
ψ = the yaw angle defined with respect to $\mathcal{F}^v$,
p = the roll rate $\dot{\phi}$ measured along $\hat{i}^b$ in $\mathcal{F}^b$,
q = the pitch rate $\dot{\theta}$ measured along $\hat{j}^b$ b in $\mathcal{F}^b$,
r = the yaw rate $\dot{\psi}$ measured along $\hat{k}^b$ in $\mathcal{F}^b$.

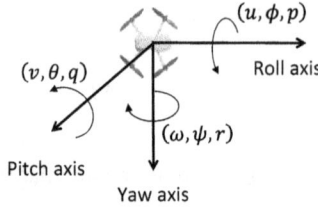

FIGURE 22.21   Definition of axes.

The state variables are shown schematically in Figure 22.21.

- The position (p_n, p_e, h) of the quadrotor is given in the inertial frame, with positive h defined along the negative z-axis in the inertial frame $\mathcal{F}^i$.
- The velocity (u, v, ω) and the angular velocity (p, q, r) of the quadrotor are given with respect to the body frame $\mathcal{F}^b$.

The Euler angles (roll ϕ, pitch θ, and yaw ψ) are given with respect to the vehicle 2-frame $\mathcal{F}^{v2}$, the vehicle 1-frame $\mathcal{F}^{v1}$, and the vehicle frame $\mathcal{F}^v$, respectively.

Quadrotor Kinematics

The state variables p_n, p_e, and $-h$ are inertial frame $\mathcal{F}^i$ quantities, whereas the velocities u, v, and ω are body frame $\mathcal{F}^b$ quantities. Therefore, the relationship between position and velocities is given by (Note $\dfrac{dx}{dt} = R v_x$, where R is Euler rotational matrix of different frames. See Equation (22.8)

$$\frac{d}{dt}\begin{bmatrix} p_n \\ p_e \\ -h \end{bmatrix} = R_b^v \begin{bmatrix} u \\ v \\ \omega \end{bmatrix} = \left[R_v^b \right]^T \begin{bmatrix} u \\ v \\ \omega \end{bmatrix} = \begin{bmatrix} c\theta c\psi & s\phi s\theta c\psi - c\phi s\psi & c\phi s\theta c\psi + s\phi s\psi \\ c\theta s\psi & s\phi s\theta s\psi + c\phi c\psi & c\phi s\theta s\psi - s\phi c\psi \\ -s\theta & s\phi c\theta & c\phi c\theta \end{bmatrix} \begin{bmatrix} u \\ v \\ \omega \end{bmatrix}$$

$$(22.13)$$

The relationship between absolute angles roll ϕ of frame $\mathcal{F}^{v2}$, pitch θ of frame $\mathcal{F}^{v1}$, and yaw ψ of frame $\mathcal{F}^v$, and the angular rates $p, q,$ and r of frame $\mathcal{F}^b$ is also complicated by the fact that these quantities are defined in different coordinate frames. The angular rates are defined in the body frame $\mathcal{F}^b$, whereas the roll angle ϕ is defined in $\mathcal{F}^{v2}$, the pitch angle θ is defined in $\mathcal{F}^{v2}$, and the yaw angle ψ is defined in the vehicle frame $\mathcal{F}^v$.

We need to relate $p, q,$ and r to $\dot{\phi}$, $\dot{\theta}$, and $\dot{\psi}$. Since $\dot{\phi}$, $\dot{\theta}$, and $\dot{\psi}$ are small and noting that:

$$R_{v2}^b\left(\dot{\phi}\right) = R_{v1}^{v2}\left(\dot{\theta}\right) = R_v^{v1}\left(\dot{\psi}\right) = I$$

We get,

$$
\begin{bmatrix} p \\ q \\ r \end{bmatrix} = R_{v2}^{b}(\dot{\phi}) \begin{bmatrix} \dot{\phi} \\ 0 \\ 0 \end{bmatrix} + R_{v2}^{b}(\phi) R_{v1}^{v2}(\dot{\theta}) \begin{bmatrix} 0 \\ \dot{\theta} \\ 0 \end{bmatrix} + R_{v2}^{b}(\phi) R_{v1}^{v2}(\theta) R_{v}^{v1}(\dot{\psi}) \begin{bmatrix} 0 \\ 0 \\ \psi \end{bmatrix} = \begin{bmatrix} \dot{\phi} \\ 0 \\ 0 \end{bmatrix} + R_{v2}^{b}(\phi) \begin{bmatrix} 0 \\ \dot{\theta} \\ 0 \end{bmatrix}
$$

$$
+ R_{v2}^{b}(\phi) R_{v1}^{v2}(\theta) \begin{bmatrix} 0 \\ 0 \\ \psi \end{bmatrix} = \begin{bmatrix} \dot{\phi} \\ 0 \\ 0 \end{bmatrix} + \begin{bmatrix} 1 & 0 & 0 \\ 0 & \cos\phi & \sin\phi \\ 0 & -\sin\phi & \cos\phi \end{bmatrix} \begin{bmatrix} 0 \\ \dot{\theta} \\ 0 \end{bmatrix}
$$

$$
+ \begin{bmatrix} 1 & 0 & 0 \\ 0 & \cos\phi & \sin\phi \\ 0 & -\sin\phi & \cos\phi \end{bmatrix} \begin{bmatrix} \cos\theta & 0 & -\sin\theta \\ 0 & 1 & 0 \\ \sin\theta & 0 & \cos\theta \end{bmatrix} \begin{bmatrix} 0 \\ 0 \\ \psi \end{bmatrix} = \begin{bmatrix} 1 & 0 & -s\theta \\ 0 & c\phi & s\phi c\theta \\ 0 & -s\phi & c\phi c\theta \end{bmatrix} \begin{bmatrix} \dot{\phi} \\ \dot{\theta} \\ \psi \end{bmatrix}
$$

$$(22.14)$$

So, we have:

$$
\begin{bmatrix} p \\ q \\ r \end{bmatrix} = \begin{bmatrix} 1 & 0 & -s\theta \\ 0 & c\phi & s\phi c\theta \\ 0 & -s\phi & c\phi c\theta \end{bmatrix} \begin{bmatrix} \dot{\phi} \\ \dot{\theta} \\ \psi \end{bmatrix}
$$

$$(22.15)$$

Inverting we get:

$$
\begin{bmatrix} \dot{\phi} \\ \dot{\theta} \\ \psi \end{bmatrix} = \begin{bmatrix} 1 & \sin\phi\tan\theta & \cos\phi\tan\theta \\ 0 & \cos\phi & -\sin\phi \\ 0 & \sin\phi\sec\theta & \cos\phi\sec\theta \end{bmatrix} \begin{bmatrix} p \\ q \\ r \end{bmatrix}
$$

$$(22.16)$$

Again, note that $p, q,$ and r angular velocities.

Equations (22.15) and (22.16) are known as attitude kinematics differential equations.

Rigid Body Dynamics

Let v be the velocity vector of the quadrotor. Newton's laws only hold in inertial frames; therefore, Newton's law applied to the translational motion is:

$$
m\frac{dv}{dt_i} = f
$$

where m is the mass of the quadrotor, f is the total applied to the quadrotor, and $\dfrac{dv}{dt_i}$ is the time derivative in the inertial frame. From the equation of Coriolis Equation (22.12), we have:

$$m\frac{dv}{dt_i} = m\left(\frac{dv}{dt_b} + \omega_{b/i} \times v\right) = f \tag{22.17}$$

where $\omega_{b/i}$ is the angular velocity of the air body frame $\mathcal{F}^b$ with respect to the inertial frame $\mathcal{F}^i$. Since the control force is computed and applied in the body coordinate system, and since ω is measured in body coordinates, we will express Equation (22.17) in body coordinates, where:

$$v^b \triangleq \begin{bmatrix} u \\ v \\ \omega \end{bmatrix}$$

$$\omega_{b/i}^b \triangleq \begin{bmatrix} p \\ q \\ r \end{bmatrix}$$

Therefore, in body coordinates, Equation (22.17) becomes:

$$m\left(\frac{d}{dt_b}\begin{bmatrix} u \\ v \\ \omega \end{bmatrix} + \begin{bmatrix} p \\ q \\ r \end{bmatrix} \times \begin{bmatrix} u \\ v \\ \omega \end{bmatrix}\right) = f$$

$$\begin{bmatrix} \dot{u} \\ \dot{v} \\ \dot{\omega} \end{bmatrix} = \frac{f}{m} - \begin{bmatrix} p \\ q \\ r \end{bmatrix} \times \begin{bmatrix} u \\ v \\ \omega \end{bmatrix} = \frac{1}{m}\begin{bmatrix} f_x \\ f_y \\ f_z \end{bmatrix} - \begin{bmatrix} p \\ q \\ r \end{bmatrix} \times \begin{bmatrix} u \\ v \\ \omega \end{bmatrix}$$

We know vector cross-product:

$$\hat{A} \times \hat{B} = \left(A_x\hat{i} + A_y\hat{j} + A_z\hat{k}\right) \times \left(B_x\hat{i} + B_y\hat{j} + B_z\hat{k}\right)$$
$$= \left(A_yB_z - A_zB_y\right)\hat{i} + \left(A_zB_x - A_xB_z\right)\hat{j} + \left(A_xB_y - A_yB_x\right)\hat{k}$$

Note: $p,q,r,u,v,w \Rightarrow$ fram $\mathcal{F}^b$;
p and $u \Rightarrow$ only $i-\text{axis}\left(\text{or } x-\text{axis}\right)$ component; so, $p_x \triangleq p; u_x \triangleq u$
q and $v \Rightarrow$ only $j-\text{axis}\left(\text{or } y-\text{axis}\right)$ component; so, $q_y \triangleq q; v_y \triangleq u$
r and $w \Rightarrow$ only $k-\text{axis}\left(\text{or } z-\text{axis}\right)$ component; so, $r_z \triangleq r; \omega_x \triangleq \omega$
r and $w \Rightarrow$ only $k-\text{axis}$ (or z-axis) component,

$$\hat{p} \times \hat{u} = \left(p_y u_z - p_z u_y \right) \hat{i} + \left(p_z u_x - p_x u_z \right) \hat{j} + \left(p_x u_y - p_y u_x \right) \hat{k} \Rightarrow 0$$

$$\hat{p} \times \hat{v} = \left(p_y v_z - p_z v_y \right) \hat{i} + \left(p_z v_x - p_x v_z \right) \hat{j} + \left(p_x v_y - p_y v_x \right) \hat{k} \Rightarrow p_x v_y \hat{k}$$

$$\hat{p} \times \hat{\omega} = \left(p_y w_z - p_z w_y \right) \hat{i} + \left(p_z w_x - p_x w_z \right) \hat{j} + \left(p_x w_y - p_y w_x \right) \hat{k} \Rightarrow -p_x w_z \hat{j}$$

$$\hat{q} \times \hat{u} = \left(q_y u_z - q_z u_y \right) \hat{i} + \left(q_z u_x - q_x u_z \right) \hat{j} + \left(q_x u_y - q_y u_x \right) \hat{k} \Rightarrow -q_y u_x \hat{k}$$

$$\hat{q} \times \hat{v} = \left(q_y v_z - q_z v_y \right) \hat{i} + \left(q_z v_x - q_x v_z \right) \hat{j} + \left(q_x v_y - q_y v_x \right) \hat{k} \Rightarrow 0$$

$$\hat{q} \times \hat{\omega} = \left(q_y w_z - q_z w_y \right) \hat{i} + \left(q_z w_x - q_x w_z \right) \hat{j} + \left(q_x w_y - q_y w_x \right) \hat{k} \Rightarrow q_y w_z \hat{i}$$

$$\hat{r} \times \hat{u} = \left(r_y u_z - r_z u_y \right) \hat{i} + \left(r_z u_x - q u_z \right) \hat{j} + \left(r_x u_y - r_y u_x \right) \hat{k} \Rightarrow r_z u_x \hat{j}$$

$$\hat{r} \times \hat{v} = \left(r_y v_z - r_z v_y \right) \hat{i} + \left(r_z v_x - r v_z \right) \hat{j} + \left(r_x v_y - r_y v_x \right) \hat{k} \Rightarrow -r_z v_y \hat{i}$$

$$\hat{r} \times \hat{\omega} = \left(r_y w_z - r_z w_y \right) \hat{i} + \left(r_z w_x - r_x w_z \right) \hat{j} + \left(r_x w_y - r_y w_x \right) \hat{k} \Rightarrow 0$$

$$\left[q_y w_z \hat{i} - r_z v_y \hat{i} \right]$$

$$\left[r_z u_x \hat{j} - p_x w_z \hat{j} \right]$$

$$\left[p_x v_y \hat{k} - q_y u_x \hat{k} \right]$$

$$\begin{bmatrix} \left(q_y w_z \hat{i} - r_z v_y \hat{i} \right) \\ \left(r_z u_x \hat{j} - p_x w_z \hat{j} \right) \\ \left(p_x v_y \hat{k} - q_y u_x \hat{k} \right) \end{bmatrix} = \begin{bmatrix} \left(q_y w_z - r_z v_y \right) \hat{i} \\ \left(r_z u_x - p_x w_z \right) \hat{j} \\ \left(p_x v_y - q_y u_x \right) \hat{k} \end{bmatrix} = \begin{bmatrix} \left(q_y w_z - r_z v_y \right) \\ \left(r_z u_x - p_x w_z \right) \\ \left(p_x v_y - q_y u_x \right) \end{bmatrix} \begin{bmatrix} \hat{i} \\ \hat{j} \\ \hat{k} \end{bmatrix}^T = - \begin{bmatrix} rv - qw \\ pw - ru \\ qu - pv \end{bmatrix} \begin{bmatrix} \hat{i} \\ \hat{j} \\ \hat{k} \end{bmatrix}^T$$

We can now rearrange as follows:

$$\begin{bmatrix} \dot{u} \\ \dot{v} \\ \dot{\omega} \end{bmatrix} = \begin{bmatrix} rv - qw \\ pw - ru \\ qu - pv \end{bmatrix} + \frac{1}{m} \begin{bmatrix} f_x \\ f_y \\ f_z \end{bmatrix} \tag{22.18}$$

where $\boldsymbol{f}^b = \left(f_x, f_y, f_z \right)^T$

For rotational motion, Newton's second law states that

$$\frac{d\mathbf{h}^b}{dt_i} = \boldsymbol{m}^b$$

where $\boldsymbol{h}$ is the angular momentum and $\boldsymbol{m}^b$ is the applied torque. Using the Equation (22.12) of Coriolis we have:

$$\frac{d}{dt_i}\mathbf{h} = \frac{d}{dt_b}\mathbf{h}^b + \omega_{b/i} \times \mathbf{h}^b = \mathbf{m}^b \tag{22.19}$$

Again, Equation (22.19) is most easily resolved in body coordinates, where $\mathbf{h}^b = \mathbf{J}\omega_{b/i}^b$, where $\mathbf{J}$ is the constant inertia matrix given by:

$$\mathbf{J} = \begin{bmatrix} \int(y^2+z^2)dm & -\int xy\,dm & -\int xz\,dm \\ -\int xy\,dm & \int((x^2+z^2)dm) & -\int yz\,dm \\ -\int xz\,dm & -\int yz\,dm & \int((x^2+y^2)dm) \end{bmatrix} \triangleq \begin{bmatrix} J_x & -J_{xy} & -J_{xz} \\ -J_{xy} & J_y & -J_{yz} \\ -J_{xz} & -J_{yz} & J_z \end{bmatrix}$$

As shown in Figure 22.22, the quadrotor is essentially symmetric about all three axes; therefore, $J_{xy} = J_{xz} = J_{yz} = 0$, which implies that:

$$\mathbf{J} = \begin{bmatrix} J_x & 0 & 0 \\ 0 & J_y & 0 \\ 0 & 0 & J_z \end{bmatrix}$$

Therefore,

$$\mathbf{J}^{-1} = \begin{bmatrix} \dfrac{1}{J_x} & 0 & 0 \\ 0 & \dfrac{1}{J_y} & 0 \\ 0 & 0 & \dfrac{1}{J_z} \end{bmatrix}$$

The inertia for a solid sphere is given by:

$$J = 2MR^2 / 5$$

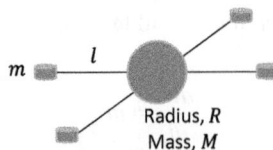

FIGURE 22.22 The moments of inertia for the quadrotor are calculated assuming a spherical dense center with mass M and radius R, and point masses of mass m located at a distance of l from the center.

Therefore,

$$J_x = \frac{2MR^2}{5} + 2l^2m$$

$$J_y = \frac{2MR^2}{5} + 2l^2m$$

$$J_z = \frac{2MR^2}{5} + 4l^2m$$

Defining $\boldsymbol{m}^b \triangleq \left(\tau_\phi, \tau_\theta, \tau_\psi\right)^T$, we can write Equation (22.19) in body coordinates:

$$\frac{d}{dt_b} \boldsymbol{h}^b + \boldsymbol{\omega}_{b/i} \times \boldsymbol{h}^b = \boldsymbol{m}^b$$

$$\frac{d}{dt_b}\left(\mathbf{J}\underset{i}{\boldsymbol{\omega}_b^b}\right) + \boldsymbol{\omega}_{b/i} \times \left(\mathbf{J}\underset{i}{\boldsymbol{\omega}_b^b}\right) = \begin{bmatrix} \tau_\phi \\ \tau_\theta \\ \tau_\psi \end{bmatrix}$$

$$\underset{i}{\dot{\boldsymbol{\omega}}_b} = \frac{1}{\mathbf{J}}\left(-\boldsymbol{\omega}_{b/i} \times \mathbf{J}\boldsymbol{\omega}_{b/i} + \begin{bmatrix} \tau_\phi \\ \tau_\theta \\ \tau_\psi \end{bmatrix}\right)$$

Since

$$\underset{i}{\dot{\boldsymbol{\omega}}_b} \triangleq \left(p, q, r\right)^T$$

We have:

$$\begin{bmatrix} \dot{p} \\ \dot{q} \\ \dot{r} \end{bmatrix} = \frac{1}{\mathbf{J}}\left(-\begin{bmatrix} p \\ q \\ r \end{bmatrix} \times \mathbf{J}\begin{bmatrix} p \\ q \\ r \end{bmatrix} + \begin{bmatrix} \tau_\phi \\ \tau_\theta \\ \tau_\psi \end{bmatrix}\right)$$

$$\hat{p} \times \hat{q} = \left(p_y q_z - p_z q_y\right)\hat{i} + \left(p_z q_x - p_x q_z\right)\hat{j} + \left(p_x q_y - p_y q_x\right)\hat{k} => p_x q_y \hat{k}$$

$$\hat{p} \times \hat{r} = \left(p_y r_z - p_z r_y\right)\hat{i} + \left(p_z r_x - p_x r_z\right)\hat{j} + \left(p_x r_y - p_y r_x\right)\hat{k} => -p_x r_z \hat{j}$$

$$\hat{q} \times \hat{p} = \left(q_y p_z - q_z p_y\right)\hat{i} + \left(q_z p_x - q_x p_z\right)\hat{j} + \left(q_x p_y - q_y p_x\right)\hat{k} => -q_y p_x \hat{k}$$

$$\hat{q} \times \hat{r} = \left(q_y r_z - q_z r_y\right)\hat{i} + \left(q_z r_x - q_x r_z\right)\hat{j} + \left(q_x r_y - q_y r_x\right)\hat{k} => q_y r_z \hat{i}$$

$$\hat{r} \times \hat{p} = \left(r_y p_z - r_z p_y\right)\hat{i} + \left(r_z p_x - r_x p_z\right)\hat{j} + \left(r_x p_y - r_y p_x\right)\hat{k} => r_z p_x \hat{j}$$

$$\hat{r} \times \hat{q} = \left(r_y q_z - r_z q_y\right)\hat{i} + \left(r_z q_x - r_x q_z\right)\hat{j} + \left(r_x q_y - r_y q_x\right)\hat{k} => -r_z q_y \hat{i}$$

$$\begin{bmatrix} \dot{p} \\ \dot{q} \\ \dot{r} \end{bmatrix} = \begin{bmatrix} \dfrac{1}{J_x} & 0 & 0 \\ 0 & \dfrac{1}{J_y} & 0 \\ 0 & 0 & \dfrac{1}{J_z} \end{bmatrix}\left(-\begin{bmatrix} \left(q_y r_z - r_z q_y\right)_x \\ \left(r_z p_x - p_x r_z\right)_y \\ \left(p_x q_y - q_y p_x\right)_z \end{bmatrix}\begin{bmatrix} J_x & 0 & 0 \\ 0 & J_y & 0 \\ 0 & 0 & J_z \end{bmatrix} + \begin{bmatrix} \tau_\phi \\ \tau_\theta \\ \tau_\psi \end{bmatrix}\right)$$

Since we have,

$$p_x \triangleq p$$
$$q_y \triangleq q$$
$$r_z \triangleq r$$

We can rearrange above equation,

$$
\begin{bmatrix} \dot{p} \\ \dot{q} \\ \dot{r} \end{bmatrix} = \begin{bmatrix} \dfrac{1}{J_x} & 0 & 0 \\ 0 & \dfrac{1}{J_y} & 0 \\ 0 & 0 & \dfrac{1}{J_z} \end{bmatrix} \left(\begin{bmatrix} 0 & r & -q \\ -r & 0 & p \\ q & -p & 0 \end{bmatrix} \begin{bmatrix} J_x & 0 & 0 \\ 0 & J_y & 0 \\ 0 & 0 & J_z \end{bmatrix} \begin{bmatrix} p \\ q \\ r \end{bmatrix} + \begin{bmatrix} \tau_\phi \\ \tau_\theta \\ \tau_\psi \end{bmatrix} \right)
\tag{22.20}
$$

Equation (22.20) can further be rewritten as:

$$
\begin{bmatrix} \dot{p} \\ \dot{q} \\ \dot{r} \end{bmatrix} = \begin{bmatrix} \ddot{\phi} \\ \ddot{\theta} \\ \ddot{\psi} \end{bmatrix} = \begin{bmatrix} \dfrac{J_y - J_z}{J_x} qr \\ \dfrac{J_z - J_x}{J_y} pr \\ \dfrac{J_x - J_y}{J_z} pq \end{bmatrix} + \begin{bmatrix} \dfrac{1}{J_x} \tau_\phi \\ \dfrac{1}{J_y} \tau_\theta \\ \dfrac{1}{J_z} \tau_\psi \end{bmatrix}
\tag{22.21}
$$

The six degree of freedom model for the quadrotor kinematics and dynamics can be summarized as follows with reference to body frame $\mathcal{F}^b$:
From Equation (22.13):

$$
\begin{bmatrix} \dot{p}_n \\ \dot{p}_e \\ \dot{h} \end{bmatrix} = \begin{bmatrix} c\theta c\psi & s\phi s\theta c\psi - c\phi s\psi & c\phi s\theta c\psi + s\phi s\psi \\ c\theta s\psi & s\phi s\theta s\psi + c\phi c\psi & c\phi s\theta s\psi - s\phi c\psi \\ s\theta & -s\phi c\theta & -c\phi c\theta \end{bmatrix} \begin{bmatrix} u \\ v \\ \omega \end{bmatrix}
\tag{22.22}
$$

From Equation (22.18):

$$
\begin{bmatrix} \dot{u} \\ \dot{v} \\ \dot{\omega} \end{bmatrix} = \begin{bmatrix} rv - qw \\ pw - ru \\ qu - pv \end{bmatrix} + \frac{1}{m} \begin{bmatrix} f_x \\ f_y \\ f_z \end{bmatrix}
\tag{22.23}
$$

From Equation (22.16):

$$
\begin{bmatrix} \dot{\phi} \\ \dot{\theta} \\ \dot{\psi} \end{bmatrix} = \begin{bmatrix} 1 & \sin\phi \tan\theta & \cos\phi \tan\theta \\ 0 & \cos\phi & -\sin\phi \\ 0 & \sin\phi \sec\theta & \cos\phi \sec\theta \end{bmatrix} \begin{bmatrix} p \\ q \\ r \end{bmatrix}
\tag{22.24}
$$

From Equation (22.21):

$$
\begin{bmatrix} \dot{p} \\ \dot{q} \\ \dot{r} \end{bmatrix} = \begin{bmatrix} \dfrac{J_y - J_z}{J_x} qr \\ \dfrac{J_z - J_x}{J_y} pr \\ \dfrac{J_x - J_y}{J_z} pq \end{bmatrix} + \begin{bmatrix} \dfrac{1}{J_x} \tau_\phi \\ \dfrac{1}{J_y} \tau_\theta \\ \dfrac{1}{J_z} \tau_\psi \end{bmatrix}
\tag{22.25}
$$

Forces and Moments

The objective of this section is to describe the forces and torques that act on the quadrotor. Since there are no aerodynamic lifting surfaces, we will assume that the aerodynamic forces and moments are negligible. The forces and moments are primarily due to gravity and the four propellers. Figure 22.23 shows a top view of the quadrotor systems. As shown in Figure 22.24, each motor produces a force F and a torque τ. The total force acting on the quadrotor is given by:

$$ F = F_f + F_r + F_b + F_l $$

The rolling torque is produced by the forces of the right and left motors as:

$$ \tau_\phi = l\left(F_l - F_r\right) $$

Similarly, the pitching torque is produced by the forces of the front and back motors as:

$$ \tau_\theta = l\left(F_f - F_b\right) $$

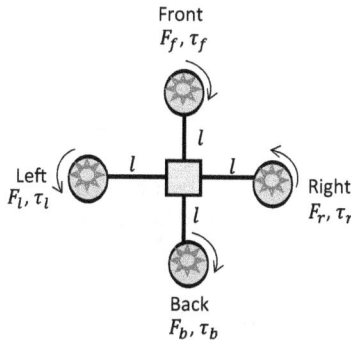

FIGURE 22.23 The top view of the quadrotor. Each motor produces an upward force F and a torque τ. The front and back motors spin clockwise, and the right and left motors spin counterclockwise.

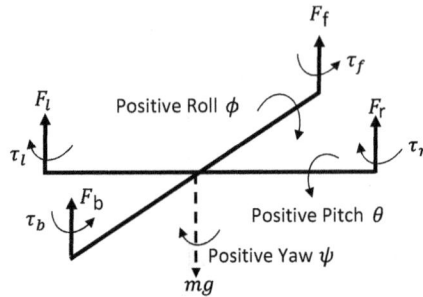

FIGURE 22.24 Definition of the forces and torques acting on the quadrotor.

Due to Newton's third law, the **drag** of the propellers produces a yawing torque on the body of the quadrotor. The direction of the torque will be in the opposite direction of the motion of the propeller. Therefore, the total yawing torque is given by:

$$\tau_\psi = \tau_r + \tau_l - \tau_f - \tau_b$$

The **lift** and **drag** produced by the propellers is proportional to the square of the angular velocity. We will assume that the angular velocity is directly proportional to the **pulse width modulation** command sent to the motor. Therefore, the **force and torque of each motor** can be expressed as:

$$F_* = \kappa_1 \delta_*$$
$$\tau_* = \kappa_2 \delta_*$$

where κ_1 and κ_2 are constants that need to be determined experimentally, δ_* is the motor command signal, and $*$ represents f, r, b, and l.

Therefore, the forces and torques on the quadrotor can be written in matrix form as:

$$\begin{pmatrix} F \\ \tau_\phi \\ \tau_\theta \\ \tau_\psi \end{pmatrix} = \begin{bmatrix} \kappa_1 & \kappa_1 & \kappa_1 & \kappa_1 \\ 0 & -l\kappa_1 & 0 & l\kappa_1 \\ l\kappa_1 & 0 & l\kappa_1 & 0 \\ -\kappa_2 & \kappa_2 & -\kappa_2 & \kappa_2 \end{bmatrix} \begin{pmatrix} \delta_f \\ \delta_r \\ \delta_b \\ \delta_l \end{pmatrix} \triangleq \mathcal{M} \begin{pmatrix} \delta_f \\ \delta_r \\ \delta_b \\ \delta_l \end{pmatrix}$$

The control strategies derived in subsequent sections will specify forces and torques. The actual motors commands can be found as:

$$\begin{pmatrix} \delta_f \\ \delta_r \\ \delta_b \\ \delta_l \end{pmatrix} = \mathcal{M}^{-1} \begin{pmatrix} F \\ \tau_\phi \\ \tau_\theta \\ \tau_\psi \end{pmatrix}$$

Note that the **pulse width modulation** commands are required to be between zero and one.

In addition to the force exerted by the motor, gravity also exerts a force on the quadrotor. In the vehicle frame $\mathcal{F}^v$, the **gravity force acting on the center of mass** is given by:

$$\mathbf{f}_g^v = \begin{pmatrix} 0 \\ 0 \\ mg \end{pmatrix}$$

However, since v in Equation (22.23) is expressed in $\mathcal{F}^b$, we must transform to the body frame to give:

$$\mathbf{f}_g^b = R_v^b \mathbf{f}_g^v$$

From Equation (22.8), we have:

$$R_v^b = \begin{bmatrix} c\theta c\psi & c\theta s\psi & -s\theta \\ s\phi s\theta c\psi - c\phi s\psi & s\phi s\theta s\psi + c\phi c\psi & s\phi c\theta \\ c\phi s\theta c\psi + s\phi s\psi & c\phi s\theta s\psi - s\phi c\psi & c\phi c\theta \end{bmatrix}$$

$$\mathbf{f}_g^b = \begin{bmatrix} c\theta c\psi & c\theta s\psi & -s\theta \\ s\phi s\theta c\psi - c\phi s\psi & s\phi s\theta s\psi + c\phi c\psi & s\phi c\theta \\ c\phi s\theta c\psi + s\phi s\psi & c\phi s\theta s\psi - s\phi c\psi & c\phi c\theta \end{bmatrix} \begin{pmatrix} 0 \\ 0 \\ mg \end{pmatrix}$$

$$\mathbf{f}_g^b = \begin{pmatrix} -mg\sin\theta \\ mg\sin\phi\cos\theta \\ mg\cos\phi\cos\theta \end{pmatrix} \tag{22.26}$$

Considering the gravity force is acting on the center of mass, the additional force showing in Equation (22.26) needs to be added in Equation (22.23). Moreover, Figure 22.24 depicts that total force $F = F_f + F_r + F_b + F_l$ is acting in opposite direction of the gravity (considering in z-axis), while forces along x-axis and y-axis, $f_x = f_y = 0$.

We can now rewrite Equation (22.23) for quadrotor as follows:

$$\begin{bmatrix} \dot{u} \\ \dot{v} \\ \dot{\omega} \end{bmatrix} = \begin{bmatrix} rv - qw \\ pw - ru \\ qu - pv \end{bmatrix} + \begin{pmatrix} -mg\sin\theta \\ mg\sin\phi\cos\theta \\ mg\cos\phi\cos\theta \end{pmatrix} + \frac{1}{m} \begin{bmatrix} 0 \\ 0 \\ -F \end{bmatrix} \tag{22.27}$$

For quadrotor, all other equations remain the same:

$$\begin{bmatrix} \dot{p}_n \\ \dot{p}_e \\ \dot{h} \end{bmatrix} = \begin{bmatrix} c\theta c\psi & s\phi s\theta c\psi - c\phi s\psi & c\phi s\theta c\psi + s\phi s\psi \\ c\theta s\psi & s\phi s\theta s\psi + c\phi c\psi & c\phi s\theta s\psi - s\phi c\psi \\ s\theta & -s\phi c\theta & -c\phi c\theta \end{bmatrix} \begin{bmatrix} u \\ v \\ \omega \end{bmatrix} \qquad (22.28)$$

$$\begin{bmatrix} \dot{\phi} \\ \dot{\theta} \\ \dot{\psi} \end{bmatrix} = \begin{bmatrix} 1 & \sin\phi \tan\theta & \cos\phi \tan\theta \\ 0 & \cos\phi & -\sin\phi \\ 0 & \sin\phi \sec\theta & \cos\phi \sec\theta \end{bmatrix} \begin{bmatrix} p \\ q \\ r \end{bmatrix} \qquad (22.29)$$

$$\begin{bmatrix} \dot{p} \\ \dot{q} \\ \dot{r} \end{bmatrix} = \begin{bmatrix} \dfrac{J_y - J_z}{J_x} qr \\[2.5ex] \dfrac{J_z - J_x}{J_y} pr \\[2.5ex] \dfrac{J_x - J_y}{J_z} pq \end{bmatrix} + \begin{bmatrix} \dfrac{1}{J_x}\tau_\phi \\[2.5ex] \dfrac{1}{J_y}\tau_\theta \\[2.5ex] \dfrac{1}{J_z}\tau_\psi \end{bmatrix} \qquad (22.30)$$

We know:

$$p = \dot{\phi};\, q = \dot{\theta};\, r = \dot{\psi}$$

So, we have:

$$\begin{bmatrix} \ddot{\phi} \\ \ddot{\theta} \\ \ddot{\psi} \end{bmatrix} = \begin{bmatrix} \dfrac{J_y - J_z}{J_x} \dot{\theta}\dot{\psi} \\[2.5ex] \dfrac{J_z - J_x}{J_y} \dot{\phi}\dot{\psi} \\[2.5ex] \dfrac{J_x - J_y}{J_z} \dot{\phi}\dot{\theta} \end{bmatrix} + \begin{bmatrix} \dfrac{1}{J_x}\tau_\phi \\[2.5ex] \dfrac{1}{J_y}\tau_\theta \\[2.5ex] \dfrac{1}{J_z}\tau_\psi \end{bmatrix} \qquad (22.31)$$

From Equation (22.30), we can also write $\tau_\phi \rightarrow \tau_x,\ \tau_\theta \rightarrow \tau_y, \tau_\psi \rightarrow \tau_z$

$$\begin{bmatrix} \ddot{\phi} \\ \ddot{\theta} \\ \ddot{\psi} \end{bmatrix} = \begin{bmatrix} \dfrac{J_y - J_z}{J_x} \dot{\theta}\dot{\psi} \\[2.5ex] \dfrac{J_z - J_x}{J_y} \dot{\phi}\dot{\psi} \\[2.5ex] \dfrac{J_x - J_y}{J_z} \dot{\phi}\dot{\theta} \end{bmatrix} + \begin{bmatrix} \dfrac{1}{J_x}\tau_x \\[2.5ex] \dfrac{1}{J_y}\tau_y \\[2.5ex] \dfrac{1}{J_z}\tau_z \end{bmatrix} \qquad (22.32)$$

Note that, in our system, the gyroscopic and aerodynamic torques are considered as external disturbances.

Deep Reinforcement Learning-Based Attitude and Altitude Control

The aerial robot used in this research study is a quadcopter [2]. Quadcopters are substantially under actuated, with six degrees of freedom (three translational and three rotational) and only four distinct inputs (rotor speeds). Rotational and translational motions are coupled to achieve six degrees of freedom. After accounting for the intricate aerodynamic effects, the resulting dynamics is highly nonlinear. As another property of quadcopters, it must be noted that, unlike conventional helicopters, the **rotor blade pitch angle in a quadcopter** does not need to be varied.

Quadcopter Coordinate Frames, Forces, and Torques

The reference coordinate frame and the coordinate frame of the vehicle body must be determined before building a mathematical model of the quadrotor, as shown in Figure 22.25.

The ground and the reference coordinate frames are both tied to $\Re_E\left\{O, \vec{I}, \vec{J}, \vec{K}\right\}$. The $\Re_B\left\{o, \vec{i}, \vec{j}, \vec{k}\right\}$ is a coordinate frame that is attached to the body of the vehicle and has its is a coordinate frame that is attached to the body of the vehicle and has its center aligned with the center of mass of the robot. In this research study, the dynamical equations governing the quadcopter were derived from the text published in [2]. The following assumptions were taken into consideration in order to determine the examined equations of the motion of the system:

- The aerial robot consists of a stiff body with a symmetrical structure.
- The geometrical center of the robot is the same as its center of gravity and mass.
- The **moment of inertia of the propellers** has been overlooked.

The dynamical model of the system could be constructed by taking into account both the translational dynamic (Newton's second law) and the rotational dynamic (Euler's rotation equations).

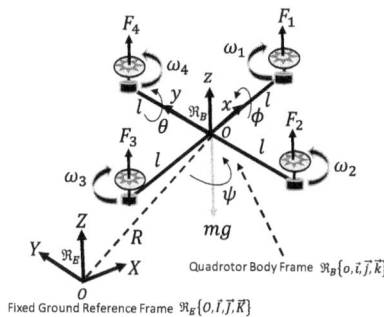

FIGURE 22.25 Coordinate frame of the quadcopter.

Translational Dynamics

The following forces acted on the system being studied:

- The total weight of the vehicle, as expressed in Equation (22.33).
- The generated thrust of rotors, which can be calculated using Equation (22.2).
- As indicated in Equation (22.3), the drag force and air friction.

$$w = \begin{bmatrix} 0 \\ 0 \\ -mg \end{bmatrix} \tag{22.33}$$

From Figure 22.25, we have:

$$F_t = R_b^e \begin{bmatrix} 0 \\ 0 \\ \sum_{i=1}^{4} F_i \end{bmatrix}$$

where R_b^e rotational matrix for transformation of the body $\mathcal{F}^b$ frame with the Earth frame $\mathcal{F}^e$. (Note that centripetal force, $F = m\dfrac{v^2}{r} = mr\dfrac{v^2}{r^2} = mr\left(\dfrac{v}{r}\right)^2 = mr\omega^2 = b\omega^2$, where ω is angular velocity and b is a trust constant.)

So, we can rewrite (for $\left[R_b^e \right]$, see page 105 of the book of Quan),

$$F_t = \left[R_b^e \right] \begin{bmatrix} 0 \\ 0 \\ \sum_{i=1}^{4} b(\omega_i)^2 \end{bmatrix} = \left[R_v^b (\phi,\theta,\psi) \right]^T = \begin{bmatrix} 0 \\ 0 \\ \sum_{i=1}^{4} b(\omega_i)^2 \end{bmatrix}$$

where ω_i is the angular velocity of the i-th propeller and b is thrust constant.

Using Equation (22.8),

$$F_t = \begin{bmatrix} c\theta c\psi & s\phi s\theta c\psi - c\phi s\psi & c\phi s\theta c\psi + s\phi s\psi \\ c\theta s\psi & s\phi s\theta s\psi + c\phi c\psi & c\phi s\theta s\psi - s\phi c\psi \\ -s\theta & s\phi c\theta & c\phi c\theta \end{bmatrix} \begin{bmatrix} 0 \\ 0 \\ \sum_{i=1}^{4} b(\omega_i)^2 \end{bmatrix} \tag{22.34}$$

$$F_t = b\sum_{i=1}^{4} (\omega_i)^2 \begin{bmatrix} c\phi s\theta c\psi + s\phi s\psi \\ c\phi s\theta s\psi - s\phi c\psi \\ c\phi c\theta \end{bmatrix} \tag{22.35}$$

Let us consider F_d represents the drag force and air friction and ξ is the position of the center of mass of the quadcopter in the flat Earth coordinates system (x, y, z). F_d is represented as follows:

$$F_d = C_d \dot{\xi} = -C_d \begin{bmatrix} \dot{x} \\ \dot{y} \\ \dot{z} \end{bmatrix}$$

where C_d is the drag coefficient. We can write C_d in matrix form, and the above equation is rewritten as follows:

$$F_d = C_d \dot{\xi} = -C_d \begin{bmatrix} \dot{x} \\ \dot{y} \\ \dot{z} \end{bmatrix} = -\begin{bmatrix} C_{dx} & 0 & 0 \\ 0 & C_{dy} & 0 \\ 0 & 0 & C_{dz} \end{bmatrix} \begin{bmatrix} \dot{x} \\ \dot{y} \\ \dot{z} \end{bmatrix} = -\begin{bmatrix} C_{dx}\dot{x} \\ C_{dy}\dot{y} \\ C_{dz}\dot{z} \end{bmatrix} \tag{22.36}$$

It is interesting to note that the total force is the summation of w, F_t and F_d. On the other hand, Newton's second law states that **Force = mass × acceleration**. So, we have the total force F:

$$F = m\ddot{\xi} = w + F_t + F_d \tag{22.37}$$

$$\ddot{\xi} = \frac{1}{m}\left(w + F_t + F_d\right) \tag{22.38}$$

In the equations, the gravity acceleration is denoted by g. In Equations (22.34) and (22.35), the Euler angles are represented by (ϕ, θ, ψ). The rotation transform matrix, the angular velocity of the i-th propeller, and the thrust constant are represented by R_b^e, ω_i, and b, respectively. In Equation (22.36), C_d is the matrix of translational drag coefficients. The position of the center of mass (ξ) in the flat Earth coordinate is defined as a 3 by 1 vector. The equation of motion that describes the translational motion of a quadcopter can be stated as follows, using Newton's second law:

$$\begin{bmatrix} \ddot{x} \\ \ddot{y} \\ \ddot{z} \end{bmatrix} = \frac{1}{m}\left(\begin{bmatrix} 0 \\ 0 \\ -mg \end{bmatrix} + b\sum_{i=1}^{4}(\omega_i)^2 \begin{bmatrix} c\phi s\theta c\psi + s\phi s\psi \\ c\phi s\theta s\psi - s\phi c\psi \\ c\phi c\theta \end{bmatrix} - \begin{bmatrix} C_{dx}\dot{x} \\ C_{dy}\dot{y} \\ C_{dz}\dot{z} \end{bmatrix}\right) \tag{22.39}$$

$$\ddot{x} = \frac{1}{m}\left(b\sum_{i=1}^{4}(\omega_i)^2 \left(c\phi s\theta c\psi + s\phi s\psi\right) - C_{dx}\dot{x}\right) \tag{22.40}$$

$$\ddot{y} = \frac{1}{m}\left(b\sum_{i=1}^{4}(\omega_i)^2 \left(c\phi s\theta s\psi - s\phi c\psi\right) - C_{dy}\dot{y}\right) \tag{22.41}$$

$$\ddot{z} = \frac{1}{m}\left(b\sum_{i=1}^{4}(\omega_i)^2(c\phi c\theta) - C_{dz}\dot{z} - mg\right) \tag{22.42}$$

Rotational Dynamics

Note the following:

Vector Cross-Product

$$\hat{A} \times \hat{B} = \left(A_x\hat{i} + A_y\hat{j} + A_z\hat{k}\right) \times \left(B_x\hat{i} + B_y\hat{j} + B_z\hat{k}\right)$$

$$= \left(A_yB_z - A_zB_y\right)\hat{i} + \left(A_zB_x - A_xB_z\right)\hat{j} + \left(A_xB_y - A_yB_x\right)\hat{k}$$

Torque = Distance × Force

$$\tau = r \times F$$

$$\begin{bmatrix} 0 \\ -l \\ 0 \end{bmatrix} \times \begin{bmatrix} 0 \\ 0 \\ -F_1 \end{bmatrix} = \begin{bmatrix} -lF_1 \\ 0 \\ 0 \end{bmatrix}$$

Rigid body rotation – gyroscopic effect – $\Sigma F_i = \mathbf{m} \cdot \mathbf{a}_i$

This tells us that the sum of the forces on a mass in the i-th direction must equal the mass of the body times the acceleration in that direction.

$$\Sigma T_x = I_x\dot{\omega}_x + \left(I_z - I_y\right)\omega_y\omega_z$$
$$\Sigma T_y = I_x\dot{\omega}_y + \left(I_x - I_z\right)\omega_x\omega_z$$
$$\Sigma T_z = I_x\dot{\omega}_z + \left(I_y - I_x\right)\omega_y\omega_z$$

A quadrotor is affected by **roll, pitch**, and **yaw** torques, as well as by an aerodynamic friction torque and the gyroscopic effect of the propeller. The torques are expressed as follows, in Equations (22.8)–(22.12):

$$\tau = \begin{bmatrix} \tau_x \\ \tau_y \\ \tau_z \end{bmatrix}$$

$$\tau_x \to \tau_\phi; \tau_y \to \tau_\theta; \tau_z \to \tau_\psi$$

From Figure 22.25, it is observed from the quadrotor geometry: Force F_2 is in the z-axis perpendicular to y-axis, but distance l is in negative direction of y-axis from the center of rotor body mass. Similarly, F_4 is perpendicular on y-axis but is in positive l distance along y-axis from the center of rotor body mass.

$$\tau_x = \begin{bmatrix} 0 \\ -l \\ 0 \end{bmatrix} \times \begin{bmatrix} 0 \\ 0 \\ F_2 \end{bmatrix} + \begin{bmatrix} 0 \\ l \\ 0 \end{bmatrix} \times \begin{bmatrix} 0 \\ 0 \\ F_4 \end{bmatrix} = \begin{bmatrix} l(F_4 - F_2) \\ 0 \\ 0 \end{bmatrix} = \begin{bmatrix} lb(\omega_4^2 - \omega_2^2) \\ 0 \\ 0 \end{bmatrix} \tag{22.43}$$

Similarly,

$$
\tau_y = \begin{bmatrix} l \\ 0 \\ 0 \end{bmatrix} \times \begin{bmatrix} 0 \\ 0 \\ F_1 \end{bmatrix} + \begin{bmatrix} -l \\ 0 \\ 0 \end{bmatrix} \times \begin{bmatrix} 0 \\ 0 \\ F_3 \end{bmatrix} = \begin{bmatrix} 0 \\ l(F_3 - F_1) \\ 0 \end{bmatrix} = \begin{bmatrix} 0 \\ lb(\omega_3^2 - \omega_1^2) \\ 0 \end{bmatrix} \tag{22.44}
$$

However, τ_z is the total torque introduced by four rotors of the quadrotor. Let us suppose $\tau_{M_i} = d\omega_i^2$ is the torque of motor M_i, d is a constant, and ω_i is the angular velocity. So, we have:

$$
\tau_z = \begin{bmatrix} 0 \\ 0 \\ \sum_{i=1}^{4} \tau_{M_i} \end{bmatrix} = \begin{bmatrix} 0 \\ 0 \\ \sum_{i=1}^{4} d(-1)^{i+1} \omega_i^2 \end{bmatrix} \tag{22.45}
$$

In addition, the aerodynamic frictional torques and gyroscopic torques of the rotors that are considered as external disturbances for the system need to be considered.

Air Frictional Torque of the Quadrotor
Our axis sequence:

$$
\phi \to x; \theta \to y; \psi \to z
$$

$$
\tau_a = C_a \begin{bmatrix} \dot{\phi}^2 \\ \dot{\theta}^2 \\ \dot{\psi}^2 \end{bmatrix} = \begin{bmatrix} C_{a_x} & 0 & 0 \\ 0 & C_{a_y} & 0 \\ 0 & 0 & C_{a_z} \end{bmatrix} \begin{bmatrix} \dot{\phi}^2 \\ \dot{\theta}^2 \\ \dot{\psi}^2 \end{bmatrix} = \begin{bmatrix} C_{a_x} \dot{\phi}^2 \\ C_{a_y} \dot{\theta}^2 \\ C_{a_z} \dot{\psi}^2 \end{bmatrix} \tag{22.46}
$$

Gyroscopic Torque of the Quadrotor Propeller
The gyroscopic effect is described by the basic equation of the gyroscope:

$$
\tau_{gp} = \Omega_r \times I_r \tag{22.47}
$$

where I_r is the rotors–propellers angular momentum vector, Ω_r is the angular velocity precession vector. In fact, each object with a spinning mass is subjected to a gyroscopic precession motion effect Ω_r. We know that the angular momentum L of a rigid body is provided as stated below:

$$
L = I\Omega \tag{22.48}
$$

where I moment of inertia and Ω is the angular velocity.

Similarly, to the angular momentum of the rigid body frame, given in Equation 22.48) the angular momentum of the propulsion unit can be expressed as follows:

$$L_r = I_r \Omega_r \tag{22.49}$$

In Equation (22.49), I_r scalar denotes simplified motor–propeller inertia, while Ω_r refers to the sum of the rotors' angular velocities (in the z-axis). The angular momentum vector (49) is obtained by multiplying the moment of inertia (I_r) by the rotor's angular velocity about the axis where the rotation occurs (Ω_r):

$$\Omega_r = \begin{bmatrix} 0 \\ 0 \\ (\omega_1 - \omega_2 + \omega_3 - \omega_4) \end{bmatrix} = \begin{bmatrix} 0 \\ 0 \\ \sum_{i=1}^{4} (-1)^{i+1} \omega_i \end{bmatrix} \tag{22.50}$$

The rotor angular velocity vector (Ω_r) contains zeros for the components in the axes x, y and the sum of the rotor's angular velocities ω_i (where i varies from 1 to 4) in the z axis. Knowing L_r and Ω_r, Equation (22.49) can be rewritten as follows:

$$\tau_{gp} = \begin{bmatrix} \omega_x \\ \omega_y \\ \omega_z \end{bmatrix} \times \begin{bmatrix} 0 \\ 0 \\ I_r \Omega_{r_z} \end{bmatrix} = \begin{bmatrix} \omega_y I_r \Omega_{r_z} \\ -\omega_x I_r \Omega_{r_z} \\ 0 \end{bmatrix} \tag{22.51}$$

Since $\hat{A} \times \hat{B} = \left(A_x \hat{i} + A_y \hat{j} + A_z \hat{k}\right) \times \left(B_x \hat{i} + B_y \hat{j} + B_z \hat{k}\right)$

$$= \left(A_y B_z - A_z B_y\right)\hat{i} + \left(A_z B_x - A_x B_z\right)\hat{j} + \left(A_x B_y - A_y B_x\right)\hat{k}$$

$$-\omega_x I_r \Omega_{r_z} \hat{j}$$

$$\omega_y I_r \Omega_{r_z} \hat{i}$$

$$0\hat{k}$$

$$\tau_x \rightarrow \tau_\phi ; \tau_y \rightarrow \tau_\theta ; \tau_z \rightarrow \tau_\psi$$

Approximation for small angles, we can assume $\dot{\phi} \approx \omega_x, \dot{\theta} \approx \omega_y, \dot{\psi} \approx \omega_z$ is valid.

$$\Omega_{r_z} \triangleq \Omega_r$$

Because Ω_r only exists in z axis only and all other components are zero. Rearranging, we have:

$$\tau_{gp} = \begin{bmatrix} \tau_{gp_x} \\ \tau_{gp_y} \\ \tau_{gp_z} \end{bmatrix} = \begin{bmatrix} \omega_y I_r \Omega_{r_z} \\ -\omega_x I_r \Omega_{r_z} \\ 0 \end{bmatrix} = \begin{bmatrix} \dot{\theta} I_r \Omega_{r_z} \\ -\dot{\phi} I_r \Omega_{r_z} \\ 0 \end{bmatrix} = \begin{bmatrix} \dot{\theta} I_r \Omega_r \\ -\dot{\phi} I_r \Omega_r \\ 0 \end{bmatrix} \tag{22.52}$$

If we rearrange the axis as follows: $\phi \to x; \theta \to y; \psi \to z$; and $\boldsymbol{I}_r \equiv \boldsymbol{J}_r$;

$$\tau_{gp} = \begin{bmatrix} \dot{\theta} \boldsymbol{J}_r \Omega_r \\ -\dot{\phi} \boldsymbol{J}_r \Omega_r \\ 0 \end{bmatrix} = \boldsymbol{J}_r \Omega_r \begin{bmatrix} \dot{\theta} \\ -\dot{\phi} \\ 0 \end{bmatrix} \qquad (22.53)$$

In Equations (22.43) and (22.44), l is the distance between the motor axis and the center of mass of the quadcopter. In Equation (22.45), $\boldsymbol{J}_r$ and Ω_r, respectively, are the inertia and rotation velocity of rotors. In Equation (22.46), Ca is a 3 by 3 matrix of aerodynamic friction coefficients.

From Equation (22.32), we have the following:

$$\begin{bmatrix} \ddot{\phi} \\ \ddot{\theta} \\ \ddot{\psi} \end{bmatrix} = \begin{bmatrix} \dfrac{J_y - J_z}{J_x} \dot{\theta}\dot{\psi} \\ \dfrac{J_z - J_x}{J_y} \dot{\phi}\dot{\psi} \\ \dfrac{J_x - J_y}{J_z} \dot{\phi}\dot{\theta} \end{bmatrix} + \begin{bmatrix} \dfrac{1}{J_x}\tau_x \\ \dfrac{1}{J_y}\tau_y \\ \dfrac{1}{J_z}\tau_z \end{bmatrix} \begin{bmatrix} \dfrac{1}{J_x}\tau_x \\ \dfrac{1}{J_y}\tau_y \\ \dfrac{1}{J_z}\tau_z \end{bmatrix}$$

Note that Equation (22.32) has not considered two more torque components: aerodynamic air friction torque τ_a of Equation (22.46) and gyroscopic torque τ_{gp} of Equation (22.53). Should we consider these two torques (let us change the inertial notations: $I_x \equiv J_x, I_y \equiv J_y, I_z \equiv J_z$; put the values of τ_x, τ_y, τ_z from Equations (22.43), (22.44), and (22.45), respectively), Equation (22.32) would look like as follows (Note that: τ_a and τ_{gp} are considered as external disturbances.

$$\begin{bmatrix} \ddot{\phi} \\ \ddot{\theta} \\ \ddot{\psi} \end{bmatrix} = \begin{bmatrix} \dfrac{I_y - I_z}{I_x} \dot{\theta}\dot{\psi} \\ \dfrac{I_z - I_x}{I_y} \dot{\phi}\dot{\psi} \\ \dfrac{I_x - I_y}{I_z} \dot{\phi}\dot{\theta} \end{bmatrix} + \begin{bmatrix} \dfrac{1}{I_x}\tau_x \\ \dfrac{1}{I_y}\tau_y \\ \dfrac{1}{I_z}\tau_z \end{bmatrix} - \begin{bmatrix} \dfrac{1}{I_x}C_{a_x}\dot{\phi}^2 \\ \dfrac{1}{I_y}C_{a_y}\dot{\theta}^2 \\ \dfrac{1}{I_z}C_{a_z}\dot{\psi}^2 \end{bmatrix} - \begin{bmatrix} \dfrac{1}{I_x}\boldsymbol{J}_r\Omega_r\dot{\theta} \\ \dfrac{-1}{I_y}\boldsymbol{J}_r\Omega_r\dot{\phi} \\ 0 \end{bmatrix} \qquad (22.54)$$

We know that Equation (22.54) is derived after applying Euler's rotation equations: Note that the following rotation Equations govern the rotating motion of the quadrotor. In the equations, I_x, I_y, and I_z are moments of inertia along the $x, y,$ and z directions, respectively. We now break down into the individual Equations (22.55), (22.56), and (22.57):

$$\ddot{\phi} = \frac{1}{I_x}\left((I_y - I_z)\dot{\theta}\dot{\psi} + lb\left(\omega_4^{\ 2} - \omega_2^{\ 2} \right) - C_{a_x}\dot{\phi}^2 - \boldsymbol{J}_r\Omega_r\dot{\theta} \right) \qquad (22.55)$$

$$\ddot{\theta} = \frac{1}{I_y}\left((I_z - I_x)\dot{\phi}\dot{\psi} + lb\left(\omega_3{}^2 - \omega_1{}^2\right) - C_{a_y}\dot{\theta}^2 + J_r\Omega_r\dot{\phi}\right) \tag{22.56}$$

$$\ddot{\psi} = \frac{1}{I_z}\left((I_x - I_y)\dot{\phi}\dot{\theta} + \sum_{i=1}^{4}d(-1)^{i+1}\omega_i^2 - C_{a_z}\dot{\psi}^2\right) \tag{22.57}$$

Note that Equations (22.56) and (22.57) differ from those of Barzergar's Equations (22.14) and (22.15). Equations (22.14) and (22.15) have some errors.

Quadrotor Dynamics Models

Having considered both translational and rotational dynamics, let us assume the following:

$$u_1 = b\sum_{i=1}^{4}\left(\omega_i\right)^2 \tag{22.58}$$

$$u_2 = lb\left(\omega_4{}^2 - \omega_2{}^2\right) \tag{22.59}$$

$$u_3 = lb\left(\omega_3{}^2 - \omega_1{}^2\right) \tag{22.60}$$

$$u_4 = \sum_{i=1}^{4}d(-1)^{i+1}\omega_i^2 \tag{22.61}$$

$$u_x = \left(c\phi s\theta c\psi + s\phi s\psi\right) \tag{22.62}$$

$$u_y = \left(c\phi s\theta s\psi - s\phi c\psi\right) \tag{22.63}$$

The entire dynamic model of the quadcopter could be stated as follows:

$$\ddot{x} = \frac{1}{m}\left(b\sum_{i=1}^{4}\left(\omega_i\right)^2\left(c\phi s\theta c\psi + s\phi s\psi\right) - C_{dx}\dot{x}\right) = \frac{1}{m}\left(u_1 u_x - C_{dx}\dot{x}\right)$$

$$\ddot{y} = \frac{1}{m}\left(b\sum_{i=1}^{4}\left(\omega_i\right)^2\left(c\phi s\theta s\psi - s\phi c\psi\right) - C_{dy}\dot{y}\right) = \frac{1}{m}\left(u_1 u_y - C_{dy}\dot{y}\right)$$

$$\ddot{z} = \frac{1}{m}\left(b\sum_{i=1}^{4}\left(\omega_i\right)^2\left(c\phi c\theta\right) - C_{dz}\dot{z} - mg\right) = \frac{1}{m}\left(u_1\left(c\phi c\theta\right) - C_{dz}\dot{z} - mg\right)$$

$$\ddot{\phi} = \frac{1}{I_x}\left((I_y - I_z)\dot{\theta}\dot{\psi} + lb\left(\omega_4{}^2 - \omega_2{}^2\right) - C_{a_x}\dot{\phi}^2 - J_r\Omega_r\dot{\theta}\right) = \ddot{\phi} = \frac{1}{I_x}\left((I_y - I_z)\dot{\theta}\dot{\psi} - C_{a_x}\dot{\phi}^2 - J_r\Omega_r\dot{\theta} + u_2\right)$$

$$\ddot{\theta} = \frac{1}{I_y}\left((I_z - I_x)\dot{\phi}\dot{\psi} + lb\left(\omega_3{}^2 - \omega_1{}^2\right) - C_{a_y}\dot{\theta}^2 + J_r\Omega_r\dot{\phi}\right) = \frac{1}{I_y}\left((I_z - I_x)\dot{\phi}\dot{\psi} - C_{a_y}\dot{\theta}^2 + J_r\Omega_r\dot{\phi} + u_3\right)$$

$$\ddot{\psi} = \frac{1}{I_z}\left((I_x - I_y)\dot{\phi}\dot{\theta} + \sum_{i=1}^{4}d(-1)^{i+1}\omega_i^2 - C_{a_z}\dot{\psi}^2\right) = \frac{1}{I_z}\left((I_x - I_y)\dot{\phi}\dot{\theta} - C_{a_z}\dot{\psi}^2 + u_4\right)a$$

$$\ddot{x} = \frac{1}{m}\left(u_1 u_x - C_{dx}\dot{x}\right) \tag{22.64}$$

$$\ddot{y} = \frac{1}{m}\left(u_1 u_y - C_{dy}\dot{y}\right) \tag{22.65}$$

$$\ddot{z} = \frac{1}{m}\left(u_1\left(c\phi c\theta\right) - C_{dz}\dot{z} - mg\right) \tag{22.66}$$

$$\ddot{\phi} = \frac{1}{I_x}\left(\left(I_y - I_z\right)\dot{\theta}\dot{\psi} - C_{a_x}\dot{\phi}^2 - J_r\Omega_r\dot{\theta} + u_2\right) \tag{22.67}$$

$$\ddot{\theta} = \frac{1}{I_y}\left(\left(I_z - I_x\right)\dot{\phi}\dot{\psi} - C_{a_y}\dot{\theta}^2 + J_r\Omega_r\dot{\phi} + u_3\right) \tag{22.68}$$

$$\ddot{\psi} = \frac{1}{I_z}\left(\left(I_x - I_y\right)\dot{\phi}\dot{\theta} - C_{a_z}\dot{\psi}^2 + u_4\right) \tag{22.69}$$

We can now rearrange $u_1, u_2, u_3,$ and u_4 in matrix form as follows (where d is the drag coefficient):

$$\begin{bmatrix} u_1 \\ u_2 \\ u_3 \\ u_4 \end{bmatrix} = \begin{bmatrix} b & b & b & b \\ 0 & -lb & 0 & lb \\ -lb & 0 & lb & 0 \\ d & -d & d & -d \end{bmatrix}\begin{bmatrix} \omega_1^2 \\ \omega_2^2 \\ \omega_3^2 \\ \omega_4^2 \end{bmatrix} \tag{22.70}$$

Taking inverse of the matrix of Equation (22.70):

$$\omega_1^2 = \frac{u_1}{4b} - \frac{u_3}{2bl} - \frac{u_4}{4d} \tag{22.71}$$

$$\omega_2^2 = \frac{u_1}{4b} + \frac{u_2}{2bl} - \frac{u_4}{4d} \tag{22.72}$$

$$\omega_3^2 = \frac{u_1}{4b} - \frac{u_3}{2bl} + \frac{u_4}{4d} \tag{22.73}$$

$$\omega_4^2 = \frac{u_1}{4b} - \frac{u_2}{2bl} - \frac{u_4}{4d} \tag{22.74}$$

These $\left(\omega_1^2, \omega_2^2, \omega_3^2, \omega_4^2\right)$ of Equations (22.71)–(22.74) are then used to calculate the current states of the quadcopter. This model is used to develop both the attitude and the trajectory controller along with implementation of the trajectory planner.

Quadrotor Control

Control Theory

A proportional–integral–derivative controller (PID controller or three-term controller) is a control loop mechanism employing feedback (Figure 22.26) that is widely

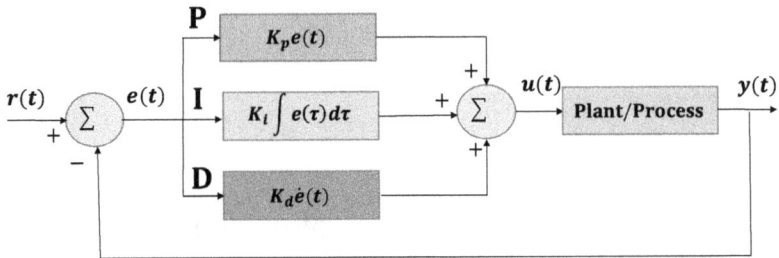

FIGURE 22.26 PID controller.

used in in industrial control systems and a variety of other applications requiring continuously modulated control. A PID controller continuously calculates an error value $e(t)$ as the difference between a desired set point (*SP*) and a measured process variable (*PV*) and applies a correction based on proportional, integral, and derivative terms (denoted by *P*, *I*, and *D* respectively), hence the name. The desired SP:

$$SP = r(t)$$

A measured process variable (*PV*):

$$PV(t) = y(t)$$

The error value:

$$e(t) = r(t) - y(t)$$

The distinguishing feature of the PID controller is the ability to use the three *control terms* of proportional, integral, and derivative influence on the controller output to apply accurate and optimal control.

In practical terms, PID automatically applies an accurate and responsive correction to a control function. An everyday example is the cruise control on a car, where ascending a hill would lower speed if constant engine power were applied. The controller's PID algorithm restores the measured speed to the desired speed with minimal delay and overshoot by increasing the power output of the engine in a controlled manner. The overall control function $u(t)$:

$$u(t) = K_p e(t) + K_i \int_0^t e(\tau) d\tau + K_d \frac{de(t)}{dt}$$

(22.75)

where K_p, K_i, and K_d, all non-negative, denote the coefficients. Let us manipulate these constants as follows to have some physical meaning:

$$K_p \rightarrow \frac{K_p}{T_i}$$

$$K_d \rightarrow K_p T_d$$

Now we will have:

$$u(t) = K_p \left(e(t) + \frac{1}{T_i} \int_0^t e(\tau) d\tau + T_d \frac{de(t)}{dt} \right) \tag{22.76}$$

where,

- $K_p T_d$ is the time constant with which the controller will attempt to approach SP.
- $\dfrac{K_p}{T_i}$ determines how long the controller will tolerate the output being consistently above or below the SP.

QUADROTOR ATTITUDE CONTROL PRELIMINARIES

Equations (22.58)–(22.70) are the equations of motion to be used in our six degree-of-freedom simulators. However, they are not appropriate for control design for several reasons. The first reason is that they are too complicated to gain significant insight into the motion. The second reason is that the position and orientation are relative to the inertial world fixed frame, whereas camera measurements will measure position and orientation of the target with respect to the camera frame.

Controlling vehicle attitude requires sensors to measure vehicle orientation, actuators to apply the torques needed to reorient the vehicle to a desired attitude, and algorithms to command the actuators based on (a) sensor measurements of the current attitude and (b) specification of a desired attitude. Once the attitude control is designed and optimized, it can be integrated with the trajectory controller. The block diagram (Figure 22.27) for attitude controller is as shown below:

A lot of different methods have been studied to achieve autonomous flight, from which three methods (both linear and nonlinear) are discussed below.

Controller Framework

A quadcopter is an under-actuated system, which means that six degrees of freedom in space are controlled by just four motors. Hence, controllers in such vehicles

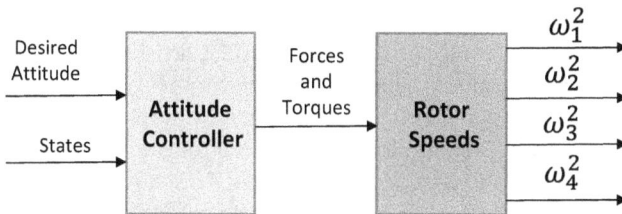

FIGURE 22.27   High-level diagram for attitude controller.

must be designed for a subset of four degrees of freedom. Furthermore, the fact must be considered that the control of x and y positions in space is influenced by changes in the pitch and roll angles. Having considered the relationships, the control of a quadrotor is normally designed for two independent subsets of coordinates. The necessity for a swashplate mechanism is eliminated with four separate rotors.

The swashplate mechanism was necessary to give the helicopter more degrees of freedom, but the same level of control be achieved by simply adding two more rotors, as implemented in the structure of quadcopters.

Even though the command is for three position coordinates (x, y, z) plus yaw angle, the control algorithm employs both roll and pitch orientation controllers. In the inertial coordinate system, the control signals of three position controllers define a force vector (thrust). Note that $u1u_x = \left(\omega_1^2 + \omega_2^2 + \omega_3^2 + \omega_4^2 \right) u_x = u1_x$ and $u_1 u_y = \left(\omega_1^2 + \omega_2^2 + \omega_3^2 + \omega_4^2 \right) u_y$. So, we could have two x- and y-component. The SPs $(u1_x, u1_y)$ transmitted to the roll and pitch controls are considered as the orientation of the vector [2].

The altitude controller and the attitude controller are the two main parts of the control architecture. The altitude controller maintains the altitude of the aerial robot at the required level. In most commercial aerial robots, the altitude controller is a PID controller with **fixed control gain** values. In this research, a proposed control architecture consisting of a model predictive controller (MPC) and a gravity compensator is proposed as a **replacement for conventional PID controllers**. MPC uses a model of the plant to make predictions about future plant outputs. It solves an optimization problem at each time step to find the optimal control action that drives the predicted plant output to the desired reference as close as possible.

The scaling factor of the compensator is adaptively adjusted during the operation of the robot, using actions generated by the reinforcement learning agent. The robot attitude controller is the second major controlling component of the system. The controller is made up of two distinct PID controller blocks.

The difference between the desired x and y position and the actual y and y location in space is measured and defined as the position error in 2D. The x-position and y-position PID controllers in the outer loop were designed to minimize the error. The control commands of the position controllers are transformed to appropriate roll and pitch SPs.

The inner loop PID controllers use the resulting roll and pitch SPs as reference inputs. The control gain values of the inner loop PID controllers are constant in our proposed control architecture, whereas the control gain values of the **outer loop PID controllers are adaptively adjusted using trained policy from the RL-based adaptation algorithm**.

The altitude controller and the attitude controller are the two main parts of the control architecture. The altitude controller, as shown in Figure 22.28, maintains the altitude of the aerial robot at the required level. In most commercial aerial robots, the altitude controller is a PID controller with fixed control gain values. In this research, a proposed control architecture consisting of an MPC controller and a gravity compensator is proposed as a replacement for conventional PID controllers. The scaling factor of the compensator is adaptively adjusted during the operation of the robot,

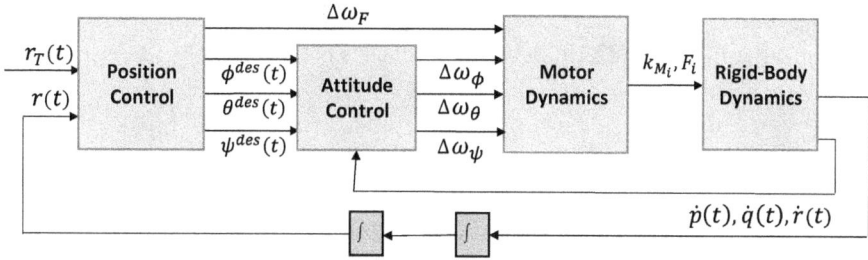

FIGURE 22.28   Control loops for position and attitude control.

using actions generated by the reinforcement learning agent. The robot attitude controller is the second major controlling component of the system. The controller is made up of two distinct PID controller blocks.

The difference between the desired x and y position and the actual x and y location in space is measured and defined as the position error in 2D. The x-position and y-position PID controllers in the **outer loop were designed to minimize the error**. The control commands of the position controllers are transformed to appropriate roll and pitch SPs. The inner loop PID controllers use the resulting roll and pitch SPs as reference inputs.

The control gain values of the **inner loop PID controllers are constant** in our proposed control architecture, whereas the **control gain values of the outer loop PID controllers are adaptively adjusted using trained policy from the RL-based adaptation algorithm**.

Attitude Control

To control the attitude of the robot, an architecture comprising of RL-based adaptive controllers is proposed in this study. The outer loop PID controllers were designed to generate $u1_x$ and $u1_y$ virtual control signals, as described in Equations (22.77)–(22.80).

$$e_x = x_{des} - x \tag{22.77}$$

$$e_y = y_{des} - y \tag{22.78}$$

Using control theory:

$$u1_x = \left(Kp_x \cdot e_x\right) + \left(Kd_x \cdot e_x^{\cdot}\right) + \left(Ki_x \int e_x dt\right) \tag{22.79}$$

$$u1_y = \left(Kp_y \cdot e_y\right) + \left(Kd_y \cdot e_y^{\cdot}\right) + \left(Ki_y \int e_y dt\right) \tag{22.80}$$

Equation (22.29) is used to convert the control commands from outer loop PIDs to the necessary roll and pitch reference values for the inner-loop PID controllers.

Motor Model

We can write Equation (22.19) in body coordinates noting that $m^b \triangleq \left(\tau_\phi, \tau_\theta, \tau_\psi \right)^T$:

$$\frac{d}{dt_b} h^b + \boldsymbol{\omega}_{b/i} \times h^b = m^b$$

$$\frac{d}{dt_b} \left(\mathbf{J} \underset{i}{\boldsymbol{\omega}_b^b} \right) + \boldsymbol{\omega}_{b/i} \times \left(\mathbf{J} \underset{i}{\boldsymbol{\omega}_b^b} \right) = \begin{bmatrix} \tau_\phi \\ \tau_\theta \\ \tau_\psi \end{bmatrix}$$

$$\underset{i}{\dot{\boldsymbol{\omega}}_b} = \frac{1}{\mathbf{J}} \left(-\boldsymbol{\omega}_{b/i} \times \mathbf{J} \boldsymbol{\omega}_{b/i} + \begin{bmatrix} \tau_\phi \\ \tau_\theta \\ \tau_\psi \end{bmatrix} \right)$$

Since

$$\underset{i}{\dot{\boldsymbol{\omega}}_b} \triangleq \left(p, q, r \right)^T$$

We have:

$$\begin{bmatrix} \dot{p} \\ \dot{q} \\ \dot{r} \end{bmatrix} = \frac{1}{\mathbf{J}} \left(-\begin{bmatrix} p \\ q \\ r \end{bmatrix} \times \mathbf{J} \begin{bmatrix} p \\ q \\ r \end{bmatrix} + \begin{bmatrix} \tau_\phi \\ \tau_\theta \\ \tau_\psi \end{bmatrix} \right)$$

Similar to Equations (22.43)–(22.45) and considering $\mathbf{J} = \mathbf{I}$,

$$\begin{bmatrix} \tau_\phi \\ \tau_\theta \\ \tau_\psi \end{bmatrix} = \begin{bmatrix} lb\left(\omega_4^2 - \omega_2^2 \right) \\ lb\left(\omega_3^2 - \omega_1^2 \right) \\ \sum_{i=1}^{4} d(-1)^{i+1} \omega_i^2 \end{bmatrix} \tag{22.81}$$

$$\begin{bmatrix} \dot{p} \\ \dot{q} \\ \dot{r} \end{bmatrix} = \frac{1}{\mathbf{I}} \left(-\begin{bmatrix} p \\ q \\ r \end{bmatrix} \times \mathbf{I} \begin{bmatrix} p \\ q \\ r \end{bmatrix} + \begin{bmatrix} lb\left(\omega_4^2 - \omega_2^2 \right) \\ lb\left(\omega_3^2 - \omega_1^2 \right) \\ \sum_{i=1}^{4} d(-1)^{i+1} \omega_i^2 \end{bmatrix} \right) \tag{22.82}$$

Each rotor has an angular speed ω_i and produces a vertical force F_i according to:

$$F_i = b\omega_i^2 \tag{22.83}$$

Experimentation with a fixed rotor at steady state shows that $b \approx 6.113 \times 10^{-8}$ $N/(r = \min^2)$. The rotors also produce a moment according to:

$$\tau_{M_i} = d\omega_i^2 \tag{22.84}$$

The constant, d, is determined to be about $d \approx 1.5 \times 10^{-9} N/(r = \min^2)$ by matching the performance of the simulation to the real system. The results of a system identification exercise suggest that the rotor speed is related to the commanded speed by a first-order differential equation:

$$\dot{\omega}_i = k_m \left(\omega_i^{des} - \omega_i \right)$$

This motor gain, k_m, is found to be about $20\,\mathrm{s}^{-1}$ by matching the performance of the simulation to the real system. The desired angular velocities, ω_i^{des}, are limited to a minimum and a maximum value determined through experimentation to be approximately 1,200 and 7,800 r/min.

Quadrotor Robot Controllers

Each robot (motor) is controlled independently by nested feedback loops, as shown in Figure 22.28. The inner attitude control loop uses onboard accelerometers and gyros to control the roll, pitch, and yaw angles and runs at approximately 1 kHz, while the outer position control loop uses the estimates of position and velocity of the center of mass to control the trajectory in three dimensions.

Similar control loops are presented in previous works. Our controllers are derived by linearizing the equations of motion and motor models (22.81)–(22.84) at an operating point that corresponds to the nominal hover state. Considering ϕ, θ, ψ angles are small and the Earth's surface is flat, we can linearize Equation (22.34) as follows: At hoover state, $R = R_0$ (see Figue 22.25), $\phi = \theta = 0, \psi = \psi_0$, and, $\dot{\phi} = \dot{\theta} = \dot{\psi} = 0$, where the roll and pitch angles are small $(c\phi = \cos\phi \approx 1, c\theta = \cos\theta \approx 1, s\phi = \sin\phi \approx \phi$, and $s\theta = \sin\theta \approx \theta)$. We can assume when angles are not small that $\dot{\phi} \approx \omega_x, \dot{\theta} \approx \omega_y, \dot{\psi} \approx \omega_z$ is valid. At this hover state, the nominal thrusts from the propellers must satisfy:

$$F_{i,0} = \frac{mg}{4}$$

The motor speeds (at hover state) are given by:

$$\omega_{i,0} = \omega_h = \sqrt{\frac{mg}{4b}}$$

Trajectory Tracking

We now present an attitude controller to track trajectories in SO(3) that are close to the nominal hover state where the roll and pitch angles are small. Note that the group SO(3) is used to describe the possible rotational symmetries of an object, as well as the possible orientations of an object in space. From Equation (22.82), substituting in the relationships between angular velocities of the rotors and forces and moments (22.83) and (22.84),

$$
\begin{bmatrix} \dot{p} \\ \dot{q} \\ \dot{r} \end{bmatrix} = \frac{1}{\mathbf{I}} \left(-\begin{bmatrix} p \\ q \\ r \end{bmatrix} \times \mathbf{I} \begin{bmatrix} p \\ q \\ r \end{bmatrix} + \begin{bmatrix} lb\left(\omega_4^2 - \omega_2^2\right) \\ lb\left(\omega_3^2 - \omega_1^2\right) \\ \sum_{i=1}^{4} d(-1)^{i+1}\omega_i^2 \end{bmatrix} \right)
$$

Using Equation (22.21), and putting $I = J$:

$$
\begin{bmatrix} \dot{p} \\ \dot{q} \\ \dot{r} \end{bmatrix} = \begin{bmatrix} \dfrac{I_y - I_z}{I_x} qr \\ \dfrac{I_z - I_x}{I_y} pr \\ \dfrac{I_x - I_y}{I_z} pq \end{bmatrix} + \begin{bmatrix} \dfrac{1}{I_x} lb\left(\omega_4^2 - \omega_2^2\right) \\ \dfrac{1}{I_y} lb\left(\omega_3^2 - \omega_1^2\right) \\ \dfrac{1}{I_z} \sum_{i=1}^{4} d(-1)^{i+1}\omega_i^2 \end{bmatrix}
\tag{22.85}
$$

$$
\begin{bmatrix} I_x\dot{p} \\ I_y\dot{q} \\ I_z\dot{r} \end{bmatrix} = \begin{bmatrix} qr(I_y - I_z) \\ pr(I_z - I_x) \\ pq(I_x - I_y) \end{bmatrix} + \begin{bmatrix} lb\left(\omega_4^2 - \omega_2^2\right) \\ lb\left(\omega_3^2 - \omega_1^2\right) \\ \sum_{i=1}^{4} d(-1)^{i+1}\omega_i^2 \end{bmatrix}
\tag{22.86}
$$

Note that the products of inertia are small (ideally, they are zero because the axes are close to the principal axes) and $I_x \approx I_y$ because of the symmetry. So, we have from Equation (22.86):

$$
\begin{bmatrix} I_x\dot{p} \\ I_y\dot{q} \\ I_z\dot{r} \end{bmatrix} = \begin{bmatrix} qr(I_y - I_z) \\ pr(I_z - I_x) \\ 0 \end{bmatrix} + \begin{bmatrix} lb\left(\omega_4^2 - \omega_2^2\right) \\ lb\left(\omega_3^2 - \omega_1^2\right) \\ \sum_{i=1}^{4} d(-1)^{i+1}\omega_i^2 \end{bmatrix}
\tag{22.87}
$$

We assume the component of the angular velocity in the z_B direction (see Figure 22.25), r is small so that the leftmost terms in Equation (22.87), which are products involving r, are small compared to the other terms. Again, we can further reduce Equation (22.87) as follows:

$$
\begin{bmatrix} I_x \dot{p} \\ I_y \dot{q} \\ I_z \dot{r} \end{bmatrix} \approx \begin{bmatrix} lb\left(\omega_4^{\,2} - \omega_2^{\,2}\right) \\ lb\left(\omega_3^{\,2} - \omega_1^{\,2}\right) \\ \displaystyle\sum_{i=1}^{4} d(-1)^{i+1}\,\omega_i^2 \end{bmatrix} \tag{22.88}
$$

The vector of desired rotor speeds can be written as a linear combination of four terms,

$$
\begin{aligned}
\omega_1^{des} &= \omega_h + \Delta\omega_F - \Delta\omega_\phi + \Delta\omega_\psi \\
\omega_2^{des} &= \omega_h + \Delta\omega_F + \Delta\omega_\phi - \Delta\omega_\psi \\
\omega_3^{des} &= \omega_h + \Delta\omega_F + \Delta\omega_\theta + \Delta\omega_\psi a \\
\omega_4^{des} &= \omega_h + \Delta\omega_F - \Delta\omega_\phi - \Delta\omega_\psi a
\end{aligned} \tag{22.89}
$$

$$
\begin{pmatrix} \omega_1^{des} \\ \omega_2^{des} \\ \omega_3^{des} \\ \omega_4^{des} \end{pmatrix} = \begin{bmatrix} 1 & 0 & -1 & 1 \\ 1 & 1 & 0 & -1 \\ 1 & 0 & 1 & 1 \\ 1 & -1 & 0 & -1 \end{bmatrix} \begin{pmatrix} \omega_h + \Delta\omega_F \\ \Delta\omega_\phi \\ \Delta\omega_\theta \\ \Delta\omega_\psi \end{pmatrix}
$$

where the nominal rotor speed required to hover in steady state is ω_h, and the deviations from this nominal vector are $\Delta\omega_F$, $\Delta\omega_\phi$, $\Delta\omega_\theta$, and $\Delta\omega_\psi$. $\Delta\omega_F$ **results in a net force along the z_B axis**, while $\Delta\omega_\phi$, $\Delta\omega_\theta$, and $\Delta\omega_\psi$ produce moments causing roll, pitch, and yaw, respectively.

Now, we linearize Equation (22.88) about the hovering operating point and write the desired angular accelerations in terms of the new control inputs:

$$
\dot{p}^{des} = \frac{4 l b \omega_h}{I_x} \Delta\omega_\phi
$$

$$
\dot{q}^{des} = \frac{4 l b \omega_h}{I_y} \Delta\omega_\theta a
$$

$$
\dot{r}^{des} = \frac{8 d \omega_h}{I_z} \Delta\omega_\psi
$$

As near the nominal hover state, $\phi \approx p$, $\theta \approx \dot{q}$, and $\dot{\psi} \approx r$, we use proportional-derivative control laws that take the form:

$$
\Delta\omega_\phi = k_{p,\phi}\left(\phi^{des} - \phi\right) + k_{d,\phi}\left(p^{des} - p\right) \tag{22.90}
$$

$$
\Delta\omega_\theta = k_{p,\theta}\left(\theta^{des} - \theta\right) + k_{d,\theta}\left(q^{des} - q\right) \tag{22.91}
$$

$$
\Delta\omega_\psi = k_{p,\psi}\left(\psi^{des} - \psi\right) + k_{d,\psi}\left(r^{des} - r\right) \tag{22.92}
$$

Substituting Equations (22.90)–(22.92) in Equation (22.89) yields the desired rotor speeds.

Position Control

Here, we present the two representative position control methods that use the roll and pitch angles as inputs via a method similar to a back-stepping approach The first, a hover controller, is used for station keeping or maintaining the position at a desired $x, y,$ and z location. The second tracks a trajectory in three dimensions.

Hover Controller

We use pitch and roll angles to control position in x_B and y_B planes, $\Delta\omega_\psi$ to control yaw angle, and $\Delta\omega_F$ to control position along z_W. We let $r_T(t)$ and $\psi_T(t)$ be the trajectory and yaw angles we are trying to track. Note that $\psi_T(t) = \psi_0$ for the hover controller. The command accelerations, $\ddot{r}_i^{des}$, are calculated from proportional-integral differential feedback of the position error, $e_i = (r_{i,T} - r_i)$ as:

$$\left(\ddot{r}_{i,T} - \ddot{r}_i^{des}\right) + k_{d,i}\left(\dot{r}_{i,T} - \dot{r}_r\right) + k_{p,i}\left(r_{i,T} - r_i\right) + k_{i,i}\int\left(r_{i,T} - r_i\right)dt = 0$$

where $\dot{r}_{i,T} = \ddot{r}_{i,T} = 0$ for hover condition.

Newton's second aw, mass × acceleration= force:

$$m\ddot{r} = \begin{bmatrix} 0 \\ 0 \\ mg \end{bmatrix} + R\begin{bmatrix} 0 \\ 0 \\ \Sigma F_i \end{bmatrix} \tag{22.93}$$

We linearize the above equation where roll and pitch angles are small $(c\phi = \cos\phi \approx 1,$ $c\theta = \cos\theta \approx 1, s\phi = \sin\phi \approx \phi,$ and $s\theta = \sin\theta \approx \theta)$:

$$\ddot{r}_1^{des} = g\left(\theta^{des}\cos\psi_T + \phi^{des}\sin\psi_T\right) \tag{22.94}$$

$$\ddot{r}_2^{des} = g\left(\theta^{des}\sin\psi_T - \phi^{des}\cos\psi_T\right) \tag{22.95}$$

$$\ddot{r}_3^{des} = \frac{8d\omega_h}{m}\Delta\omega_F \tag{22.96}$$

These relationships are inverted to compute the desired roll and pitch angles for the attitude controller, from the desired accelerations, as well as $\Delta\omega_F$.

$$\phi^{des} = \frac{1}{g}\left(\ddot{r}_1^{des}\sin\psi_T - \ddot{r}_2^{des}\cos\psi_T\right) \tag{22.97}$$

$$\theta^{des} = \frac{1}{g}\left(\ddot{r}_1^{des}\cos\psi_T + \ddot{r}_2^{des}\sin\psi_T\right) \tag{22.98}$$

$$\Delta\omega_F = \frac{m}{8d\omega_h}\ddot{r}_3^{des} \tag{22.99}$$

The position control loop for the hover controller runs at 100 Hz, while the inner attitude control loop runs at 1 kHz. There is the usual trade-off in optimizing the control gains between speed of response and stability. Experimental results show (see the representative trial in Figure 22.4(a) and (b)) for a tightly optimized stiff controller the horizontal positioning errors that are within 2 cm, and the error in the vertical direction is always less than 0.6 cm. However, this set of gains leads to a relatively small basin of attraction. By optimizing the gains for a softer response, we can increase the size of this basin of attraction. We can experimentally characterize this basin by perturbing the quadrotor from the hover state and measuring the response of the hover controller. We found the robot to be quite robust if we used a softer controller, allowing it to recover from disturbances as large as 1:5 m (three body lengths) in the horizontal direction and 2:0 m (four body lengths) in the vertical direction, pitch or roll angle errors of 60°, and velocity errors of up to 3:0 m = s.

Let us consider: Newton's second law – mass × acceleration = force:

$$\ddot{r} = \begin{bmatrix} \ddot{x} \\ \ddot{y} \\ \ddot{z} \end{bmatrix}$$

$$m\ddot{r} = \begin{bmatrix} 0 \\ 0 \\ mg \end{bmatrix} + \begin{bmatrix} c\theta c\psi & s\phi s\theta c\psi - c\phi s\psi & c\phi s\theta c\psi + s\phi s\psi \\ c\theta s\psi & s\phi s\theta s\psi + c\phi c\psi & c\phi s\theta s\psi - s\phi c\psi \\ -s\theta & s\phi c\theta & c\phi c\theta \end{bmatrix} \begin{bmatrix} 0 \\ 0 \\ \sum F_i \end{bmatrix}$$

$$= \begin{bmatrix} 0 \\ 0 \\ mg \end{bmatrix} + \sum F_i \begin{bmatrix} c\phi s\theta c\psi + s\phi s\psi \\ c\phi s\theta s\psi - s\phi c\psi \\ c\phi c\theta \end{bmatrix}$$

We linearize the above equation where roll and pitch angles are small $\left(c\phi = \cos\phi \approx 1, c\theta = \cos\theta \approx 1, s\phi = \sin\phi \approx \phi, \text{and } s\theta = \sin\theta \approx \theta \right)$:

$$m\ddot{r} = \begin{bmatrix} 0 \\ 0 \\ mg \end{bmatrix} + \sum F_i \begin{bmatrix} \theta c\psi + \phi s\psi \\ \theta s\psi - \phi c\psi \\ 1 \end{bmatrix}$$

$$m\ddot{x} = \sum F_i \left(\theta c\psi + \phi s\psi \right)$$

$$m\ddot{y} = \sum F_i \left(\theta s\psi - \phi c\psi \right)$$

$$m\ddot{z} = mg + \sum_i F_i$$

$$\alpha \triangleq \sum F_i$$

$$m\ddot{x} = \alpha \left(\theta c\psi + \phi s\psi \right)$$

$$m\ddot{y} = \alpha \left(\theta s\psi - \phi c\psi \right)$$

$$\begin{bmatrix} \dfrac{m}{\alpha} \ddot{x} \\ \dfrac{m}{\alpha} \ddot{y} \end{bmatrix} = \begin{bmatrix} c\psi & s\psi \\ s\psi & -c\psi \end{bmatrix} \begin{bmatrix} \theta \\ \phi \end{bmatrix}$$

$$\begin{bmatrix} \theta \\ \phi \end{bmatrix} = \dfrac{\dfrac{m}{\alpha}\begin{bmatrix} \ddot{x} \\ \ddot{y} \end{bmatrix}}{\begin{bmatrix} c\psi & s\psi \\ s\psi & -c\psi \end{bmatrix}} = \begin{bmatrix} c\psi & s\psi \\ s\psi & -c\psi \end{bmatrix}^{-1} \left(\dfrac{m}{\alpha}\begin{bmatrix} \ddot{x} \\ \ddot{y} \end{bmatrix} \right)$$

Note:
Well, for a 2×2 matrix the inverse is:

$$\begin{bmatrix} a & b \\ c & d \end{bmatrix}^{-1} = \frac{1}{ad - bc}\begin{bmatrix} d & -b \\ -c & a \end{bmatrix}$$

In other words: **swap** the positions of a and d, put **negatives** in front of b and c, and **divide** everything by $ad - bc$.

$$\begin{bmatrix} c\psi & s\psi \\ s\psi & -c\psi \end{bmatrix}^{-1} = \frac{1}{-c\psi c\psi - s\psi s\psi}\begin{bmatrix} -c\psi & -s\psi \\ -s\psi & c\psi \end{bmatrix} = \frac{1}{c\psi^2 + s\psi^2}\begin{bmatrix} c\psi & s\psi \\ s\psi & -c\psi \end{bmatrix}$$

$$\begin{bmatrix} \theta \\ \phi \end{bmatrix} = \frac{m}{\alpha}\left(\frac{1}{c\psi^2 + s\psi^2}\right)\begin{bmatrix} c\psi & s\psi \\ s\psi & -c\psi \end{bmatrix}\begin{bmatrix} \ddot{x} \\ \ddot{y} \end{bmatrix} = \frac{m}{\alpha}\begin{bmatrix} c\psi & s\psi \\ s\psi & -c\psi \end{bmatrix}\begin{bmatrix} \ddot{x} \\ \ddot{y} \end{bmatrix}$$

$$c\psi^2 + s\psi^2 = 1$$

- **Conditions for hovering**

$$mg = F_1 + F_2 + F_3 + F_4$$

All moments $\neq 0$

$$\frac{m}{\alpha} = \frac{m}{\Sigma F_i} = \frac{m}{mg} = \frac{1}{g}$$

$$\begin{bmatrix} \theta \\ \phi \end{bmatrix} = \frac{1}{g}\begin{bmatrix} c\psi & s\psi \\ s\psi & -c\psi \end{bmatrix}\begin{bmatrix} \ddot{x} \\ \ddot{y} \end{bmatrix}$$

$\ddot{x} =$ desired acceleration toward $x - $ axis $\equiv u1_x$

$\ddot{y} =$ desired acceleration toward $y - $ axis $\equiv u1_y$

Therefore, our desired SP for acceleration is $(u1_x, u1_y)$
Now we can write,

$$\begin{bmatrix} \theta_d \\ \phi_d \end{bmatrix} = \frac{1}{g}\begin{bmatrix} \cos\psi_{des} & \sin\psi_{des} \\ \sin\psi_{des} & -\cos\psi_{des} \end{bmatrix}\begin{bmatrix} u1_x \\ u1_y \end{bmatrix} \qquad (22.100)$$

As indicated in Equations (22.101)–(22.103), three PID controllers were implemented in the inner-loop PID control block to provide manipulated variables for robot attitude control.

$$u_2 = \left(Kp_2\left(\phi_d - \phi\right)\right) + \left(Kd_2\left(\dot{\phi}_d - \dot{\phi}\right)\right) + \left(Ki_2\int\left(\phi_d - \phi\right)dt\right) \qquad (22.101)$$

$$u_3 = \left(Kp_3\left(\theta_d - \theta\right)\right) + \left(Kd_3\left(\dot{\theta}_d - \dot{\theta}\right)\right) + \left(Ki_3\int\left(\theta_d - \theta\right)dt\right) \qquad (22.102)$$

$$u_4 = \left(Kp_4\left(\psi_d - \psi\right)\right) + \left(Kd_4\left(\dot{\psi}_d - \dot{\psi}\right)\right) + \left(Ki_4\int\left(\psi_d - \psi\right)dt\right) \qquad (22.103)$$

TABLE 22.1

Control Gain Values of the Inner-Loop PID Controllers

PID Control	K_p	K_i	K_d
Roll (ϕ)	0.021	0.001	0.003
Pitch (θ)	0.014	0.030	0.001
Yaw (ψ)	0.002	0.070	0.013

The optimal control gains for the inner loop PID controllers were obtained, based on several trials and errors. The obtained gain coefficients are listed in Table 22.1. The control gains of outer-loop PID controllers are actively estimated and adjusted by the RL agent.

Altitude Control

In this study, the proposed altitude controller utilizes a linear model predictive controller and a gravity compensator in its controlling architecture. The gravity compensator is responsible for alleviating the effect of forces that arise from fluctuations in the weight of the robot. **The scaling factor of the compensator is adaptively updated using the RL policy.** The proposed algorithm aimed to mitigate the impact of disturbances arising from changes in the weight of the robot on the performance of the aerial robot in trajectory tracking and altitude stabilization.

Model predictive control is an advanced method of process control that is used to control a process while satisfying a set of constraints. The main advantage of MPC is the fact that it allows the current timeslot to be optimized, while keeping future timeslots in account. This is achieved by optimizing a finite time-horizon, but only implementing the current timeslot and then optimizing again, repeatedly, thus differing from a linear–quadratic regulator (LQR). Also, MPC has the ability to anticipate future events and can take control actions accordingly. PID controllers do not have this predictive ability. MPC is nearly universally implemented as a digital control, although there is research into achieving faster response times with specially designed analog circuitry.

The MPC is based on an iterative, finite-horizon robot model optimization. The present states of the quadcopter are sampled at time t, and a cost-minimizing control strategy for a relatively short time horizon in the future $\left[t, t+T\right]$ is computed (using a numerical minimization technique). At each control interval, model predictive control solves an optimization problem, a quadratic program (QP). Until the next control interval, the solution generated a sequence of manipulated variables to be applied to the robot. A series of online optimizations are run to estimate possible state trajectories that would arise from the present states.

Furthermore, the solution identifies a cost-minimizing control strategy (by the solution of Euler–Lagrange equations) from at t until time ran out at $t+T$. Although the MPC computes a series of manipulated variables, only the first step of the computed control strategy is applied to the quadcopter, after which the updated states of the robot are sampled again and the computations are repeated using the updated states, resulting in computation of fresh control inputs and a new anticipated state route. Model predictive control is considered as a multivariable controlling algorithm incorporating the following components:

- A dynamics model of the system under control

A cost function J.

- An optimization mechanism
  - The optimal manipulated variable (uMPC) is computed by minimizing the cost function J using the optimization algorithm

A typical cost function in the **MPC algorithm is made up of four terms**, each of which focuses on a different element of controller performance (k represents the current control interval):

$$J\left(d_k\right) = J_{RT}\left(d_k\right) + J_u\left(d_k\right) + J_{\Delta u}\left(d_k\right) \tag{22.104}$$

$$d_k = \left[u_{MPC}(k \mid k)u_{MPC}(k+1 \mid k)...u_{MPC}(k+p-1 \mid k)\right] \tag{22.105}$$

where,

- d_k signifies optimal control inputs that are obtained by solving a quadratic programming (QP) problem, as indicated in Equation (22.104).
- In Equation (22.104), $J_{RT}\left(d_k\right)$ is reference tracking cost.
  - Here, $J_u\left(d_k\right)$ and $J_{\Delta u}\left(d_k\right)$ are representations of manipulated variable tracking and manipulated variable move suppression, respectively.
- The last cost term in Equation (22.104), $J_{\Delta u}\left(d_k\right)$, decreases the control effort, thereby reducing the energy consumption of the actuators (e.g., the dc motors of the quadrotor).

The MPC cost function can be formulated as follows:

$$J\left(d_k\right) = \sum_{i=0}^{p-1}\left[e_{RT}^T\left(k+i\right)Q_{eRT}\left(k+i\right)\right] + \left[e_u^T\left(k+i\right)Re_u\left(k+i\right)\right] + \Delta u^T\left(k+i\right)R_{\Delta u}\left(k+i\right)$$

$$\tag{22.106}$$

In Equation (22.35),

- Q is a $\left(n_{RT} \times n_{RT}\right)$ weight matrix (n_{RT} is the number of plant output variables).
- Here, R and $R_{\Delta u}$ $\left(n_u \times n_u\right)$ are **positive-semi-definite weight matrices** (n_u represents the number of manipulated variables).
- In the cost function, p is the prediction horizon (Note: How far ahead the model predicts the future. When the prediction horizon is well matched to the lag between input and output, the user learns how to control the system more rapidly, and achieves better performance), which can be adjusted according to the controller performance and the processing power of the hardware.

In Equation (22.106), e_{RT}, e_u, and Δu can be computed, using Equations (22.107)–(22.109).

$$e_{RT}\left(k+i\right) = r(k+i+1\,|\,k) - y(k+i+1\,|\,k) \tag{22.107}$$

$$e_u\left(k+i\right) = u_{des}(k+i+1\,|\,k) - u_{MPC}(k+i\,|\,k) \tag{22.108}$$

$$\Delta u\left(k+i\right) = u_{MPC}(k+i\,|\,k) - u_{MPC}(k+i-1\,|\,k) \tag{22.109}$$

The **reference input value** (or reference values) given to the controller at the i-th prediction horizon step is specified as $r(k+i\,|\,k)$. Similarly, the value (or values) of n_{RT} outputs variables of the plant, sampled at the ith prediction horizon step, is defined as $y(k+i\,|\,k)$.

In the equation, $u_{des}(k+i\,|\,k)$ reflects the value (or values) of n_u **desired control inputs** corresponding to $u_{MPC}(k+i\,|\,k)$. In the proposed MPC control architecture, in this chapter, there is one manipulated variable $u_{MPC}(k+i\,|\,k)$.

In addition, in this chapter, $y(k+i\,|\,k)$ **is the altitude (z position) of the robot**, while $r(k+i\,|\,k)$ is the **desired altitude for the robot**. In order to reduce the controller effort and alleviate the effects of arising fluctuations in the weight of the aerial robot, this study proposed to use a gravity compensator after the MPC controller. The **final altitude control input** u_1 is defined as:

$$u_1 = u_{MPC}(k\,|\,k) - g\left(\text{nominal mass of the robot} + a_1\right) \tag{22.110}$$

In Equation (22.110), a_1 **is the scaling factor of the compensator that is actively estimated by the RL policy**. In order to evaluate the performance of the proposed control architecture in controlling an aerial robot, a dynamics model of a commercial aerial robot named **Parrot** was chosen and implemented in the simulator environment as the aerial robot under control.

This research study employs a linear model predictive control. Hence, a linear state-space model of the quadcopter is required. Equations (22.111) and (22.112) describe the standard form of a linear time-invariant (LTI) state-space model, which has p inputs, q outputs, and n state variables.

$$\dot{x} = Ax + Bu \tag{22.111}$$

$$y = Cx + Du \tag{22.112}$$

where $x, y,$ and u are the state vector, the output vector, and the input vector, respectively. In the equation, $A, B,$ and C are $(n \times n)$ the state matrix; $(n \times p)$ the input matrix; and $(q \times n)$ the output matrix, respectively. It should be noted that D is a $(q \times p)$ feedforward matrix. The D matrix is zero in this study, as there is no direct feedthrough. The linear model was computed around the operational point in the state-space model using the MPC Designer app in the MPC toolbox. The toolbox included in the MATLAB software provides control blocks for developing not only linear model predictive control but also nonlinear and adaptive model predictive controllers. In addition, the MPC Designer app included in the software package facilitates the design of an LMPC by automating the processes for plant linearization and tuning the controller parameters. The state-space parameters of the linearized plant can be found in the MPC object (MPCobj) generated by the MPC Designer app. The Simulink MPC Designer linearizes each block in the model independently, then combines the outputs of the individual linearized models to produce the linearized model of the whole plant. Deep Reinforcement Learning for Parameter Estimation and Tuning

The proposed parameter estimation approach in the study utilizes a DRL algorithm to actively estimate and adjust the parameters both in PID controllers and the compensator. A reinforcement learning agent was developed in the Simulink environment to construct an adaption topology for actively estimating the tuning parameters of the designed controllers. In the first stage, the reinforcement learning algorithm interacted with the dynamics model of the robot (in the simulator environment) to learn the appropriate tuning rules. During the operation time of the robot, the trained policy is used to actively adjust the gain values of the controllers. In this study, the RL algorithm is the Deep Deterministic Policy Gradient MATLAB Simulations. The details of using RL for solving the quadrotor attitude and altitude control described in [2] and MATLAB simulations results are not included here for the sake of brevity.

SUMMARY

We have described unmanned robot attitude and altitude control in details: Quadrotor drone, reference frames, rotational matrices, rotation formula, Euler angles, Euler angles rotation, coordinate frames, equation of Coriolis, rotor kinematics and dynamics, forces, and momentum. A new deep reinforcement learning-based adaptive controller for controlling an aerial robot has been described. To interact with the robot dynamics model and learn the right policy for actively adjusting the controller, the proposed adaptive control method leveraged a deep deterministic policy gradient algorithm. A linear model predictive controller and an adaptive gravity compensator gain have been used in the control system for the robot altitude controller. The performance of the proposed control architecture has been compared to that of traditional PID controllers with fixed settings in the Simulink environment (results are not shown for the sake of brevity). Experiments in a simulated environment demonstrated that the presented control algorithm outperforms ordinary PID controllers in terms of trajectory tracking and altitude control.

REFERENCES

1. Beard, R.W., "Quadrotor Dynamics and Control," Brigham Young University, February 19, 2008.
2. Ali Barzegar, A. and Lee, D. J., "Deep Reinforcement Learning-Based Adaptive Controller for Trajectory Tracking and Altitude Control of an Aerial Robot," *Applied Science* May 2022, 2022(12), 4764. https://doi.org/10.3390/app12094764; https://www.mdpi.com/journal/applsci

23 Unmanned Robot Network with Swarm Capabilities

OVERVIEW

In swarm robotics, multiple robots [1] collectively solve problems by forming advantageous structures and behaviors like the ones observed in natural systems, such as swarms of bees, birds, or fish. However, the step to industrial applications has not yet been made successfully. The real-world swarm applications that apply actual swarm algorithms solving all the problems related to complex swarm behaviors are described here. Swarm robotics behaviors can be categorized into spatial organization, navigation, decision-making, and miscellaneous. One of the key functions of swarm robotics is whether the control is centralized or distributed. Swarm robotics architecture is another area that needs to be explored based on the types of real-world applications that may span from terrestrial (e.g., unmanned ground vehicles (UGVs), to aerial (e.g., unmanned aerial vehicles (UAVs), to aquatic (e.g., unmanned underwater vehicles (UUVs), to outer space, for example, unmanned space vehicles (USVs).

SWARM ROBOTICS BEHAVIOR

Swarms typically consist of many individuals, simple, and homogeneous or heterogeneous agents. They traditionally cooperate without any central control and act according to simple and local behavior. Only through their interactions a collective behavior emerges that can solve complex tasks. These characteristics lead to the main advantages of swarms: adaptability, robustness, and scalability. Swarms can be considered as a kind of quasi-organism that can adapt to changes in the environment by following specific behaviors, e.g.:

- Pursuing a specific goal
- Aggregating or dispersing in the environment
- Communicating (direct, indirect)
- Memorizing (local states, morphologies)

In swarm robotics, multiple robots – homogeneous or heterogeneous – are interconnected, forming a swarm of robots. Since individual robots have processing, communication, and sensing capabilities locally on-board, they are able to interact with each other and react to the environment autonomously. Despite the large number of swarm algorithms, the step to industrial applications has not been mastered successfully,

DOI: 10.1201/9781003499480-23

TABLE 23.1
Swarm Behavior Types

Swarm behavior	Spatially organization behaviors	Aggression
		Pattern formation
		Chain formation
		Object clustering and assembling
		Self-assembling and morphologies
	Navigation behaviors	Collective exploration
		Coordinated motion
		Collective transport
		Collective localization
	Collective decision-making	Consensus achievement
		Task allocation
		Collective fault detection
		Collective perception
		Synchronization
		Group size regulation
	Other behaviors	Collision avoidance
		Self-healing
		Self-reproduction
		Human-swarm interaction

yet. In our research work on real-world applications, we noticed that oftentimes industry applications use the term "swarm," but typically do not implement swarm algorithms. They rather use parts of swarm algorithms and implement them using centralized control. The taxonomy of swarm behaviors is given in Table 23.1.

In most swarm algorithms, individuals perform according to local rules, and the overall behavior emerges organically from the interplay of the individuals of the swarm. Translated to the swarm robotics domain, individual robots exhibit a behavior that is based on a local rule set which can range from a simple reactive mapping between sensor inputs and actuator outputs to elaborate local algorithms. Typically, these local behaviors incorporate interactions with the physical world, including the environment and other robots.

Each interaction consists of reading and interpreting the sensory data, processing this data, and driving the actuators accordingly. Such a sequence of interactions is defined as basic behavior that is repeatedly executed, either indefinitely or until a desired state is reached. In the following subsections, we classify and list the basic swarm behaviors which are adapted and expanded with additional swarm robotic behaviors, including collective localization, collective perception, synchronization, self-healing, and self-reproduction. The behaviors are explained from a high-level view describing the task of individual robots and the resulting global objective achieved by the swarm. We do not detail the sensing and actuation part which is specific to each robotic platform.

Taxonomy

The taxonomy of swarm behaviors is given in Table 23.1. It is based on the classification which are extended by several categories [1]. We first give an overview of the taxonomy, and the additional behavior categories are provided. For these behaviors, we also give the basic principles.

Spatial Organization

These behaviors allow the movement of the robots in a swarm in the environment in order to spatially organize themselves or objects:

- **Aggregation** moves the individual robots to congregate spatially in a specific region of the environment. This allows individuals of the swarm to get spatially close to each other for further interaction.
- **Pattern formation** arranges the swarm of robots in a specific shape. A special case is chain formation where robots form a line, typically to establish multihop communication between two points.
- **Self-assembly** connects the robots to establish structures. They can either be connected physically or virtually through communication links. A special case is morphogenesis where the swarm evolves into a predefined shape.
- **Object clustering and assembly** lets the swarm of robots manipulate spatially distributed objects. Clustering and assembling objects are essential for construction processes.

Navigation

These behaviors allow the coordinated movement of a swarm of robots in the environment:

- **Collective exploration** navigates the swarm of robots cooperatively through the environment to explore it. It can be used to get a situational overview, search for objects, monitor the environment, or establish a communication network.
- **Coordinated motion** moves the swarm of robots in a formation. The formation can have a well-defined shape, e.g. a line, or be arbitrary as in flocking.
- **Collective transport** by the swarm of robots enables to collectively move objects which are too heavy or too large for individual robots.
- **Collective localization** allows the robots in the swarm to find their position and orientation relative to each other via establishment of a local coordinate system throughout the swarm.

Decision-Making

These behaviors allow the robots in a swarm to take a common choice on a given issue:

- **Consensus** allows the individual robots in the swarm to agree on or converge toward a single common choice from several alternatives.
- **Task allocation** assigns arising tasks dynamically to the individual robots of the swarm. Its goal is to maximize the performance of the entire swarm system. If the robots have heterogeneous capabilities, the tasks can be distributed accordingly to further increase the system's performance.

- **Collective fault detection** within the swarm of robots determines deficiencies of individual robots. It allows to determine robots that deviate from the desired behavior of the swarm, e.g. due to hardware failures.
- **Collective perception** combines the data locally sensed by the robots in the swarm into a big picture. It allows the swarm to make collective decisions in an informed way, e.g. to classify objects reliably, allocate an appropriate fraction of robots to a specific task, and to determine the optimal solution to a global problem.
- **Synchronization** aligns frequency and phase of oscillators of the robots in the swarm. Thereby, the robots have a common understanding of time which allows them to perform actions synchronously.
- **Group size regulation** allows the robots in the swarm to form groups of desired size. If the size of the swarm exceeds the desired group size, it splits into multiple groups.

Miscellaneous

There are further behaviors of swarm robots that fit neither of the abovementioned categories:

- **Self-healing** allows the swarm to recover from faults caused by deficiencies of individual robots. The goal is thus to minimize the impact of robot failure on the rest of the swarm to increase its reliability, robustness, and performance (see also collective fault detection above).
- **Self-reproduction** allows a swarm of robots either to create new robots or replicate the pattern created from many individuals. The goal is to increase the autonomy of the swarm by eliminating the need of a human engineer to create new robots.
- **Human-swarm interaction** allows humans to control the robots in the swarm or receive information from them. The interaction can happen remotely, e.g. through a computer terminal or proximal in a shared environment, e.g. through visual or acoustic clues.

DETAILED DESCRIPTION OF ADDITIONAL SWARM BEHAVIOR CATEGORIES

In the following, we describe the additional categories of basic swarm behaviors with which we extended the taxonomy by, namely collective localization, collective perception, synchronization, self-healing, and self-reproduction.

COLLECTIVE LOCALIZATION

Collective localization allows the robots in the swarm to find their position and orientation relative to each other via establishment of a local coordinate system throughout the swarm.

Basic Principles

The approaches given below are engineered from simultaneous localization and mapping along with landmarks.

Methodologies

There are two approaches which originate from the multi-robot research domain. First, creating a map of the environment and localizing relative to it. This approach is called simultaneous localization and mapping (SLAM). It can use and merge different sources of information, such as range sensors or visual sensors. Second, using stationary landmarks with known positions and localizing relative to them. To avoid relying on external information, other robots can be used as landmarks. The robots can move alternatingly through the environment while keeping precise localization information. If the initial positions of the robots are known, then also absolute localization is possible. The dead-reckoning approach where robots use odometry for localization is another possibility but introduces an accumulating error which renders it useless for most scenarios.

Applications

A mapping algorithm where multiple robots can localize in a globally fused map is developed. It requires that the approximate initial positions of the robots are known by all other robots. It uses an incremental expectation maximization approach which allows robots to localize themselves in maps created by other robots. Experiments demonstrate that the robots can localize robustly in real time in large-scale environments using low-end computers. In the follow-up work, the requirement of known initial positions is relaxed, if robots share an overlapping part of their explored maps. It employs the concept of information filters that represents the robot positions by Gaussian Markov random fields. The robots can identify the correct alignments between different local maps by maximizing the correspondence of similar-looking landmark configurations.

A belief-based approach for collaborative multi-robot localization has been proposed. It fuses localization information from different sources, such as odometry, environment measurements, and mutual robot detections by combining visual and range sensors. This allows improvement in the robots' belief of the world, which is by learning about the detection model from data using a maximum likelihood estimator. Experiments demonstrate that a team of robots is superior in localization compared to single robots with a relatively small communication overhead.

The method of cooperative positioning using robots as landmarks. There are two groups of robots that move alternatingly while using the other, stationary group as localization reference. An increased number of robots also increases the redundancy of position information. With the weighted least square method, this redundancy decreases the localization error. The authors perform experiments where the robots use range sensors to measure their respective positions. Results show that this localization method performs better than the dead reckoning method including environments with uneven terrain.

Another method where robots visually observe each other to improve the dead reckoning localization has been developed. They have proposed two algorithms based on triangulation and trapezoidation for small- and large-scale environments, respectively. They perform experiments with two robots where one robot carries markers and the other a camera. The results demonstrate that joint localization leads to much more robust localization than odometry alone. When more robots are used, the localization precision is increased.

A maximum likelihood method combined with a distributed numerical optimization has been proposed to eliminate the need for landmark robots to be stationary. It combines range measurements with odometry. Robots observe each other's motion and exchange this information to create a graph consisting of their positions and respective observations. Experiments with four robots demonstrate that this method can localize with adequate precision and is robust to changes in the environment and to flawed odometry. Furthermore, robots can infer the position of other robots they have never seen before. Another method has applied the robot-landmark approach to a large swarm of 1,024 Kilobots. Kilobot is a small mobile robot that can operate in groups of dozens to more than 1,000 units. The robots mimic how ants and other insects coordinate their swarm behaviors. There are four pre-localized seed robots which define the coordinate system. The other robots localize relative to these seed robots using trilateration of infrared signals. The robots were able to self-assemble and let the swarm morph to a given shape.

Collective Perception

Collective perception combines the data locally sensed by the robots in the swarm into a big picture. It allows the swarm to make collective decisions in an informed way, e.g., to classify objects reliably, to allocate an appropriate fraction of robots to a specific task, or to determine the optimal solution to a global problem.

Basic Principles

Many social insects are able to get a global view using only local information. Examples are honeybees that assess the current global workload balancing by evaluating simple cues like queuing delays and ants that use pheromone trails to find shortest paths in large environments.

Methodologies

For collectively determining the type of object observed, the predominant approach is classification of the object among a set of predefined models. Sometimes, the mobility of the robots is used to improve the perception of individual robots. The robots use explicit communication to propagate their findings and achieve consensus. The way the robots exchange the information is an important aspect. They must add contextual information that allows the other robots to correctly interpret the data. Furthermore, the information can be simply forwarded and thus spread in the swarm or modified in order, e.g., to measure the distance to a specific location. There are also approaches from other research domains, such as camera networks, but the agents are typically stationary and often centrally controlled.

Applications

A strategy has been proposed where sensing agents collect, analyze, and categorize data, enrich it with contextual information, and forward it to synthesizing agents. The latter are then able to use the different aspects observed by the sensing agents to perceive events using an eigenspace method. Using simulation experiments, the authors demonstrate that events can be detected reliably using only the first few eigen values.

Another scheme has developed a swarm of micro-robots for a collective classification task. Based on evidence theory, the swarm has to identify the geometries of objects in space by exchanging data from infrared depth sensing while having limited communication capabilities. Experiments show that a wrong belief of an individual quickly converges to the correct belief after exchanging only a few messages.

A simple model of a hexagonal grid world has been proposed in which a swarm has to differentiate between differently shaped objects. They show that an increased number of agents leads to an overproportional decrease of object detection time. Another model presents an approach for cooperative gesture recognition with a robot swarm. Each robot processes and classifies camera images locally. Using a distributed consensus protocol, the robots exchange their opinions over a low-bandwidth wireless channel to find a common decision by exploiting the different viewpoints and mobility of the agents. The approach is evaluated through simulation and physical experiments on 13 robots. The results show that the recognition accuracy of the system scales effectively with the number of agents and is robust to communication failures.

Another proposal has developed a method that allows a swarm of robots with different types of low-resolution sensors to collectively classify objects. Each robot processes the sensor data locally and exchanges its estimation. Using the naive Bayes classifier together with the information received from other robots, the swarm can robustly classify objects. The more diverse the sensors are, the better the results.

A theoretic framework has been proposed which employs mobility to improve the information sensed by the swarm using the Kalman-consensus filter. It is employed to track a target with a swarm of agents. Each agent tries to improve its sensing while avoiding collisions with the others. Simulations show that this solution can effectively track linear and non-linear maneuverable targets.

Another method has been presented for simultaneous coverage of surfaces with a swarm of robots. This method assigns robots to different viewpoints to allow effective 3D reconstruction of objects. Simulation results show that this method can coordinate the robots while minimizing the mission duration and maximizing the coverage quality.

A different model compares different communication strategies for collective perception of a swarm, namely the hop-count strategy and the trophallaxis-inspired strategy. The presented solutions allow a swarm to collectively compare sizes of target areas which are too large for individual robots to perceive. Simulations show that the robots can aggregate in the target areas while their numbers are proportional to the size of the target area. A new model tackles a similar problem in which robots should distinguish good and bad spots using models of chemical reaction networks. They perform experiments with five robots and demonstrate that robots with limited sensing capabilities can collectively achieve good performance.

Synchronization

Synchronization aligns frequency and phase of oscillators of the robots in the swarm. Thereby, the robots have a common understanding of time which allows them to perform actions synchronously.

Basic Principles

During courtship, males of certain animal species synchronize their behavior. In some firefly species, the phase difference between the blinking of male and female flashing period is important for mating. Hence, the fireflies synchronize by influencing their flashing phase. Likewise, bush crickets synchronize or alternate using chirps by altering their chirp periods in response to other chirps. Another proposal has developed a model that describes the spatiotemporal wave patterns observed from myxobacteria cells. There are many more examples of coupled oscillating systems, e.g. pacemaker cells in the heart or clapping of spectators in a theater.

Methodologies

The oscillators are synchronized to the same frequency with the phases being aligned among the robots in the swarm. Two approaches exist, either the oscillators continuously influence each other to adjust phase and frequency or they are pulse coupled, meaning that they regularly fire a signal corresponding to their current phase. The latter one is mostly used as it requires fewer interactions between the robots. Robots interact through either acoustic, visual signals, or radio communication.

Applications

A model employs synchronized oscillators as a communication and navigation system. In a synchronized system, robots at a target area increase their frequency and thereby produce phase waves in the swarm that can be used by the robots to perform wave-front navigation, i.e., travel toward higher frequencies. The authors analyze the robustness of the system by simulating up to 300 robots. The results show that this communication system is robust to changes in signal strength, signaling period length, and communication obstacles. Nevertheless, the signaling period is an important parameter to be fit to the scenario. They conclude that pulse-coupled oscillator synchronization is especially suited for swarms of robots as it has low hardware requirements in terms of communication range and processing power.

Another model applies synchronization to detect faulty robots in the swarm. Using pulse-coupled oscillators and visual signaling, the swarm can determine malfunctioning robots when their phases are not aligned to the rest of the swarm. The authors develop a discrete model that can be applied to robots. Simulations with 100 robots show that robots synchronize faster when they are mobile, synchronization time is linearly proportional to the swarm size where denser networks synchronize faster, and synchronization is robust to communication obstacles but decreases in performance. Experiments with ten physical robots confirm the simulation results, despite the inherent latencies associated with the sensor and actuator systems.

In contrast to the bio-inspired approaches, the model synthesizes synchronization strategies using artificial evolution. These strategies perform phase coupling between robots to allow synchronous movement. Simulations with up to 96 robots show that the strategies scale well and are mainly limited by collision avoidance behaviors. This is confirmed through experiments with up to three robots where sensor and actuator noise is introduced.

Another model synchronizes robots in a swarm to determine the network topology and detect changes. They develop a strategy for estimating the degree of oscillator coupling in the swarm and synchronizing them continuously. Applying this strategy to formation control allows three simulated robots to move in a formation. Simulations with five robots show that the network topology can be detected reliably.

A noble concept of swarmalators has been applied to robots. *Swarmalators* are generalizations of phase oscillators that swarm around in space as they synchronize in time. This concept couples the oscillator phase with spatial location in such a way that they mutually influence each other. They modify the original model considering the discrete nature of robots. Simulations of 100 robots and experiments with 10 robots show that the spatiotemporal patterns can be performed in stationary and dynamic scenarios.

A case study has been performed to analyze how motion and sensing capabilities influence the synchronization capabilities of a robot swarm. By altering the field of interaction (e.g., camera field of view) and the speed at which the robots travel, the emergence of synchrony can be influenced. The robot speed influences the time until synchrony is reached, whereas a narrow field of interaction results in a low degree of synchronization. Furthermore, high robot densities limit the synchronization possibility due to signal occlusion and robot collisions.

Self-Healing

Self-healing allows the swarm to recover from faults caused by deficiencies of individual robots. The goal is to minimize the impact of robot failure on the rest of the swarm to increase its reliability, robustness, and performance. After detecting the fault, appropriate countermeasures must be taken.

Basic Principles

The immune system of vertebrates shows how biological systems protect complex organisms against diseases. This serves as inspiration for the artificial immune system (AIS). An overview of how AISs and swarm intelligence relate has been described. They point out many similarities and conclude that both systems are complementary tools for solving complex engineering problems. Regeneration in biological systems allows animals to self-heal their body, e.g. salamanders, starfish, and lizards can regenerate lost limbs. Another prominent example is the morphallaxis, i.e., tissue regeneration of Cnidarian hydra [1]. It exhibits what is sometimes referred to as scalable self-healing: When the hydra is dissected, each part can self-heal into a fully functional and independent hydra where its size is proportional to the number of cells.

Mythologies

There are two ways to tackle the problem of self-healing. First, healthy robots can aid the faulty robots in recovering. It requires an explicit failure management routine which is typically inspired by the immune system. Second, the swarm can adjust its behavior while ignoring failing robots. This does not require any special handling of the failure case. It is typically inspired by biological regeneration.

Applications

As self-healing is a relatively challenging topic for swarm robotics, only a few embodied studies exists and most work is done through simulation experiments. Dai et al. (2006) present a model for detecting and healing software components of swarm robots. It is part of the NASA autonomous nano technology swarm (ANTS) concept mission. This model is only partly distributed as it relies on a central cyber disease library. Each robot runs one or more virtual neurons as background processes. They monitor certain system variables, such as CPU, memory, and network usage. In case of anomalies, it freezes the process in question and reports its behavior to a higher-level controller. It can perform further diagnosis, e.g. by assigning more neurons, and generate a prescription based on the cyber disease library. The prescription is applied by the executor process which reports back results. This allows the cyber disease library to learn and improve prescriptions. In case the prescription does not work, further escalation steps are possible, such as killing the faulty process or even rebooting the whole machine. A simulation case study shows that a memory leaking process is successfully detected and eventually killed. The results show that the system becomes more reliable and robust against failures and failure propagation. Even though faulty processes degrade the overall system performance, the performance improves compared to systems without the self-healing properties.

A model has been developed for self-healing to overcome hardware failures. They present a solution that is inspired by granuloma formation, a process of containment and repair found in the immune system. They apply it to a swarm of robots performing flocking and taxis toward a beacon. When a robot has a discharged battery and loss of mobility, it would anchor the whole swarm which would then fail to reach the beacon. The proposed solution allows energy sharing between healthy and faulty robots. The faulty robots signal their need of help to the other robots within range. Depending on the required energy and the energy available at the healthy robots, a varying number of robots surround the discharged robots to share energy. Other robots ignore this robot cluster and regard it simply as an obstacle, continuing their mission. Simulation experiments with ten robots show that the granuloma formation algorithm works well even when half of the robots in the swarm are experiencing low energy levels. Other algorithms are compared where only the nearest healthy robots perform the healing. They fail to heal the swarm when three or more robots have a discharged battery.

A broad body of research is directed toward pattern formation and morphogenesis in self-healing. The SHAPEBUGS approach where agents evenly disperse within a predefined shape has been proposed. First, the agents perform trilateration to establish a common coordinate system. This is aided by allowing a few agents to know their initial position. Then the agents use the contained gas model to move in a way that they are equally spaced within the desired shape. In case of agent failure, the other agents simply adapt their positions to again reach an equilibrium density. The agent model contains a proximity sensor and wireless communication to exchange positions. It furthermore requires a compass to determine the global orientation of the agents. Simulations show that the swarm can self-repair and restabilize in cases of agent death or displacement. Furthermore, it can overcome large degrees of sensor and movement errors of the agents.

Another model relaxes some of these assumptions and still achieve similar results. The model differs in that the robots build compact shapes. Thereby, the size of the shape varies rather than its density. The model requires only a single sensor for local information and communication. The shape is given to the robots as a potential field where the robots aim to move to its center while avoiding collisions. The scale of the shape is calculated as a function of the estimated swarm size.

Additionally, each robot changes its color to a color that is predefined depending on the position. Thereby, the robots can form colored patters or displays. When properly synchronized, they can even show time varying patterns. Simulation results show that the robots can perform scalable self-healing by recreating the desired shape for varying swarm sizes. In later work, this model has been demonstrated on a large swarm of physical robots, as described above.

A model has been proposed where the robot swarm builds shapes by arranging on the boundaries of a polygon. They propose an external compilation process that uses the polygon to derive a set of parameterized local rules to be executed by a swarm of homogeneous, stateless robots. By attaching physically to each other, the robots can communicate directionally. The agents stay connected as long as they are communicating. By replying with predefined messages, the state of the system is "externalized" in the circulating messages. By attaching to each other, the agents grow the edges of the polygon, while randomly wandering robots "diffuse" into the interior of the polygon by replacing boundary robots that in turn move into the polygon interior. Simulation experiments show that the swarm can build simple polygons. It can heal from failures due to communication faults, such as dropping messages. Robots that do not communicate anymore drop out of the shape and are replaced by new ones. If the shape is broken into two, the swarm creates two shapes with the original size.

SELF-REPRODUCTION

Self-reproduction allows a swarm of robots either to create new robots or to replicate the pattern created from many individuals. In the first case, the robots produce identical copies of themselves. The goal is to increase the autonomy of the swarm by eliminating the need of a human engineer to create new robots. In the second case, the robots copy a structure consisting of many individual robots. Existing approaches are not fully autonomous yet and typically require at least the building blocks to be provided to the swarm. In contrast to self-reconfiguration of formed patterns, the goal of self-replication is to assemble a functional robot from passive components.

Basic Principles

All biological organisms possess the ability to reproduce, either sexually or asexually.

Methodologies

The theory of self-reproducing machines already exists for several decades, which is the idea of an automaton model for self-reproduction. The research in this direction followed the general idea of template-replicating systems, i.e., to create a new robot according to an existing model. Other approaches follow the evolutionary design strategy. The existing approaches assume the robot hardware to be modular to have base building blocks. The finer the modules, the more difficult the process, but the more flexibly a new robot can be created.

Applications

A model has evolved the design for simple electromechanical systems through simulation experiments. The building blocks are bars and actuators that are connected through joints and controlled by artificial neurons. The performance of the evolved systems is measured in terms of the distance it can locomote. The best performing designs are fabricated using rapid, additive manufacturing technology. This process allows robots to design new robots with minimized human intervention. The manufactured robots can locomote similarly to the simulation models.

A fully autonomous, self-replicating robot built from Lego parts has been presented. It assembles a new robot from four preassembled subsystems that hold together using magnets and shape-constraining blocks. The controller of the replica is already preprogrammed with the same program as the original. Experiments show that the original robot, which is guided by lines on the ground, can detect the subsystems and assemble them into a fully functional replica.

Another approach presents the design of a modular robot cube, called Molecube. These cubes can attach to each other using electromagnets and hence form complex patterns. These cubes have one actuated degree of freedom to control the shape of the assembled pattern. The authors present experiments where the spatial patterns and corresponding controllers are both manually created and automatically designed with artificial evolution. The results show that the swarm of Molecubes reproduces identical copies of its pattern, both in simulation and physical experiments. The only human interaction is by providing enough Molecubes as building material. The authors conclude that the more units are involved, and the simpler and more homogeneous they are, the more information is being reproduced by the system itself, as compared with information preexisting in the parts and environment.

ARTIFICIAL INTELLIGENCE-BASED SWARM ROBOTICS

Automatic Design Methods

Automatic methods use intelligent algorithms [2] to produce the behaviors for all system entities without the developer's explicit interior. The most used approaches are evolutionary algorithms like particle swarm optimization (PSO) and reinforcement learning (RL) algorithms, which are a field of machine learning. Furthermore, it is helpful to depend on artificial intelligence algorithms to reduce developers' computations and efforts. This chapter will illustrate the most used algorithms in automatic design methods for swarm robotics systems and show the modifications on these algorithms and how they affect the collective behavior of the swarm.

PARTICLES SWARM OPTIMIZATION

The PSO is one of the well-regarded algorithms in the literature of optimization and has been widely used in various science and industry fields. The PSO mimics the navigation and foraging of a flock of birds or schools of fishes. Table 23.1 shows a mapping between PSO terms and swarm robotics system for applying PSO to the system. Synchronous particle swarm optimization (SPSO) means updating *pBest*, *gBest* after evaluating fitness value for all particles. In another way, we can update these values for every particle after its fitness evaluation. This method is called asynchronous particles swarm optimization (ASPSO). It is better to use it when some information is missed for neighbors. After applying the two methods SPSO and APSO on the same swarm system, it is noticeable that there are similarities in results. Still, APSO is slower than SPSO for convergence.

However, it is not a disadvantage because it allows the particles to discover the search space. In addition to that, APSO is preferred for swarm systems. It will enable efficient usage of the robot's PSO is considered an efficient evolutionary algorithm to plan the path for swarm robots; for example, fuzzy logic is used to update the velocity of particles, making the system more reliable and faster to reach the target. Another modification has been implemented to find the shortest path. It used another evolutionary algorithm called bacterial foraging Optimization Algorithm (BFOA) (Tables 23.2 and 23.3).

Particles search randomly in an area for the best solution, every particle is initialized by a random one, and then they will search for the best by updating their solutions during every iteration; every particle has two basic values that should be recorded during the process:

- *pBest*: The best fitness value particle has now
- *gBest*: The best fitness value among the swarm

TABLE 23.2
Mapping between PSO and Swarm Robotics System

Swarm robotic search	PSO
Signal detection	Fitness evaluation
Relative localization	Absolute localization
Path planning	Particle update
Continuous control	Iteration
Local communication based on physical distance	Global communication-based fixed neighborhood

---

TABLE 23.3
Reward Sets

Reward Set	Des. rd	Lan. rl	Fel. rf	Col. rc
A	5	5	1	−1
B	5	1	5	−1
C	5	1	1	−5

After finding previous values, particles update the velocity and the position according to the following equations, which illustrate the producers of PSO:

$$v_{i,t+1}^d = \omega^* v_{i,t}^d + c_1^* \text{rand}_i() * \left(p_{i,t}^d - x_{i,t}^d \right) + c_2^* \text{rand}_i() * \left(p_g^d - x_{i,t}^d \right)$$

$$x_{i,t+1}^d = x_{i,t}^d + v_{i,t+1}^d$$

$v_{i,t+1}^d$: Velocity of particle i at $t + 1$ in d dimension
$v_{i,t}^d$: Velocity of particle i at t in d dimension
$p_{i,t}^d$: The best fitness value particle has now
p_g^d: The best fitness value among the swarm
Ω: Inertial weight
c_1: Cognitive learning factor
c_2: Social learning factor
$x_{i,t}^d$: Position of particle I at t moment

PSO has three repetitive steps: Calculation of the fitness value for each particle i, update *pBest* and *gBest*, and update the velocity for each particle. Figure 23.1 depicts the PSO flowchart.

Reinforcement Learning

As mentioned before, according to the developments in machine learning and deep learning field, it is efficient to apply these techniques in swarm robotic systems, especially RL and deep neural networks, to reduce the number of needed computations. Many approaches are used to make the robot learn the behavior in its swarm by trail-and-error by receiving negative or positive feedback from the environment to evaluate the behavior. For example, as shown in Figure 23.2, each robot gets a state from the environment by sensing s_t.

The process defined by S, A, P, r, γ, T

Policy π: $S \times A, R \geq 0$
Expected reward $E_r = \sum_{t=0}^{T} \gamma^t r \left(s_t, a_t \right)$
s, S: State
a, A: Action
P: Probabilities of transition
r, R: Reward
γ: Discount factor
T: Number pf the taken actions during one episode

According to this state, the robot will do an action a_t to move the environment to a new state, s_{t+1}. The action will be evaluated by giving the robot reward or punishment R. The robot will learn its behavior by taking the behavior which achieves the highest cumulative rewards. So, RL aims to maximize the expected cumulative rewards. In other words, the aim is to map between states and actions. This is called Policy π. Moreover, it is deterministic, which means there is one action for every

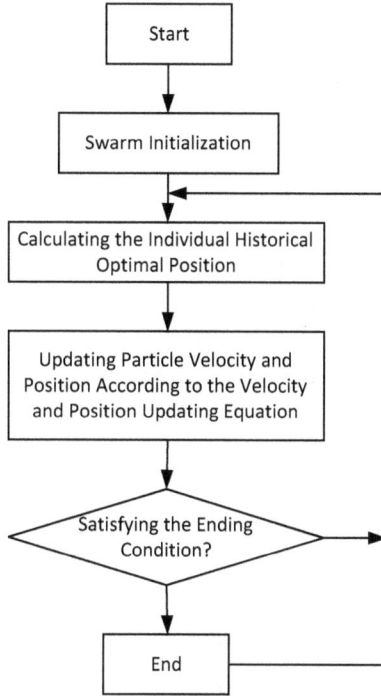

FIGURE 23.1 PSO flow start.

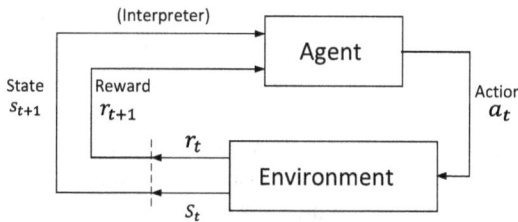

FIGURE 23.2 Reinforcement learning model.

state, or it is stochastic. Instead of one action for each state, there is a distributed probabilities function of actions.

There are two ways to obtain optimal policy: policy-based methods and value-based methods. For policy-based methods, there is a function policy that maps directly between actions and states. In general, the neural network is used to produce policy functions. It takes states as input and actions as output. This algorithm is called policy gradient (PG), as shown in Figure 23.3.

In value-based methods, a value function calculates the Q value for each state action and then chooses actions that lead to the highest value states. It is common to use a neural network to train the robot to calculate the Q value, so the neural network takes the state as input, and the output is the Q value. The RL algorithm chooses the

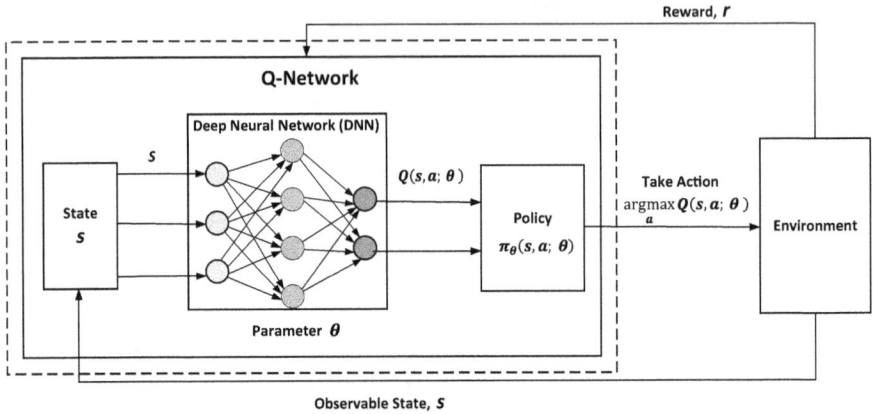

FIGURE 23.3 Policy gradient architecture.

best actions according to calculated Q values. This architecture is called deep Q network DQN, as shown in Figure 23.3.

DEEP REINFORCEMENT LEARNING IN SWARM ROBOTICS SYSTEMS

Foraging task means that swarm of robots should transport items from sources to a sink. DQN is used to learn swarm of robots to perform foraging task. This approach has compared the results after adding many modifications in the architecture of DQN, like freezing the target network. To know how freezing is done depends on the error function during the learning process. The error function is used in the neural network by calculating the square of the difference between the target and Q values. These two values change simultaneously, so it will cause a high variance, and so it is better to copy the parameters for the current network to the target network every specific time step. Another modification is to use two neural networks to reduce the overestimation. It is called double DQN DDQN; also, it is possible to use future rewards in the N-step Q network NSQ. This shows that freezing the target network and adding NSQ with DDQN will maximize the expected reward and improve and fasten the learning process. Another technique used to enhance the performance of the swarm robotic system is to use shared memory for all robots, as shown in Figure 23.4.

All robots store their experiences in the memory, and all of them can access this memory and invest it during the training process. It is noticeable that increasing the number of robots who share this memory will increase learning speed and reduce the variance. The most crucial parameter in RL algorithms is the reward choosing R affects the swarm's collective behavior. For example, the swarm behavior has been studied for a swarm consisting of blue and yellow robots. Every group should learn how to reach its destination without any collision. The study was conducted with three sets of R, as shown in Table 23.3, and the environment has two cases, free and with obstacles.

- *Des. rd*: When the robot arrives at its destination.
- *Lan. rl*: The captured image by camera contains more pixels related to the destination than the previous one.

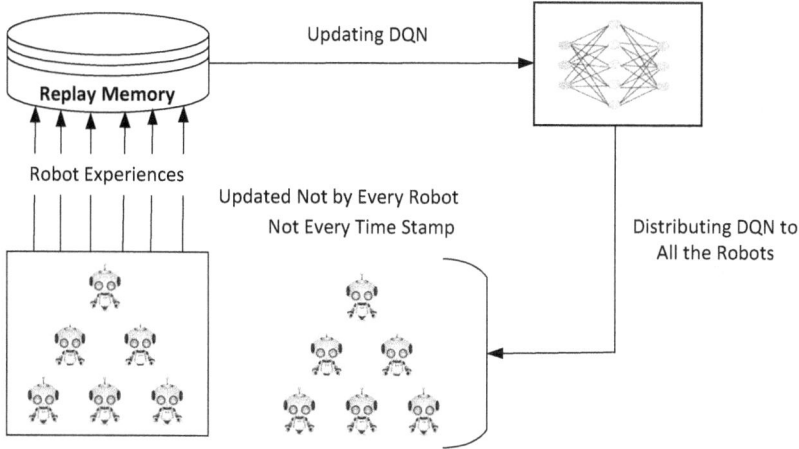

FIGURE 23.4   Sharing memory for swarm robots.

- *Fel. rf*: The captured image contains more pixels, indicating that the robot is with his group.
- *Col. rc*: Punishment for collision with other robots or with obstacles.

The study shows that when robots take severe punishment, it will affect the path planning to reach their destination because robots will focus on avoiding obstacles more than getting their goal. Although, in general, the R-value should be chosen to motivate the swarm entities to reach their destination, lots of studies neglect the punishment by adding R equal to zero. Many improvements are added to traditional RL algorithms like the sequential Q-learning algorithm based on knowledge sharing to increase the number of robots. The multi-agent deep deterministic policy algorithm with a proposed experience samples the optimizer for exploration task. Reinforcement Learning (RL) is used for formation control of swarm robots. Bilateral Biased Neighbors-Sharing Cooperative Reinforcement Learning (BBNSCRL) method is presented to accelerate the learning process by integrating the neighboring robots' knowledge with local knowledge.

SUMMARY

Research on swarm algorithms is a relatively young topic. Despite the large number of swarm algorithms, the transition to industry and industrial production, not to mention daily use, has not been made successfully. Nevertheless, several steps toward swarm applications have already been taken. The main objective of this chapter is to motivate future research and engineering activities by providing a comprehensive list of existing platforms, projects, and products as a starting point for applied research in swarm robotics.

We have classified the basic swarm behaviors and present a comprehensive overview of current research platforms and industrial applications. While this demonstrates the possibility of integrating basic swarm behaviors in current applications, it also shows that many applications of swarm robotics cannot fully exploit the advantages offered by distributed swarm architectures due to systems with only few agents or central control. Swarm algorithms build upon self-organized swarm behaviors, e.g. observed in natural swarm systems, such as insect colonies or flocks of birds that can handle extremely diverse and dynamic environments. The same holds for robot swarms. They are meant to operate in the physical world, which typically faces continual dynamic changes and must cope with events and external conditions that are hard to predict or model. Besides huge potential for applications in areas like logistics, agriculture, and inspection, one suitable working environment for swarms are places that are unsuitable for humans, including places that are hard to reach, dangerous, or dirty. Applications in these environments could help to better observe, understand, and exploit the advantages of swarm behaviors: adaptability, robustness, and scalability.

In addition to industrial applications, we have also surveyed different research hardware platforms dedicated to swarm robotic experiments. On the one hand, this overview allows us to choose an appropriate research platform for implementing and testing swarm algorithms in laboratory environments. On the other hand, it shows that there is a huge potential in research to transform these platforms from pure prototyping platforms to productive, industrial robotic systems that can perform in the real world. This might require shifting from the current simplified robot models and controls to a trade-off between simplicity of design and capability of solving complex tasks in a reliable way, e.g., from reduced resource consumption to a more intensive usage of sensor data and information sharing.

We have also introduced swarm engineering term by focusing on automatic design methods that depend on artificial intelligence to produce the required behavior for each swarm robot, making it learn from the environment. In general, automated methods have two directions, evolutionary methods like PSO, bacterial foraging, and others, and RL, which is more reliable and needs less data to make swarm learn. Lots of modifications were applied to the two approaches. For example, fuzzification of some parameters of PSO algorithms enhances the learning speed. Furthermore, they were using deep learning in RL to reduce the number of computations. Other enhancements were found in the neural network architecture to solve the problems of variance and overestimation and illustrate how modifications affect the swarm behavior. Most of them give a good speed of learning process and a reliable swarm system, but there is no precise formula to model swarm. It still depends on the experiment. Further, this study can be extended for the development of swarm robotics algorithms to reduce the problems and obstacles to achieving the swarm's collective behavior.

REFERENCES

1. Schranz, M. et al., "Swarm Robotic Behaviors and Current Applications," *Frontiers in Robotics and AI* April 2020, 7, Article 36. https://doi.org/10.3389/frobt.2020.00036
2. Iskandar, A. and Kovács, B., "A Survey on Automatic Design Methods for Swarm Robotics Systems," *Carpathian Journal of Electronic and Computer Engineering* 2021, 14(2), 1–5. https://doi.org/10.2478/cjece-2021-0006

24 Digital Twins

OVERVIEW

The first practical definition of a digital twin (DT) originated from NASA to improve the physical model simulation of spacecraft in 2010. Initially, product drawings and engineering specifications used in DT have progressed from handmade drafting to computer-aided drafting/computer-aided design to model-based systems engineering and strict link to signal from the physical counterpart. Formally, we define digital twin as follows:

> A digital twin (DT) is a mathematical model with an updating mechanism that generates data which are indistinguishable from its physical counterpart. DT is defined as a virtual representation of a physical asset enabled through data and the multi-physical, multiscale, and probabilistic simulators for real-time prediction, optimization, monitoring, controlling, and improved decision-making based on the feedback.

The DT allows us to solve several complex problems of various kinds. The difference between the DT and the conventional mathematical models is that it allows us to display the current state of the investigated object or process very accurately. At the same time, the DT can not only operate with the reality that is supposed to exist objectively but also work with augmented reality and with an altered reality.

Since information is granular, the DT representation is determined by the value-based use cases it is created to implement. So, the DT can and does often exist before there is a physical entity. The virtual representation of the object or system that spans its lifecycle, is updated from real-time data, and uses simulation, even using machine learning (ML) and reasoning to help decision-making. DT is a true embodiment of a cyber-physical system (CPS). DT technology is used to create virtual models of CPS such as computing, internet of things (IoT), cloud nodes, edge nodes, radio access network (RAN), and sensors, of communications networks, and roads, bridges, and trains of transportation systems. These DT s are used to simulate the behavior of these systems in different scenarios, such as traffic congestion, weather conditions, and maintenance events. It is a living, intelligent, and evolving model which can optimize the processes and continuously predicts future statuses (e.g., defects, damages, and failures) through the closed-loop optimization between DT and surrounding environment.

DIGITAL TWINS APPLICATIONS

DT applications are in health, meteorology, manufacturing and process technology, education, cities, transportation, energy sector, and other areas.

DOI: 10.1201/9781003499480-24

Digital Twin Categories

The technologies required for DT can be divided into two categories, one is a statistical model driven by data and the other is a mechanical model that integrates multiscale knowledge and data. The numerical model is used to calculate the structural performance, while the analytical model is used for structural analysis. Note that artificial intelligence (AI) is the key enabling technology for DTs where physical processes are integrated, for example, with the network and computer domains by analyzing a set of AI agents at the application and infrastructure levels. For example, real data generated by a DT for robotics system can be analyzed and validated by a set of AI agents at the application and infrastructure levels.

In some cases, DTs are commonly divided into three subtypes: digital twin prototype (DTP), digital twin instance (DTI), and digital twin aggregate (DTA) [1]. The DTP consists of the designs, analyses, and processes that realize a physical product. The DTP exists before there is a physical product. The DTI is the DT of each individual instance of the product once it is manufactured. The DTI is linked with its physical counterpart for the remainder of the physical counterpart's life. The DTA is the aggregation of DTIs whose data and information can be used for interrogation about the physical product, prognostics, and learning. The specific information contained in the DTs is driven by use cases. The DT is a logical construct, meaning that the actual data and information may be contained in other applications.

Digital Twin Characteristics

DT technologies have certain characteristics that distinguish them from other technologies: connectivity, servitization, and homogenization, intelligent reprogrammable, digital traceability, and modularity [1].

Connectivity

For DT developed for any object, for example network, product, or system, there remains an inherent thread that shows how the internal things that can be defined or conceptualized as technologies of the object. First and foremost, the technology enables connectivity between the physical component and its digital counterpart. The basis of DTs is based on this connection; without it, DT technology would not exist. For example, the connectivity created by IoT, cloud nodes, edge nodes, RAN, and sensors of communications networks on the physical product obtain data and integrate and communicate this data through various integration technologies. Similarly, DT technology enables increased connectivity between organizations, products, and customers. For example, connectivity between partners and customers in a supply chain can be increased by enabling members of this supply chain to check the DT of a product or asset. These partners can then check the status of this product by simply checking the DT.

Servitization

Servitization is the process by which organizations add value to their core corporate offerings through services. For example, DT developed for the communications network shows what kinds of application services are provided to their customers.

Customers can then check how can they subscribe to those services checking the DT. In another the example, the manufacturing of the engine could be the core offering of an organization; others can then add value by providing a service of checking the DT of the engine and can offer maintenance services.

Homogenization

DTs can be further characterized as a digital technology that is both the consequence and an enabler of the homogenization of data. Due to the fact that any type of information or content can now be stored and transmitted in the same digital form, it can be used to create a virtual representation of the product (in the form of a DT), thus decoupling the information from its physical form. Therefore, the homogenization of data and the decoupling of the information from its physical artifact have allowed DTs to come into existence. However, DTs also enable increasingly more information on physical products to be stored digitally and become decoupled from the product itself.

As data is increasingly digitized, it can be transmitted, stored, and computed in fast and low-cost ways. According to Moore's law, computing power will continue to increase exponentially over the coming years, while the cost of computing decreases significantly. This would, therefore, lead to lower marginal costs of developing DTs and make it comparatively much cheaper to test, predict, and solve problems on virtual representations rather than testing on physical models and waiting for physical products to break before intervening.

Another consequence of the homogenization and decoupling of information is that the user experience converges. As information from physical objects is digitized, a single artifact can have multiple new affordances. DT technology allows detailed information about a physical object to be shared with a larger number of agents, unconstrained by physical location or time. With the DT, not only the factory manager, but everyone associated with factory production could have that same virtual window not to only a single factory but also to all the factories across the globe, if needed.

Reprogrammable and Smart

A DT enables a physical product to be reprogrammable in a certain way with the use of AI, and predictive analytics, and smart technologies embedded in the physical products. Furthermore, the DT is also reprogrammable in an automatic manner. A consequence of this reprogrammable nature is the emergence of functionalities. If we take the example of an engine again, DTs can be used to collect data about the performance of the engine and, if needed, adjust the engine, creating a newer version of the product. Also, servitization can be seen because of the reprogrammable nature as well. Manufactures can be responsible for observing the DT, adjusting, or reprogramming the DT when needed, and they can offer this as an extra service. Similar is the case for DT of the communications network, as we described earlier.

Digital Trace Making

Another characteristic that can be observed, is the fact that DT technologies leave digital traces. These traces can be used by engineers, for example when a machine malfunctions, to go back and check the traces of the DT, to diagnose where the problem occurred. These diagnoses can in the future also be used by the manufacturer of these machines, to improve their designs so that these same malfunctions will occur less often in the future. Similar is the case for DT of the communications network operations and maintenance.

Modularity

In the sense of the manufacturing industry, modularity can be described as the design and customization of products and production modules. By adding modularity to the manufacturing models, manufacturers gain the ability to tweak models and machines. DT technology enables manufacturers to track the machines that are used and notice possible areas of improvement in the machines. When these machines are made modular, by using DT technology, manufacturers can see which components make the machine perform poorly and replace these with better fitting components to improve the manufacturing process. Similar is the case for DT for the operations and maintenance of different functional entities of the communications network.

AI INTEGRATION OF DIGITAL TWINS

Industry 4.0 concept envisions industrial systems with the ability to make decentralized and autonomous decisions using CPSs. The rapid advancements in information and communication technology (ICT) are transforming the industrial sector toward a full digitalization and integration concept. Consequently, the industrial world can improve the productivity and logistics and lower production costs [2]. CPSs are the main linchpin for Industry 4.0 to move toward a fully automated industrial infrastructure that relies on real-time capabilities, distributed control systems, virtualization, service orientation, and modularity [2]. DT, which represents a virtual physical asset through data and simulator, truly embodies the cyber-physical integration combining any industrial process achieved through closed-loop feedback mechanisms. Fifth-generation/sixth-generation (5G/6G) networks of ICT are architected to simultaneously support different types of service profiles in the shared infrastructure, such as enhanced mobile broadband (eMBB+), massive machine type communication (mMTC+), and ultra-reliable low-latency communication (URLLC+). Together with edge and fog computing, they provide a communication link with low end-to-end (E2E) latency, low jitter, and localization awareness to industrial services. Still, by themselves these technologies cannot efficiently manage automation or compute the best decisions to achieve dynamic adaptation.

In this sense, cyber space mirrored through DTs arises as the perfect playground for the development of AI agents. Moreover, ML is a strong candidate to implement such agents, as an alternative to heuristic or decision tree-based solutions, among others. DTs provide the tools for transferring the domain expertise of specialized personnel into raw data in cyber space, which can later be used to train and cross-validate different ML algorithms used in AI agents. These agents not only develop expertise in specific tasks but also extend and optimize them beyond human capability due to the volume of data they can handle to make decisions. Ultimately, smarter and more accurate DTs can be devised where autonomy is achieved through AI-controlled

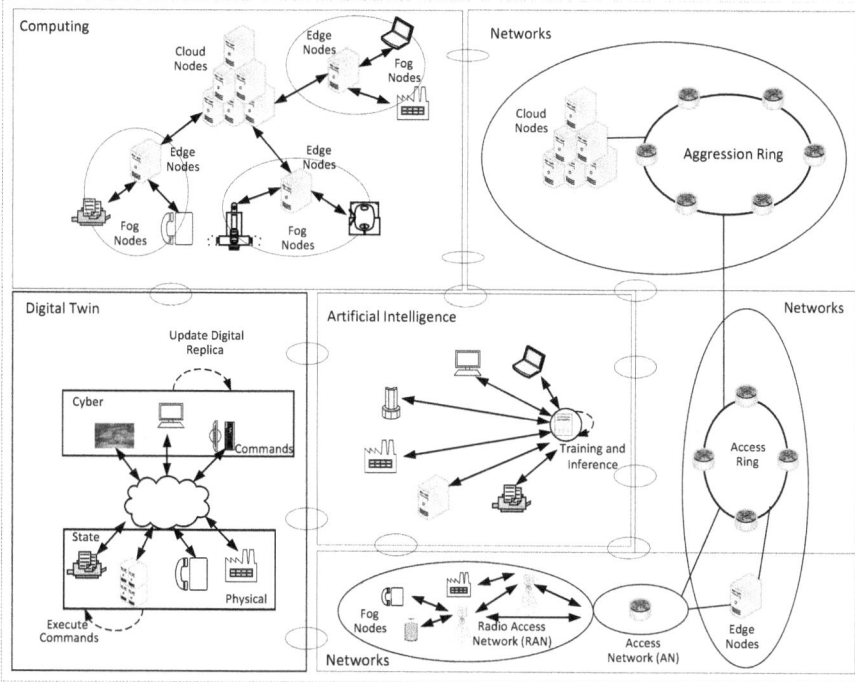

FIGURE 24.1 Generalized concept of digital twin for computing and network.

processes that operate in all types of environments and conditions [2]. Figure 24.1 shows a generalized concept of DT for computing and networks.

Computing

In Industry 4.0, physical objects are composed by either low-performance and constrained hardware or hardware tailored to a specific task. Due to the development of virtualization, software components of the physical object are represented as modular virtualized functions whose execution is outsourced into more powerful computing resources. Cloud-based solutions were initially exploited for implementing such concepts, by providing the elastic and powerful computing capabilities required to support the DT. However, cloud providers cannot ensure the performance of the network between the physical object and its digital replica, worsening with their network distance and the number of providers in between.

As a result, cloud-based DTs suffer from time-varying network delay, unpredictable jitter, limited bandwidth, and data loss. These drawbacks prevent time-sensitive tasks, including real-time remote control, being fully supported by the cloud computing substrate. To overcome the shortcomings of cloud computing, edge and fog computing emerged as a natural extension. While edge computing provides computing capabilities near the physical objects via static substrates, fog computing also integrates volatile, constrained, or mobile resources (including the physical objects). Figure 24.1 shows a generalized view that the computing architecture consists of cloud nodes, edge nodes, and fog nodes.

By exploiting edge and fog computing, the DT can off-load time-sensitive processing from the physical object, which in turn contributes toward further optimizations of the hardware costs. Additionally, new algorithms for efficient data filtering, envisioning privacy, and security improvements can be applied, and the data can be restricted within a trusted private infrastructure. Finally, due to the proximity, edge-based DTs can use the available radio network information to adapt physical objects' operations or to optimize resource allocation to improve the quality of experience (QoE). However, it requires to optimally allocate computing resources for DTs in the cloud-to-thing continuum, to satisfy key performance indicators (KPIs) such as latency and security requirements.

Networks

The underlying network infrastructure of the DT comprises different dynamic and heterogeneous topologies. It can be divided into three segments, as shown in Figure 24.1:

- Aggregation ring
- Access ring
- RAN
- Access network (AN)

The aggregation ring resides far from the physical objects, relying on wired connectivity to connect cloud-based DTs that are suitable for human-scale responsive services and delay-tolerant tasks (e.g., monitoring). The access rings are closer to the physical objects, interconnecting multiple RANs and ANs. The access rings are locally present and expose radio network information (e.g., radio channel) to edge-based DTs, namely for time-sensitive tasks (e.g., remote manipulation). Finally, the RAN/AN is in the vicinity of the factory floor, providing connection to the physical objects using both wired and wireless connectivity.

Different radio access technologies (RATs) are available (e.g., Wi-Fi, LTE, 5G, and 6G), differing in their capabilities with respect to latency, range, data rate, power profile, and scalability. Wired technologies are most suitable for fulfilling the communication requirements of DTs. Due to their limitations in terms of flexibility, mobility, and high-density connections, wireless technologies are becoming more appealing in the RAN/AN. However, the critical processes within industrial environments are sensitive to radio frequency (RF) interference, requiring RATs to be interference-free, to work on licensed bands, and to provide an extremely controlled environment.

Industry 4.0 claims 5G/6G as a key enabler to fulfill the communication requirements set by DT, not only through radio enhancements but also by employing network slicing and virtualization as core features. At the same time, Wi-Fi 6E appears as another candidate for Industry 4.0, with trials already showcasing its capability to sustain the presence of interference and noise, and to meet the stringent requirements of most use cases. The key is to build DTs that benefit from optimal use of heterogeneous RAT resources and overcome radio interference problems. The challenges demand DTs that satisfy the expected real-time and secure performance. Furthermore, DTs should also tackle the problems derived from its integration with the network and computing infrastructure. AI agents are strong candidates to handle such challenges, as they can benefit from ML algorithms to exploit existing data sources with context information at both the application and infrastructure levels.

AI Agent for Digital Twins

AI agents for DT can be trained for cloud, edge, and fog computing (Figure 24.1). AI agents will use appropriate ML/deep learning (DL) (ML/DL) algorithms to meet DT requirements such as movement prediction, task learning, risk reduction, predictive maintenance, heterogeneous network (HetNet) selection, dynamic scaling, privacy, security, and intrusion detection.

> **Movement prediction**: Remotely controlling a physical object over a wireless channel via its DT can be prone to unpredictable RF interference that introduces high jitter and packet loss. Consequently, the remote operator experiences lagging behavior that breaks the real-time control of the physical object and creates an unsafe environment. A solution that recovers from such unpredictable behaviors, keeping the remote control uninterrupted, is required. AI stands out as a strong candidate that can forecast future movements providing extra reliability in case movement commands are lost. In this context, *movement prediction* uses historic commands to predict future ones using ML time series algorithms like temporal convolutional network (TCN), gated recurrent unit (GRU), long short-term memory (LSTM). Whenever the next command is lost or does not arrive on time, movement prediction triggers the forecasting of such a command to keep the remote control uninterrupted (as showcased later). To prevent packet loss and high latencies, movement prediction has to be deployed in the fog, executing when a failure occurs, or in the edge, piggybacking predictions with every real command.
>
> **Task learning**: In industrial scenarios, there are still highly complex and dynamic tasks that require human expertise/presence. Traditional agents based on finite state machines are not suitable for automating such tasks, as they cannot react in unforeseen situations, such as the appearance of unpredictable obstacles. *Task learning* AI agents based on IL and RL algorithms are potential solutions to overcome such situations, as they are designed to learn and afterward interact with a dynamic environment. Task learning is first trained through observations of human-based operations, in a trial-and-error fashion, and its behavior is validated in the simulated environment, which includes unexpected and random situations. Finally, its runtime deployment is envisioned on the factory floor (i.e., fog or edge) to ensure secure, reliable, and low latency execution of the task. For example, task learning can introduce generalization and adaptability to drive a lift truck for package delivery. Thus, the DT can learn how to act robustly upon introduction of obstacles in the path, or changes in the shapes or position of packages.
>
> **Risk reduction**: As remote control mechanisms emerge and physical objects become more autonomous, safety plays a critical role in the design of a DT. When considering human-machine collaboration scenarios, failures of

either humans or machines may pose a safety risk. Factory floors equipped with surveillance cameras could reuse them to perform image segmentation and pattern recognition to identify and mitigate dangerous situations. In recent years, it has been proved that AI solutions based on CNN algorithms achieve the best performance on computer vision-related tasks. Hence, *risk reduction* uses CNN algorithms to identify dangerous situations analyzing a video stream, helping the DT to act preventatively, such as blocking the physical object or adapting its operation. Since fast countermeasures are required, its runtime deployment is best fit in the edge or, in scenarios with higher degree of autonomy, in the fog. For example, based on a real-time video stream, risk reduction can detect that a human operator is in dangerous proximity to an operational industrial machine, and use this information to block the machine.

Predictive maintenance: Industrial physical objects have always been held to a higher reliability and predictability standard than any general-purpose systems. Industrial companies consider unplanned downtime and emergency maintenance caused by failures a major challenge. For preventing eventual failures, the future state of a given component must be forecast and classified to verify if it requires maintenance. ML-based solutions provide high accuracy to solve both prediction and classification problems. Thus, a predictive maintenance AI agent is a suitable candidate to preemptively detect failures or repair needs by using combinations of algorithms including autoregressive integrated moving average (ARIMA), LSTM, logistic regression (LR), and support vector machine (SVM). Predictive maintenance checks of the available sensor data might lead to failure situations, and, if so, it schedules the maintenance of the physical object. Since this AI agent is not performing time-sensitive operations, it can be deployed anywhere from the edge up to the cloud. For example, if historical data reports high vibrations upon the break of a screw, predictive maintenance can forecast future vibrations (e.g., LSTM) and decide if maintenance is required (e.g., SVM).

Dynamic scaling: With the recent development of virtualization technologies, smart factories benefit from having DTs coexisting under the same cloud-to-thing continuum. During the lifetime of a given application, adequate scaling of resources is required so that DT-related KPIs (e.g., latency) are satisfied without deteriorating the performance of others. Such a problem is analyzed in the existing literature as an NP-hard problem, that is, optimal scaling polices cannot be found in feasible runtimes. Consequently, AI solutions based on Markov decision processes can be used to find near-optimal scaling policies in feasible times using algorithms based on ARIMA. Dynamic scaling follows scaling policies learned with the algorithms, training with data such as resource consumption, date and time, task, and number of instances and sessions. Dynamic scaling can then compute scaling decisions to fulfill KPIs and service-level agreements (SLAs). The runtime deployment of this AI agent is most suitable on the network side (i.e., edge or cloud), depending on inference time and network latency toward the orchestrator. For example, whenever a new robotic arm is added on the factory floor, dynamic scaling increases the allocation of virtual central processing unit (vCPUs) to the virtual instance in charge of holding its digital replica, allowing its processing delay to stay below a threshold.

Privacy, security, and intrusion detection: By employing DTs in an industrial environment, huge volumes of network traffic are distributed in the cloud-to-things continuum to create the digital factory. This makes the detection and diagnosis of security breaches and intrusions very challenging and complex for the infrastructure operators and their tenants. Performing an exhaustive analysis of all the network traffic would take a vast amount of time, which is infeasible to detect intrusions or security breaches early. ML algorithms, like principal component analysis (PCA), K-means, and auto-encoders, are ideal solutions to shrink traffic volume and speed up traffic inspection. Privacy, security, and intrusion detection use these algorithms to detect malicious traffic and consequently block remote control of physical objects through their DTs. Moreover, federated learning and transfer learning appear as ML approaches boost collaborative training across different industrial players, which, by not centralizing the training data, retain the privacy and locality of private data. The edge and cloud are candidate locations to deploy this AI agent, depending on whether on-site security operations are required or not.

Heterogeneous network selection: In an industrial environment comprising multiple RATs, the challenge of being always best connected arises, directly affecting the design and performance of DTs. RAT selection is traditionally solved by applying rules derived from the network infrastructure with prior domain knowledge and experience of experts. However, applying this type of RAT selection to DTs is often complex to manage on dynamic and heterogeneous industrial environments. A HetNet selection AI agent that uses ML algorithms, for example reinforcement learning (RL), artificial neural network (ANN), and fuzzy logic, appears as a tool to mitigate the challenges. It exploits the locally available radio context information to select the best RAT for each physical object on the factory floor and, if required, the best handover candidate. The radio context information is defined by European Telecommunications Standards Institute (ETSI) multi-access edge computing (MEC), and provided by different radio information services, such as radio network information service (RNIS) and wireless local-area network (WLAN) access information service (WAIS). Based on such information, Het-Net selection detects when, for example, an automated guided vehicle (AGV) will be out of coverage and lose the connection to the point of attachment. The DT can use this information to preemptively transfer state information to the new point of attachment and to instruct the AGV to change its RAT to seamlessly move around the factory floor. Since this AI agent depends on locally available information, its preferable deployment location is the edge or fog.

DIGITAL TWIN FOR 6G NETWORK

The 6G network needs to support distributed mobile users connected over multi-tier base stations (BSs), which is termed as "heterogeneous networks" (Het-Nets), dynamically. It becomes critical for the 6G Het-Nets to achieve both diverse quality of service (QoS) requirements of all users and entire network performance. Figure 24.2 provides a reference architecture of 6G Digital Twin Network (DTN) based on the International Telecommunication Union-Telecommunication (ITU-T) recommendation Y.3090.

Accordance to the ITU-T recommendation Y.3090, a reference to 6G DTN architecture may consist of three layers: 6G physical network layer, 6G twin layer, and 6G network application layer. Figure 24.2 provides an illustration of the reference 6G DTN architecture. Note that the significantly increased complexity of matching the irregularly distributed users and BSs as well as highly dynamic network traffic often cause unbalanced spatial-temporal loads for multitier BSs during user association. They are emerging as a powerful technology for design, diagnosis, simulation, what-if-analysis, and AI/ML/ DL-driven real-time optimization and control of the 6G wireless networks.

Despite the great potential of what DTs can offer for 6G, realizing the desired capabilities of 6G DTNs requires tackling many design aspects including data, models, and interfaces. The foundation of 6G DTNs is built upon a set of design principles shaped by various use case requirements, including reliability & latency, scalability, agility, generalizability, security, and interpretability [3].

Reliability & latency: Trustworthiness is at the core of DTN's fundamental characteristics. Rendering high degree of stability throughout the lifecycle of a DTN requires robustness in its operational infrastructure, including proactive handling of latency critical adverse situations involving human error, equipment failure or malicious attacks with real-time response, capability of ensuring robust data collection, storage, modeling, and information exchange between the DTN and other network entities, high degree of availability, and capabilities for disaster recovery (such as backup provision, and ability to restore critical historical states/data points).

Scalability: Scale of a DTN can vary over a wide range, depending on the dimension and complexity of its physical counterpart. A physical network may grow or shrink in scale and the DTN should automatically adjust itself. Such flexible scalability is an essential attribute of DTNs across various network domains.

Agility: Serving various network applications on-demand requires a DTN to have sufficient flexibility in its functionalities as needed. Ability for cross domain interaction, information exchange and service cooperation between multiple DT entities are some of the important attributes of a DTN towards meeting needs of various operational stages of a physical network.

Generalizability: Applicability of DTNs across network equipment of various network applications and topologies requires a certain degree of compatibility. Specifically, data collection, storage, modeling, and interfaces need to

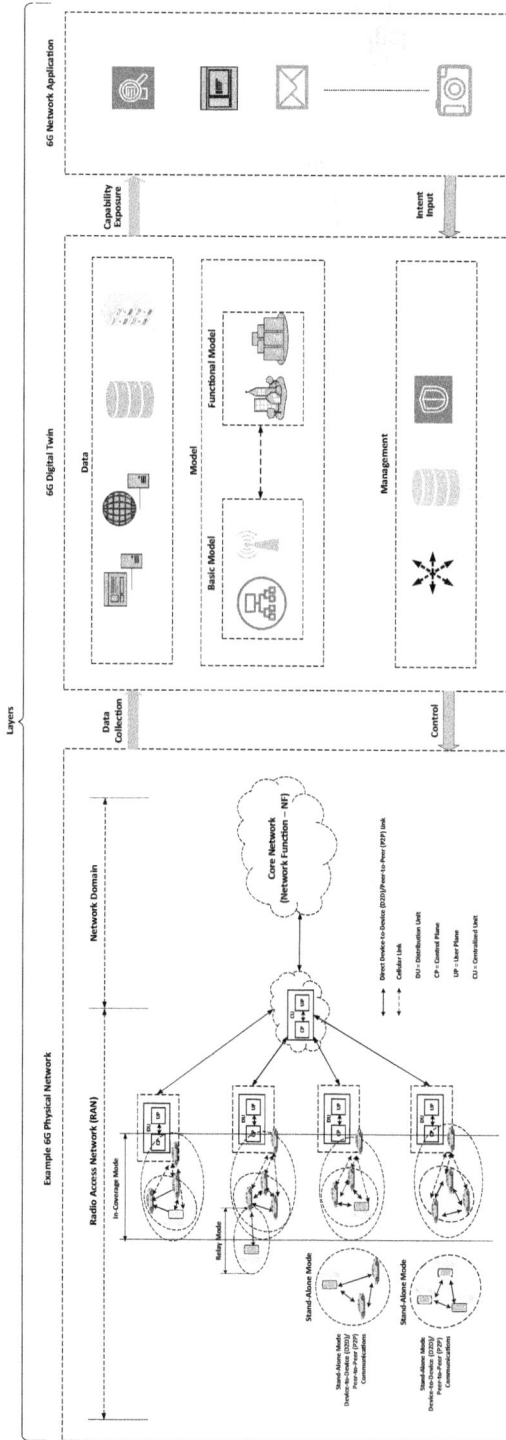

FIGURE 24.2 6G Digital Twin Reference Architecture

be generalizable to support multi-vendor, multi-standard interoperability and ensure backward compatibility between updated and older versions of DTNs.

Security: Guarantee of adequate protection from potential attacks as well as defense mechanism for threat prevention should be imprinted in the DNA of a robust DTN. A comprehensive security and privacy consideration for DTN elements including data, models, interactive interfaces, and overall network infrastructure is warranted to ensure integrity of a DTN.

Interpretability: Augmenting DTNs with easy-to-use asset management services aids user-friendly management of overall lifecycle of the twin entities like network equipment, links, traffic flows, etc. Ability to observe discernable changes in the functionalities of a DTN through display tools improves interpretability of the DTN's behavior with associated events like instance creation, aggregation, update, and termination.

We describe the architectural aspects of 6G DTNs that enable interactive virtual-real mapping and control. The 6G physical network layer refers to the target real-world 6G network, consisting of physical network elements and their operating environment. Depending on the need, a 6G DTN may focus on different parts of the target 6G physical network (e.g., a radio cell, a radio access network, a transport network, and a core network) or the entire end-to-end 6G network. The 6G physical network layer exchanges data and control messages with the 6G twin layer.

The 6G twin layer is the core of a 6G DTN. It consists of three domains: data domain, model domain, and management domain. Data domain is a data repository subsystem responsible for collecting data from the 6G physical network to build accurate and up-to-date models in the twin. The functions of the data repository subsystem may include collection, storage, service, and management of data. Model domain is a service-mapping subsystem comprising of models to represent the real-world objects in the 6G physical network based on the collected data. There are two types of models: basic and functional. Basic models represent network elements and topology to describe the target physical network. Functional models refer to analytical models for extracting insights from the DTN. Examples include network planning, traffic analysis, fault detection, network emulation, and prediction. Management domain is a DT entity management subsystem responsible for the management function of the 6G twin layer. Examples include model management (e.g., model creation, configuration, update, and monitoring), and security management (e.g., authentication, authorization, encryption, and integrity protection).

The 6G network application layer includes various network applications that exploit the capabilities exposed by the 6G twin layer. The network applications provide intent to the 6G twin layer which then emulates the required services and, if needed, sends control messages to the 6G physical network layer. Examples of the network applications include network operations, administration, and maintenance (OAM), network optimization, and network visualization.

There will be diverse 6G DTNs with different capabilities to meet the demands of a wide range of use cases. It is vital to develop common components, generally applicable frameworks, standardized interfaces, and universal platforms and tools to enable interoperability and extensibility of the 6G DTNs. These developments will gradually evolve into unified tools to build the diverse 6G DTNs on universal platforms hosting data, models, and management functions.

FUNDAMENTAL BUILDING BLOCK

With a reference 6G DTN architecture in place, the next question is how we can realize it. The inception of DTN begins with the blending of three fundamental building blocks: data, models, and interfaces. The degree of reliability with which a DTN can mirror its physical counterpart depends on the quality of data upon which its models are built, and the accuracy of the models. Interfaces around a DTN, on the other hand, enable its seamless interaction with the underlying physical network and associated network applications.

Data

Data is the basis for building a 6G DTN. Detailed, up-to-date data enables the construction of high-accuracy models in the DTN. With AI/ML at its core, the efficacy of a 6G DTN relies fundamentally on both the quality as well as the quantity of data collected from different parts of the physical network infrastructure.

What-to-collect (data type): In general, the collected dataset needs to be comprehensive enough to constitute a holistic representation of the physical network. At the same time, proper analysis of specific network topology to determine required data type is crucial to avoid redundant data collection, which may result in unnecessary cost of compute and storage resources. Some of the data type examples include data representing operational and provisional status of a physical network, network telemetry, performance measurement, data from network services/applications and lifecycle management (LCM) of network entities, and user data. Striking the right balance between data quality and quantity is essential in building a sustainable and trustworthy DTN.

When-to-collect (time/frequency): To swiftly adjust to the dynamic variation of a live network, the DTN requires constant monitoring, performance reassessment, and timely adjustment. Collection of data to feed a DTN, therefore, cannot be a one-time event. The frequency and pattern of the time series of data collection depend primarily on the data type. For example, data related to faults/events are usually collected in an event-triggered manner, whereas network performance specific data are usually collected at a regular interval, with potential aperiodic trigger enablement depending on pre-defined conditions/intents. Some types of data that statistically vary over shorter periods of time (e.g., port/interface statistics) require higher collection frequency. On the other hand, some other types of data that are critical to network's operational health (e.g., flow status and link failures) need to be gathered in a time sensitive manner.

How-to-collect (mechanism/tool): Building a comprehensive DTN requires network data with diverse characteristics, which, in turn, necessitates usage of various data collection mechanisms and tools. The state of physical devices may be collected by the associated sensors that send periodic or event-triggered updates to data collectors. The state of physical sites may be captured by using lasers and drones besides connected sensors. Integrated sensing and communication utilizing terahertz (THz) or optical frequency bands can perform high-resolution sensing, localization and tracking, and 3D imaging to collect data for building and updating the models of wireless propagation environments. Data collection capabilities can be embedded in the 6G physical network to collect status about network elements, topology, traffic, etc. Technologies enabling simultaneous and collaborative data collection from heterogeneous network nodes in a performant and time-synchronized manner are critical in establishing a high-quality data collection process to serve as the solid foundation of an ultra-reliable DTN.

Where-to-store (repository): Collection of massive amounts of network data comes with the burden of complex storage requirements. To that end, construction of a unified data repository for myriad network data becomes one of the key prerequisites for creating a successful DTN. The salient features of a hyperconverged data repository include support for heterogeneous database, provision for efficient data services (such as data augmentation and federation, historical data reporting snapshot and rollback, and fast database query), and assurance of high data availability, among others.

How-to-access (retrieval): Acceleration is the key for efficient operations (e.g., data ingestion, search, query, and visualization) on the DTN data repository which has large-volume data and fast data streams ingestion from the physical network. To this end, it is crucial to exploit the massive parallel capabilities provided by GPU and the corresponding tools such as *Dask* to accelerate the three-phase data processing - extract, transform, and load. This is particularly important when the database operations need to be real-time to synchronize the DTN with its physical counterpart, derive lightning-fast data analytics, and achieve immediate closed-loop decisions.

How-to-maintain (management): Long term stability, adaptability, and performance of a DTN depend on the efficacy of the underlying data management framework, including the LCM of data collection, storage, maintenance, and retrieval. Data management can either be an integral part of the overall DTN management framework or can be an independent entity in the twin layer. Automation is arguably the most desired feature of data management, along with other capabilities like secure handling, storage, deposition and destruction of sensitive network data (e.g., data related to user privacy), traceability and accurate records of historical data transactions, and guarantee of data integrity (i.e., accuracy, completeness, and consistency).

Models

DTs of a physical network can be modularized at different levels, depending on the use case requirement. For example, DTNs can be of individual physical network domains (like radio access/core/transport network), or of a combination of collocated network entities/functionalities (like DT of cell site/near edge/far edge/cloud), or at the extreme case, a single DTN for the entire end-to-end physical network. The degree of modularity, in turn, determines specific aspects of a DTN. For example, a DTN responsible for a cell site can be operating at a high-level of detail using high-fidelity ray tracing employing multi-bounce specular reflection, 3D diffraction, and diffuse scattering to enable site-specific optimizations in the base station's physical and medium-access control layers. A DTN in the cloud would be responsible for end-to-end service analysis and network operation including resource management and, for example, AI/ML/DL approaches for network slicing.

Arguably, the most challenging design aspect of these multimodal DTNs is their models that represent accurate digital versions of physical entities constituting the real network peers. The generic requirements of a typical DTN model are manifold. The crux of a DTN model meeting the diverse set of requirements is built upon two pillars: physically accurate modeling and its functional manifestation using AI-powered simulation, visualization and control.

Physically accurate modeling: High-fidelity models are fundamental building blocks for 6G DTNs. The models need to accurately capture the up-to-date characteristics of their physical counterparts, including geometries, materials, properties, lighting, behaviors, and rules, among others. One key requirement of the models for the 6G DTN is the need of physically accurate simulation of radio wave propagations in true-to-reality environments. This is particularly pertinent considering that 6G will utilize THz technology with ultra-wide signal bandwidth, integrate sensing with radar or lidar technology, use extremely large antennas, and incorporate RIS. Therefore, it is vital to accurately model the effects of radio wave reflection, diffraction, scattering, and multipath propagation in the environment. Traditional software ray tracing techniques may have difficulty in performing efficient, physically accurate simulation of the radio wave propagations in the complex 6G systems. Ray tracing at-scale is extremely computationally intensive. At least two components are required to bring radio frequency (RF) ray tracing at-scale to reality in a DTN:

1. Programmable hardware accelerators supporting the mathematical operations at the heart of ray tracing and
2. optimized libraries for the ray tracing pipeline.

An example of a ray tracing hardware acceleration (RTX) is the ray tracing cores found in the many hundreds of streaming multiprocessors that comprise the NVIDIA Ampere class GPU and that are capable of performing many trillions of floating-point math operations per second. There are several ray tracing frameworks available including DX12, Vulkan, and OptiX. Take OptiX for example. Based on compute unified device architecture (CUDA), it is an application framework for achieving optimal ray tracing performance on a GPU that takes advantage of RTX hardware accelerators. OptiX enables the developers to concentrate on wave propagation rather than on low-level ray tracing optimization. Creating the geometry for the DT at interactive rates has become feasible based on capture using instant neural graphics primitives.

AI-powered simulation, visualization, and control: Interactive virtual-real mapping and control is a key distinguishing characteristic of 6G DTN, compared to the traditional link, system, or network simulators. However, real-time interaction and control is challenging due to the large amount of data, stringent latency requirements, and computation-intensive simulations. AI-powered functions are key to achieve an autonomous feedback loop between the 6G physical network and its DTN. As one example, AI techniques can leverage historical and real-time data to create and evolve models in the 6G twin layer to constantly run simulations for anomaly detection and network optimization, leading to intelligent decision making for the target 6G physical network. AI models can also be used to constantly run predictive "what-if" simulations in the 6G twin layer to determine in advance the consequences of hypothetical actions, which allows for impact analysis of changes, prevention of outage events, and finding optimal settings for future operations. In addition, AI techniques can create photorealistic 3D graphics to visualize the 6G DTN and illustrate findings, facilitating the involvement of humans in the decision-making processes.

Interfaces

Open and standard interfaces between twin layer and physical network layer/network application layer are essential towards proliferating a multi-vendor, interoperable DTN ecosystem. In general, the guiding principles of interface protocols around DTNs include extensibility, backward compatibility, easy usage and accessibility, capability of handling high concurrency of dataflow, and mechanisms to ensure secure and reliable communication pathway between DTNs and other network entities.

> *Network-bound interfaces*: As the bridge between a DTN and its underlying physical network, network-bound interfaces provide efficient information exchange between the endpoints, including both control and user plane data. Network-bound interfaces carry various types of network data (e.g., provisional and operational data) collected via different methods such as on-demand, subscription-based, event-triggered, direct measurement, and indirect observation. The interfaces support versatile data requirements from DTN models, resulting in varying data transfer speed requirements ranging from real-time (~milliseconds) to near-real time (~second) and to non-real time (~minute). It is imperative to support management and control signaling through these interfaces for the DTN to configure data collection protocol, time/frequency of collection, and other configurations of network elements in the physical network. Realization of network-bound interfaces using low-latency interconnects is crucial in enabling real-time interactive mapping between the DTN and its physical embodiment. Another important aspect of network-bound interfaces is 'decoupling,' i.e., making the interfaces agnostic of the type of information being transported and specificities of physical entities at its endpoints. Decoupling enables the DTN to use a harmonized interface towards the physical network, without requiring data specific or network entity/topology specific customization.
>
> *Application-bound interfaces*: Delivering requirements (related to network management, optimization, and protocol validation) or intents of network

applications to a DTN is fulfilled through application-bound interfaces. Application-bound interfaces support the capability to convey the DTN's features, including common data models and their performance to authenticated, third-party network applications. Speed, latency, and bandwidth requirements for application-bound interfaces are less stringent and less demanding than network-bound interfaces, which makes these interfaces realizable in relatively lightweight ways with reduced complexity. Application-bound interfaces provide a layer of abstraction towards the DTN, implying that the differences between network applications are transparent to the DTN, and the same set of interfaces serves seamless interaction between the endpoints, without requiring application-specific tailoring.

Intra/inter-DTN interfaces: Depending on the use case, scalability and flexibility requirements of a DTN drive the degree of modularity, resulting in either a single DTN model or multiple DTN models. For the latter, inter-DTN interfaces enable efficient communication between various DTN models that are either co-located or distributed. For a DTN built upon federated learning-based ML models, both communication between distributed model objects (during training) and aggregation of trained objects can be facilitated through intra-DTN interfaces. While standardization of inter-DTN interfaces between multiple DTN models is crucial for multi-vendor implementation, intra-DTN links between multiple model objects leading to a DTN model ensemble can be proprietary.

SUMMARY

We have defined the DT and have taken an example for DT modeling of computing and networks. We have explained how to use AI/ML/DL. AI agents, with the help of ML/DL algorithms, open the range of opportunities to enable optimizations in terms of reliability, robustness, and performance in the DT. The fast growth of network scale towards 6G and the stringent performance requirements of diverse use cases call for innovative tools and platforms. The unique capabilities of DTNs make them a powerful technology for the design, analysis, diagnosis, simulation, and control of 6G wireless networks.

Through AI, DTs are evolving into powerful, dynamic, and automated tools to explore and monitor the whole industrial environment through, for example, an immersive digital world without temporal or spatial constraints.

Altogether, there are several challenges to cope with industrial environments, like the creation and validation of virtual models, the need for expertise from different engineering fields (e.g., robotics, networking, and software), and real-time access, connection, and synchronization to production data. The latter aspect will be a driving factor for the future 5G/6G networks. DT can be used in all disciplines, and scope of modeling the DT is also huge and requires interdisciplinary skill sets. The innovations and applications of DT technology in wireless networks are as vast as imagination. It will be exciting to see how DT technology will improve lives and transform industries on the path to 6G and beyond.

REFERENCES

1. https://en.wikipedia.org/wiki/Digital_twin
2. Groshev, M. et al., "Toward Intelligent Cyber-Physical Systems: Digital Twin Meets Artificial Intelligence," *IEEE Communications Magazine*, August 2021, # 3.
3. Lin, X., et al., "6G Digital Twin Networks: From Theory to Practice," URL: https://safe. menlosecurity.com/doc/docview/viewer/docND8B3C5AAC5A17cfe5b3596c4e6f54b2 e7fcf9e2b593d1558d1c6a809194fe793524ef539391d

HOLOGRAPHIC COMMUNICATIONS

OVERVIEW

It is expected that the 6G communication systems will be endogenously holography-capable and intelligent and have a programmable radio propagation environment, which will offer unprecedented capabilities for high spectral and energy efficiency, low latency, and massive connectivity [1]. The holographic multiple-input multiple-output (HMIMO) antenna system is expected to actualize holographic radios with reasonable power consumption and fabrication cost. The HMIMO is facilitated by ultra-thin, extremely large, and nearly continuous surfaces that incorporate reconfigurable and sub-wavelength-spaced antennas and/or metamaterials. Such surfaces comprising dense electromagnetic (EM) excited elements are capable of recording and manipulating impinging fields with utmost flexibility and precision, as well as with reduced cost and power consumption, thereby shaping arbitrary-intended EM waves with high energy efficiency.

The powerful EM processing capability of HMIMO opens the possibility of wireless communications of holographic imaging level, paving the way for signal processing techniques realized in the EM-domain, possibly in conjunction with their digital-domain counterparts. However, despite the significant potential, the studies on HMIMO communications are still at an initial stage, its fundamental limits remain to be unveiled, and a certain number of critical technical challenges need to be addressed.

We have described that Massive MIMO and ultra-Massive MIMO systems achieve the critical beamforming functionality by using the beam-space model and depicting the spatial domain with beams in specific angular directions, which is considered as a low-dimensional approximation. The approximation optimality is achieved based on a set of ideal assumptions as follows:

- Perfect calibration with predefined antenna array geometry with perfect calibrations
- Considering propagation without mutual coupling and near-field scattering, which is not valid with larger aperture shaped in arbitrary geometries and/ or covered with dense antenna elements

With emerging 6G technologies, new technologies are emerging while compensating the shortcomings of existing architectures with a completed EM field characterization and a full manipulation of the EM wave are considered. Holography, an innovative technology, is capable of recording and reconstructing the amplitude and phase of wavefronts. As a result, holography that enlarges the EM wave manipulation freedom has great potential to enable holographic radios, satisfying the emerging 6G's extreme requirements.

On the other hand, with the emergence and development of metamaterials and metasurfaces, feasible solutions are provided for supporting the realization of holography in EM wave recording and reconstruction [1] with their broad applications on wireless communications. Note that metamaterials indicate a class of artificial composite materials capable of interacting with incident EM waves in various expected effective electric and/or magnetic responses not found in nature. The design structure and employed material define the EM properties of the metamaterial, yielding expected EM responses and enabling desired EM functionalities. In principle, metamaterials can realize arbitrary values of permittivity and permeability, thus enabling them to manipulate EM waves.

Metasurfaces are developed as a two-dimensional (2D) equivalent of volumetric metamaterials, whose meta-atoms form an ultrathin planar structure that can be readily fabricated. Without following the propagation phase accumulating principle of metamaterials, metasurfaces utilize the abrupt phase and amplitude discontinuity of EM waves occurring at the interfaces of meta-atoms. As such, spatially varying EM waves with desired amplitude, phase, and/or polarization can be fully achieved by properly arranging the meta-atoms. Metasurfaces can also be further integrated with a programming capability, programmable (reconfigurable or dynamic) metasurfaces can be formulated.

Holographic MIMO

The HMIMO communications have been formally defined as follows:

> Holographic wireless communication—it is the physical process of realistically and completely restoring the three-dimensional (3D) target scene transmitted by the transceiver ends with the help of new holographic antenna technology and wireless Electro-Magnetic (EM) signal technology, and at the same time realize 3D remote dynamic interactions with people, objects and their surrounding environment.

[1]

The features have been summarized in Table 24.1.

The fundamental changes of HMIMO are from the physical hardware perspective in comparison with the massive MIMO. Massive MIMO antenna arrays appear as a spatially discrete aperture whose inter-element spacing follows half a wavelength condition, which, although simplifies transceiver designs without considering mutual coupling of antenna elements, sacrifices a large amount of spatial information. However, HMIMO surfaces are considered as almost spatially continuous apertures, the spacing of which between antenna elements are much smaller than half a wavelength of incident EM waves.

Consequently, HMIMO surfaces can form sharp beams with weak sidelobes. More importantly, the nearly continuous aperture can control and record almost

TABLE 24.1
Holographic MIMO Features

Holographic MIMO

New Features	Qualitative Changes
Nearly spatially continuous aperture	• Unprecedented EM wave control
Electrically extremely lager aperture	• Near-field holographic-imaging-level spatial
Intelligently reconfigurable and tunable	multiplexing
Low-cost and low-power consumption	• High energy efficiency

continuous phase changes of the wavefront and manipulate EM waves with an unprecedented flexibility. On the other hand, HMIMO surfaces implement amplitude and phase tuning through a totally different hardware structure, replacing a large amount of costly and power-hungry RF devices by either utilizing the holographic-based leaky-wave antennas (LWAs) or the photonic tightly coupled antenna arrays (TCAs). As a result, it facilitates communication signal processing in the analog domain based on reconfigurable and simplified hardware with reduced size, weight, cost, and power consumption.

This also facilitates fabrications of electrically extremely large HMIMO surfaces to combat high path-loss in high frequencies (e.g., THz bands). The differences in hardware structure as well as the quantitative changes of antenna elements (from sparse to dense) and aperture size (from small to extremely large) brought by HMIMO surfaces, inevitably cause qualitative changes in HMIMO communications, which constitutes the second aspect of revolutions. Firstly, the distinct hardware structure in HMIMO corresponds to a dedicated working mechanism that is different from massive MIMO. Therefore, the unique hardware structure and working mechanism necessitate new mathematical models for system depiction, which should capture the essence and comply with the physical constraints.

Another qualitative change emerges as HMIMO antenna elements become more and more dense, and it creates a nearly spatially continuous aperture. Moreover, mutual coupling between antenna elements, considered harmful to communication systems and mitigated in Massive MIMO antenna arrays configured in half a wavelength spacing, cannot be neglected in HMIMO communications. HMIMO mutual coupling can offer super-directivity enabling significantly large antenna array gains. In addition to mutual coupling, spatially continuous apertures allow signal processing to be shifted from conventional digital domain to future EM-domain. Consequently, new analysis and design ideas from electromagnetism will be introduced to revolutionize existing wireless communication frameworks to (hybrid digital-) EM-domain ones, paving the way for realizing high flexibility, high spatial resolution, low latency wireless communications. For example, communication models and channel models can be characterized in the EM-domain. Shannon's information theory that ignores the underlying physical phenomena of EM wave propagation needs to be augmented with EM theory.

Lastly, the remaining qualitative change of HMIMO is induced by the extremely large aperture sizes of HMIMO surfaces. Distinct to massive MIMO communications, always considering far-field scenarios, HMIMO can naturally transform the far-field (i.e., more than one full wavelength) region to the near-field region (i.e., the Fresnel region – less than one full wavelength) as the aperture size increases significantly, which enables holographic near-field communications. Compared with massive MIMO far-field communications that are angle-aware, HMIMO nearfield communications can discriminate not only the angle of an object but also its distance. This leads to a totally distinct near-field channel model, and the conventional angle-aware beamforming transforms to the distance-angle-aware HMIMO beam focusing. This will bring significant benefits in communication performance, such as broadening the degree of freedom (DOF) of communication systems.

Taking advantage of the unprecedented flexibility in EM wave manipulation as well as the near-field communications, we expect that HMIMO communications will realize the holographic imaging-level radios with ultra-high pixel density and extremely large spatial multiplexing, which can be made possible because the nearly infinite number of antennas in HMIMO constitute the asymptotic limit of massive MIMO, assuming very large capacity. It is worth noting that above-described HMIMO are mainly emphasized for HMIMO surfaces being active transceivers. They can also be operated as passive reflectors, coinciding with the reconfigurable intelligent surfaces (RISs) or the intelligent reflecting surfaces (IRS) that are deployed at positions between transceivers, the wireless environment treated conventionally as a random process can thereby be transformed to the smart radio environment that can be intelligently software programmable, enhancing communications performance.

Basic Principle of Holography

Recoding and reconstruction are the two main basic steps used for the realization of the holography mainly, as depicted in Figure 24.3. In the recording process, the recording media can be such as holographic plates, charge coupled device, or complementary metal oxide semiconductor cameras, etc. are employed to track intensities of the hologram (interference pattern) formed by interfering a known reference wave with a desired object wave.

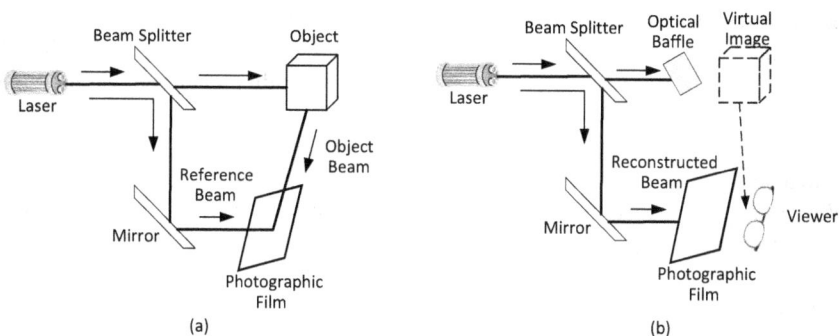

FIGURE 24.3 Schematic view of optical holography: (a) recording and (b) reconstruction.

In the reconstruction process, the employed recording medium is illuminated by a replica of the reference wave, thereby reconstructing the object wave perfectly. The main idea of this realization is that the interference between two coherent waves presents their phase differences that can be used for object wave reconstruction subsequently. We illustrate this idea in the following equations. First, we denote the object wave $\mathcal{O}$ as:

$$\mathcal{O} = |\mathcal{O}| e^{j\theta} \tag{24.1}$$

with $|\mathcal{O}|$ being the wave amplitude and θ being the wave phase. Likewise, we present the reference wave $\mathcal{R}$:

$$\mathcal{R} = |\mathcal{R}| e^{i\varnothing} \tag{24.2}$$

Assuming an intensity-sensitive recording media (alternatively, it only records the phase information), we thus have this intensity of the hologram represented as:

$$\mathcal{I} = |\mathcal{O} + \mathcal{R}|^2 = |\mathcal{O}|^2 + |\mathcal{R}|^2 + \mathcal{O}\mathcal{R}^* + \mathcal{O}^*\mathcal{R} \tag{24.3}$$

where $*$ indicates the complex conjugate operation. One can directly find that the last two terms in (24.3) include the phase difference between the reference and object waves, critical for object wave reconstruction. Retaining the intensity of the hologram by recording media and illuminating the recording media with a replica of the reference wave, we have the following expression:

$$\mathcal{I}\mathcal{R} = |\mathcal{O}|^2 \mathcal{R} + |\mathcal{R}|^2 \mathcal{R} + \mathcal{O}|\mathcal{R}|^2 + \mathcal{O}^* |\mathcal{R}|^2 \tag{24.4}$$

One can see from the last two terms of (24.4) that it is possible to completely reconstruct the object wave that possesses both the intensity and phase information.

Holography is primarily realized via the optical technology, where coherent light sources are produced by a laser. The recording process of optical holography can be found in Figure 24.3a. It can be seen that the laser beam propagates to a beam splitter that divides the single input into two coherent output light beams. On the one hand, the first output beam is controlled to propagate to an object such that the desired object beam is scattered by the object. On the other hand, the second output beam is guided to a mirror for the purpose of reflecting the beam as a reference beam toward the photographic film to superpose with the reflected object beam and thus form a hologram. In this process, the hologram is recorded by the photographic film. With a successful recording process, one can reconstruct the object beam through the experiment setup shown in Figure 24.3b. It is noted from the figure that the object is removed and an extra-optical baffle is added for blocking the first output light beam.

In this reconstruction process, the laser produces a replica of the beam in the recording process. This beam then propagates along the same path as in the recording process from the beam splitter to the mirror, generating a reconstruction beam identical to the reference beam accordingly. The reconstruction beam illuminates the hologram recorded on photographic film such that the object beam is reconstructed based on light diffraction. From the side of the viewer, a virtual 3D image of the original object is generated. In the optical holography roadmap, different schemes, such as (compressive) optical scanning holography and the phase-shifting holography, each with a unique hardware structure, have been presented.

CONVENTIONAL HOLOGRAPHY

Conventional optical holography requires a vast amount of complicated optical components for recording and reconstruction. It also needs a real object for different hologram acquisition, which is inefficient and limits the generalization for imaging of different objects. With the assistance of computer technology and spatial light modulators (SLMs), computer-generated holography (CGF) was invented for mitigating the problems encountered in optical holography. Instead of recording and reconstructing via a hardware-dependent optical manner, CGH performs both recording and reconstruction through numerical calculations. This working mechanism not only reduces the requirement for complicated hardware but also facilitates imaging for different objects without requiring real objects. The schematic of CGH is depicted in Figure 24.4.

Basically, the hologram design methods of CGH can be divided into two main groups:

- Wavefront-based methods
- Ray-based methods

Through simulating the wave diffraction process, wavefront-based methods numerically calculate the 3D wave fields of a given object/scene, as well as its 2D distribution on the hologram plane. In this group, the following are the most widely adopted approaches:

- Point cloud model
- Polygon-based model
- Layer-based representation of 3D objects/scenes, all utilizing 3D positional information

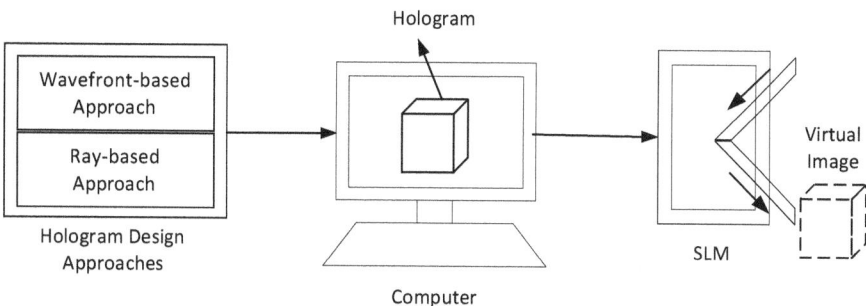

FIGURE 24.4   Schematic view of computer-generated holography.

Differently, capturing incoherent 2D images of a given 3D object/scene based on the transformation from ray-based representations to wavefront-based holographic information, ray-based methods generate the hologram accordingly. This group consists of two typical categories: the holographic stereogram and the multiple viewpoint projection holography.

With arbitrarily computer-generated holograms illuminated by the reconstruction beam, high-fidelity virtual 3D images can be reconstructed. This kind of holography is primarily for image processing. However, we will be focusing primarily on EM holographic communications using HMIMO antennas.

EM HOLOGRAPHY

We know that holography can realize 3D imaging as mentioned in optical holography and CGH. This kind of imaging can be interpreted as wave field reconstruction of a given object, which can be generalized to the EM region.

Combining holography with antenna technologies, novel holographic wireless communications can be achieved based on the EM holography. In this regard, we embrace two typical schemes for realizing the EM holography:

- Holographic LWA-based EM holography
- Photonic TCA-based EM holography

HOLOGRAPHIC LWA-BASED EM HOLOGRAPHY

In such type of communication systems, an EM wave source used for reference wave generation replaces the role of the laser in optical holography,

- An EM antenna aperture, referred to as the HMIMO surface, plays the role of photographic film.
- One or more communication nodes take the place of the object, correspondingly.

It is noteworthy that the EM wave source can be located externally or integrated internally into an HMIMO surface, which is one of the main differences compared to optical holography structures. Take the typical integration case for example, we present the schematic of EM holography with recording and reconstruction processes in Figure 24.5, complying with those shown in optical holography. An HMIMO surface commonly consists of a substrate with one or more feeds integrated inside and antenna elements printed on its surface. The substrate serves as a waveguide, allowing reference waves to propagate along it.

The feeds are used for launching reference waves coming from RF chains. The antenna elements are designed to construct various holograms with corresponding explicit textures or to approach holograms via different tuning mechanisms without presenting a specific texture. With configured holograms, specific radiations, with radiating signals leaked from the reference waves, can be realized.

During the recording process, the hologram should be designed and constructed by the HMIMO surface. By backpropagating the object wave from a given direction

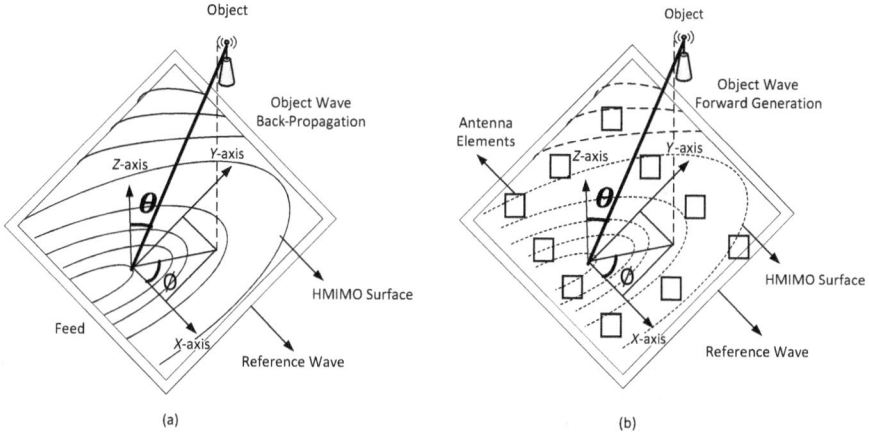

FIGURE 24.5   LWA-based holography: (a) recording and (b) reconstruction.

to the HMIMO surface, a certain hologram is obtained as a superposition of object and reference waves, as depicted in Figure 24.4a. To explicitly show this process, we describe the reference wave excited by a point source within a lossless substrate as:

$$E_{rw} = A_r e^{-i\beta_r d_r(x,y)} \tag{24.5}$$

where A_r denotes the amplitude of reference wave, a constant in the lossless substrate; β_r represents the wavenumber of reference wave; $d_r(x, y)$ indicates the distance between the (x, y)-coordinate of the HMIMO surface and the point source (assumed as the original coordinate), expressed as:

$$d_r(x, y) = \sqrt{x^2 + y^2}$$

Additionally, denote by $d_0(x, y)$ the distance between the object and the (x, y)-coordinate, by β_0 the free space wavenumber, as well as by $\varnothing$ and θ the azimuth and elevation angles of the object, respectively. The phase variation caused by $d_0(x, y)$ can be described through a projection of the free space wavenumber on x and y axis, respectively. We express this process as:

$$\left[\beta_0 \sin\theta\cos\varnothing, \beta_0 \sin\theta\sin\varnothing\right]\begin{bmatrix} x \\ y \end{bmatrix} = \beta_0\left(x\sin\theta\cos\varnothing + y\sin\theta\sin\varnothing\right)$$

Based on the result, the field distribution of object wave on the (x, y)-coordinate is:

$$E_{obj} = A_0 e^{i\beta_0 d_0(x,y)} = A_0 e^{i\beta_0(x\sin\theta\cos\varnothing + y\sin\theta\sin\varnothing)} \tag{24.6}$$

where A_0 indicates the wave amplitude. At any given (x, y)-coordinate, the hologram can be formulated based on (24.5) and (24.6) as:

$$E_{int} = E_{obj}E_{rw}^*$$

which is detailed as follows:

$$E_{int} = A_0 A_r e^{i\left[\beta_0(x\sin\theta\cos\varnothing + y\sin\theta\sin\varnothing) + \beta_r\sqrt{x^2+y^2}\right]} \tag{24.7}$$

By properly designing the HMIMO surface for capturing the hologram, one can realize the recording process of EM holography. It is noteworthy that HMIMO surfaces can be implemented through various hardware forms in fixed or tunable mechanisms, which will be detailed in the next section.

In reconstruction process, with the hologram implemented by the HMIMO surface based on (24.7), an object wave toward the $(\varnothing, \theta)$-direction can be forward generated once the reference wave is excited and travels along the surface, as demonstrated in Figure 24.3b. The mapping rule of an HMIMO surface from reference wave to object obeys the LWA that follows the wave diffraction principle.

PHOTONIC TCA-BASED EM HOLOGRAPHY

In photonic TCA-based EM holography communication systems, the EM holography is realized through an optical domain processing with the assistance of photonic TCA-based HMIMO surfaces that realize a holographic RF-optical mapping capable of achieving a domain transformation from RF signals to optical beams and vice versa. The mapping is enabled by electro-optic modulators (EOMs) and uni-traveling-carrier photodetectors (UTC-PDs), responsible for transformations from electrical to optical signals in reception stage and oppositely from optical to electrical signals in transmission stage, respectively. Each EOM, connected to one antenna element, upconverts the received RF signal to optical regime to be propagated by an optical-fiber bundle. Additionally, each UTC-PD is bonded to adjacent antenna elements enabled by flip-chip technology. UTC-PDs can convert optical beams to electrical signals with high power, large bandwidth, and high converting efficiency, making them feasible to directly drive the antenna elements with a very large bandwidth, i.e., ≥ 40 GHz.

LOCALLY MAXIMUM PHASE LINES OF HOLOGRAM

The primary holographic design is achieved via the locally maximum phase lines of hologram for its straightforward and easy interpretation. It should be first noted that the hologram generated by the reference and object waves is indeed an interference wave with a certain period. This hologram includes a certain number of locally maximum phase lines within a dedicated distance, each of which exists within one period.

This design methodology enables antenna elements to be located at the locally maximum phase lines of the hologram, forming a surface pattern consisting of a series of curve/line-shaped strip gratings. The resulting surface pattern allows the generation of an object wave with a specific direction. To explicitly show this methodology, we assume a reference wave propagating in the y-direction $E_{rw} = A_r e^{-i\beta ry}$, and an object wave steering toward a given direction, $\theta = \theta_s$ and $\varnothing = \pi/2$. Based on (24.7), the hologram is reduced to:

$$\hat{E}_{int} = A_r A_0 e^{i(\beta_r + \beta_0 \sin\theta_s)y} \tag{24.8}$$

whose wavenumber is given by $(\beta r + \beta_0 \sin\theta s)$, equivalent to wavelength of:

$$\frac{2\pi}{\left(\beta_r + \beta_0 \sin\theta_s\right)}$$

Within each wavelength of the hologram, the antenna element is placed on the locally maximum phase line (namely, the peak value achieved by coordinates). All locally maximum phase lines of the hologram within the aperture compose the surface pattern constructed by a series of curve/line-shaped strip gratings.

From such a design methodology, one can see that the phase information between strip gratings is inevitably lost, lowering the performance of holographic mapping between reference wave and object wave. We can alternatively interpret this from a sampling perspective that the antenna elements take a small number of specific samples of the reference wave, incurring the result of under-sampling. Another noticeable question is that the HMIMO surfaces based on this design methodology are not reconfigurable. Each surface pattern only corresponds to a specific wave radiation, such that one should change the surface pattern to manipulate the radiated wave.

MACROSCOPIC SURFACE IMPEDANCE-BASED APPROACH

Deploying a large amount of sub-wavelength conductive patches over the surface, the reference wave can be densely sampled. The relatively small sizes of antenna elements and adjacent distances compared with wavelengths of both reference and object waves enable the scatter properties to be described using macroscopic effective surface impedance. The macroscopic effective surface impedance is defined as the ratio of electric field E_x to magnetic field H_y near the surface averaged over the unit cell, which can be expressed as [1]:

$$Z = \int_{cell} \frac{E_x}{H_y} ds$$

It is related to substrate permittivity and thickness, as well as the period and gap between adjacent antenna elements. For instance, high impedance values are achieved with high substrate permittivity, thick substrate, large period, and small gap. Under a desired surface impedance, the antenna elements can be engineered in various shapes.

Holographic architecture realized by the macroscopic surface impedance-based approach is established on collecting a large amount of specifically designed antenna elements for obtaining the expected surface impedance that is determined by the hologram created via the interference between reference and object waves. To be more specific, the surface impedance of (x, y)-coordinate is given by:

$$Z(x,y) = i\left[X + M\mathfrak{N}\left(E_{\mathrm{obj}}E_{rw}^*\right)\right] \tag{24.9}$$

where X is the real-valued average impedance value, and M denotes the real-valued modulation depth that spans the entire available range of impedance values Equation (24.9) builds a mapping between the surface impedance and the hologram. Combining this mapping with the relation between surface impedance and geometry parameters of substrate and antenna elements, one can seamlessly correspond the hologram to geometry parameters. With a selected substrate (permittivity and thickness are fixed), the hologram is implemented as a surface pattern that is constructed by many patches with specific periods and gaps, displayed as concentric ellipses, spiral or concentric circles. These textured surface patterns correspond to different radiation properties, determining the radiation directions. Holographic design via macroscopic surface impedance can be dated back to the sinusoidally modulated reactance surfaces whose modal surface impedance is modulated sinusoidally for realizing an expected radiation.

It should be emphasized that the above surface impedance is designed for radiating a single object wave. When multiple object waves are required to be radiated simultaneously, the surface impedance is designed as a division or a superposition of multiple textured surface patterns on one shared aperture. For the division case, the surface impedance design of each division follows from (24.9), indicating a corresponding object wave. While for the superposition case, the surface impedance design is guided by the following expression:

$$Z(x,y) = i\left[X + \frac{M}{K}\sum_{k=1}^{K}\mathfrak{N}\left(E_{\mathrm{obj}}E_{rw}^*\right)\right] \tag{24.10}$$

where K is the number of object waves. In this case, the HMIMO surfaces allow multibeam radiations simultaneously, where each object wave is excited by one reference wave. Taking the surface impedance approach for holographic design can be boiled down to configuring the geometrical parameters of substrate and antenna elements, which impose a limitation on the realization of reconfigurable HMIMO surfaces, namely once the geometry parameters are designed, the radiating properties are determined accordingly. Tunability can be achieved by introducing special materials, such as graphene. Bridging the connection between conductivity of graphene and surface impedance, the radiation properties are tunable with an external DC control on graphene's conductivity.

GEOMETRIC POLARIZABLE PARTICLE-BASED APPROACH

An alternative holographic design is achieved from a geometric perspective via using the polarizable particles (dipoles)-based approach. It is feasible as an antenna element is small with respect to the free space wavelength; thus, its radiation field can be well described by that of a dipole. This approach describes the generated radiations as a weighted sum of far-field patterns of all dipoles. Each antenna element corresponds to a weight that follows a specific constraint and is configured by the holographic principle. For example, assume the far-field approximation of a radiated wave from an 1D microstrip line HMIMO surface as one microstrip line of the aperture such as curve/line-shaped strip gating, slitted square/ellipse-shaped patches, coffee bean, grain of rice, double π, double anchor, circle-shaped patches with cross slot, square hexagonal-shaped loop-wire units, cross-shaped patch, C-shaped element, and slot shaped of the company CELEC.

Under these settings, the radiation wave at a distance $do(x, y)$ can be expressed as:

$$E_{\mathrm{rad}} = \frac{A_r \omega^2}{4\pi d_o\left(x,y\right)} e^{-i\beta_0 d_o\left(x,y\right)} \times \cos\theta \sum_{n=1}^{N} \alpha_n\left(\omega,x,y\right) e^{-i\beta_r \, d_r\left(x,y\right)} e^{-i\beta_0 d_r\left(x,y\right)\sin\varnothing}$$

$$= \frac{A_r \omega^2}{4\pi d_o\left(x,y\right)} e^{-i\beta_0 d_o\left(x,y\right)} AF\left(\theta,\varnothing\right)$$

where,

$$AF\left(\theta,\varnothing\right) \triangleq \cos\theta \sum_{n=1}^{N} \alpha_n\left(\omega,x,y\right) e^{-i\beta_r \, d_r\left(x,y\right)} e^{-i\beta_0 d_r\left(x,y\right)\sin\varnothing} \qquad (24.11)$$

and ω denotes the operating frequency, $\alpha_n(\omega, x, y)$ represents the weight of the *nth* antenna element (located at the (x, y)-coordinate) at frequency ω, and $AF(\theta, \varnothing)$ is defined as the array factor. It is emphasized that β_r and $dr(x, y)$ are the same as in (24.5). Comparing the object wave in (24.6) with the radiation wave in (24.11), we should model the array factor $AF(\theta, \varnothing)$ to be some constant ensuring $E_{\mathrm{obj}} = E_{\mathrm{rad}}$, which means that the HMIMO surface radiates the desired wave to the object. As such, the weights of all antenna elements satisfying the requirement can be obtained as:

$$\alpha_n\left(\omega,x,y\right) = e^{i\beta_r \, d_r\left(x,y\right)} e^{jd_r\left(x,y\right)\sin\varnothing} \qquad (24.12)$$

One can see from (24.12) that the weights require full control over the phase of each antenna element, which cannot be satisfied due to the inherent constraints faced by each antenna element. Each antenna element can be considered as a resonant electrical circuit scattering as a dipole, and each weight (overlooking the index n and its coordinates (x, y)) has the Lorentzian form expressed as [1]:

$$\alpha\left(\omega\right) = \frac{F\omega^2}{\omega_0^2 - \omega^2 + i\omega\gamma} \qquad (24.13)$$

where F is the real-valued oscillator strength, ω_0 denotes the resonance frequency, and $\gamma = \dfrac{\omega_0}{2Q_m}$ describes the damping factor with Q_m being the quality factor of the resonator. The amplitude and phase of the weight are coupled through the connection $|\alpha(\omega)| = \dfrac{F\omega|\cos\psi|}{\gamma}$ with $\tan^{-1}\left(-\dfrac{\gamma\omega}{\omega_0^2-\omega^2}\right)$, which limits the weight range to a restricted subset compared with the independent control over amplitude and phase. The antenna element can be tuned in two possible ways, namely by either shifting the resonance frequency or changing the damping factor. Depending on the selected tuning case, the weights can be configured in three forms:

- Amplitude-only
- Binary amplitude
- Lorentzian-constrained phase

First, in the amplitude-only case, the antenna element is near resonance such that $\alpha(\omega) = -\dfrac{F\omega}{\gamma}$. By adjusting the oscillator strength F or the damping factor γ, amplitude tuning of the weight can be achieved without changing its phase. The amplitude-only weight is deduced from (24.12) by taking its real part in the following formulation [1]:

$$\alpha_n(\omega, x, y) = X_n + M_n \cos\left[\beta_r\, d_r(x,y) + \beta_0 d_r(x,y)\sin\varnothing\right] \qquad (24.14)$$

where X_n and M_n are real-valued positive variables. Additionally, the binary amplitude case is applicable to antenna elements tuned between "ON" and "OFF" states, which can be realized by toggling the resonance frequency between a valid value within the operating frequency and an invalid value outside the operating frequency. In this case, each antenna element achieves only two amplitudes, given by [1]:

$$\alpha_n(\omega, x, y) = X_n + M_n\Theta\cos\left[\beta_r\, d_r(x,y) + \beta_0 d_r(x,y)\sin\varnothing\right] \qquad (24.15)$$

where $\Theta \in \{0, 1\}$ enables $\alpha_n(\omega, x, y)$ to be an offset square wave. Finally, the Lorentzian-constrained phase case depicts the inherent coupling between amplitude and phase faced by the weight. The Lorentzian resonator limits the range of phase within $[0, \pi]$, losing the untouchable set $(0, 2\pi]$. Building the following weight function expressed as:

$$\alpha_n(\omega, x, y) = \frac{i + e^{i\left(\beta_r\, d_r(x,y) + \beta_0 d_r(x,y)\sin\varnothing\right)}}{2} \qquad (24.16)$$

the phase of $\alpha_n(\omega, x, y)$ is ensured to have a range of $[0, \pi]$, and the amplitude satisfies the constraint of:

$$|\alpha_n(\omega, x, y)| = \left|\cos\frac{\beta_r\, d_r(x,y) + \beta_0 d_r(x,y)\sin\varnothing}{2}\right|$$

Tuning Mechanisms of HMIMO Surfaces

Nontuning mechanism is fixed in designs and fabrication process, but paradigm is shifting to the dynamically reconfigurable tuning capabilities such as lumped element tuning, liquid crystal tuning, graphene tuning, optical tuning, and others.

- Lumped element tuning: Positive-intrinsic-negative (PIN) diode is an example of this tuning presenting two states "ON" and "OFF" using direct current (DC) bias voltage for reconfiguring HMIMO surfaces.
- Liquid crystal (LC) tuning: The LCs also demonstrate an excellent tuning capability under external stimuli, such as electric or magnetic bias field, which controls the permittivity tensor of LCs. The tunability is utilized for achieving reconfigurable HMIMO surfaces.
- Graphene tuning: The excellent electronic properties of graphene allow a realization of dynamically controllable conductivity via electric or magnetic manners, enabling a feasible way for implementing reconfigurable devices, particularly for HMIMO surfaces.
- Optical tuning: In optical tuning, the reconfigurability of HMIMO surfaces can be implemented via either utilizing photosensitive semiconductors, such as silicon, or driving dedicated devices, such as photodiodes.
- Others: Some emerging tuning mechanisms are under study for HMIMO surfaces reconfigurations by controlling a piezoelectric actuator for varying the mechanical separation between antenna elements and a ground layer.

HMIMO Surface Fabrication Methodologies

Photolithography, electron beam lithography, focused-ion-beam lithography, interference lithography, self-assembly lithography and nanoimprint lithography are primarily used for HMIMO surfaces.

HMIMO Surfaces Aperture Shapes

2D planar HMIMO surfaces is commonly used, where square/rectangle-shaped apertures and circular/hexagon-shaped apertures are currently typical forms employed, although 1D microstrip line is the simplest aperture shapes.

HMIMO Surfaces Functionalities

EM wave polarization and EM wave steering are two most important functionalities for HMIMO surfaces.

EM Wave Polarization

EM wave polarization indicates the oscillation orientation of an HMIMO surface transverse wave. Both fixed and reconfigurable HMIMO surface wave polarizations are now possible. The polarization control is achieved by dividing the HMIMO surface into different regions and changing the phase of surface impedance modulation of one region relative to others, capable of realizing arbitrary linear and circular polarization. Alternatively, an orthogonally discrete unit cell with four working states has been proposed, for realizing linearly and circularly polarized waves via matching surface impedance along two orthogonal directions. The polarization functionality can achieve a transformation of the oscillation orientation via different designs or tuning mechanisms.

EM Wave Steering

EM wave steering, which defines the process of generating a single beam or multiple beams in different directions by constructively or destructively adding the phases of EM waves, is being demonstrated as single-beam HMIMO radiation surface wave. The single beam is generated based on strip grating (discrete dipole) radiations that are constructively superposed in-phase in the desired direction and destructively canceled in remaining directions. The reconfigurability has further achieved facilitated by PIN diodes and enabled by LCs. Beyond realizations of single-beam radiation, the steering capability for supporting multibeam radiations attracts significant interest for the promising application potential in supporting future communication systems.

HMIMO SURFACES PROTOTYPES

The HMIMO surface theory has been validated by development of many HMIMO surface prototypes. These include two $\frac{55.5\,mm}{49.0\,mm} \times 30\,mm$ apertures with 20 and 12 strip gratings (antenna elements), respectively, for achieving a fixed direction radiation at 60 GHz with more than 20 *dBi* maximum gain achieved. Several initial commercial products (39 GHz CPE beamformer, 28 GHz RAN beamformer, 28 GHz repeater beamformer, and 14 GHz A2G beamformer) have been released. These products highlight the potential of HMIMO surfaces in promoting communication systems with a tremendous reduction in cost, size, weight, and power consumption.

HMIMO SURFACE-AIDED COMMUNICATION PROTOTYPES

A 256-element aperture as RF chain-free transmitter and space-down-conversion receiver was first integrated into a communication system to form an E2E MIMO prototype working at 4.35 GHz. Using a 16-quadrature amplitude modulation (QAM), an experimental 2×2 MIMO-16QAM transmission with 20 megabit-per-second data rate is achieved, validating the great potential to enable the cost-effective and energy-efficient systems. They demonstrated various E2E communication prototypes, including binary frequency-shift keying/phase-shift keying/QAM transmitters, pattern modulation system, multichannel direct transmission system, and space-/frequency division multiplexing system.

HMIMO COMMUNICATIONS NETWORKING

The enormous physical advances of HMIMO surfaces described earlier increase the possibility of turning HMIMO communications into reality. However, the corresponding theoretical foundations of HMIMO communications are still under development. A fundamental transformation of HMIMO surfaces design is the shifting from the conventional digital domain to the EM domain. The new domain opens new possibilities for wireless communications for development of a fundamentally a new theory combining both new emerging EM wave sampling and existing EM information theory. One challenge is that the channel model of HMIMO systems faces an underlying shift as the dense packing of nearly infinite small antenna elements and the large area deployments of HMIMO surfaces. They will not only unveil the fundamental limits of HMIMO communications and facilitate the ultimate performance analysis on HMIMO systems but also lay the foundation to develop critical enabling technologies.

Channel Modeling

The channel model of HMIMO systems faces an underlying shift as the dense packing of nearly infinite small antenna elements and the large area deployments of HMIMO surfaces. According to the physical aspects of HMIMO surfaces presented previously, they can be mathematically modeled as uniform planar array (UPA) apertures with small antenna element spacing, or spatially continuous planar surface (CPS) apertures. UPA for HMIMO has an antenna element spacing (denoted by Δ) much smaller than half a wavelength (denoted by λ), i.e., $\Delta \ll \lambda/2$. The rectangular shape of apertures as our indicative example and demonstrate the UPA and CPS for HMIMO. The UPA-based aperture corresponds to the spherical-wave propagation channel model, and the CPS-based aperture is in line with the tensor Green's function-based channel.

DOF

The DOF represents the number of communication modes of EM waves, which reveals that the number of independent data streams can be transmitted simultaneously by the wireless propagation media. Many studies have been exploring the optimal communication capability of an HMIMO system along with their limitations. Since the mathematical modeling of HMIMO transceivers shifts from the conventional discrete antenna array to the continuous EM surface, and the far-field region tends to be near-field, the DOF of HMIMO systems varies accordingly, which needs to be examined.

Conversely, another model put an emphasis on a far-field isotropic scattering environment and analyzed the spatial DOF of a point-to-point HMIMO system. They examined the DOF for 1D linear, 2D planar, and 3D volumetric apertures, respectively, based on the newly derived 4D Fourier planewave series expansion of the channel response. The results provided a guideline of discrete antenna elements spacing to achieve the DOF, which is $2L_x/\lambda$ for 1D linear apertures, $\pi L_x L_y/\lambda^2$ for 2D planar apertures, and $2\pi L_x L_y/\lambda^2$ for 3D volumetric apertures, where L_x and L_y are lengths of the surface. The work has then extended to a more general study on the relation between DOF and Nyquist sampling under arbitrary propagation conditions.

The work has been then extended to a more general study on the relation between DOF and Nyquist sampling under arbitrary propagation conditions. This was performed by modeling the EM wave propagation to a corresponding linear system, for which multidimensional sampling theorem and Fourier theory are applied for analysis.

The study showed that the DOF per unit of area is the Nyquist samples per square meter for large antenna surfaces. With further consideration in the presence of evanescent waves, the spatial DOF for near-field HMIMO communications has been studied using the Fourier plane-wave series expansion. It mainly focused on the isotropic scattering environment while is capable of being extended to the non-isotropic case. The study revealed that the evanescent waves can be further exploited to add extra-spatial DOF and increase the system capacity.

The DOF has been explored through a specially developed joint spatial-temporal correlation model for isotropic scattering environment. Therein, they noticed that the spatial DOF decreases with an increase of the number of antennas of HMIMO transceivers, which seems counterintuitive. An additional explanation of this anomalous phenomenon has been presented via utilizing the power spectrum of the spatial correlation function. With a particular emphasis on the existence of mutual coupling, this study presented an analysis on the effective spatial correlation as well as the eigenvalue structures of the spatial correlation matrix in terms of different antenna element intervals, by leveraging a specially designed metric that indicates the interelement coupling strength. The corresponding results revealed the connection between eigenvalues and evanescent waves, which are potentially beneficial for near-field communications.

System Capacity

The intrinsic system capacity for emerging HMIMO communication systems is yet to be developed. Since HMIMO exhibits strong mutual coupling effect due to numerous of antenna elements and behaves differently in near-field and far-field regions, the fundamental limits require to be uncovered and novel performance evaluation methods should be developed. Some studies were performed based on channel types: line-of-sight (LOS) and non–line-of-sight (NOLS) channel capacity.

The near-field capacity limit of point-to-point HMIMO systems has been analyzed in an LOS propagation environment considering a more fundamental EM perspective. They applied a tensor Green's function-based channel model. A tight upper bound was derived based upon an EM-domain analysis framework, where it revealed that the capacity limit grows logarithmically with the product of transmit element area, receive element area, and the combined effects of $\frac{1}{d_{mn}^2}, \frac{1}{d_{mn}^4}$, and $\frac{1}{d_{mn}^6}$ over all transmit and receive antenna elements, where d_{mn} indicates the distance between each transmit and receive element.

In the case of NLOS HMIMO communications, the channel between HMIMO transceivers includes scatterers that influence the wave propagation. The system capacity of a point-to-point NLOS HMIMO system has been investigated based on the 4D Fourier plane wave representation of HMIMO channels with arbitrary spatially stationary scattering. Particularly, the wavenumber domain channel was established, instead of using the conventional spatial domain channel model, to capture the essence of the physical channel and used to evaluate the system capacity for rectangular volumetric arrays. On this basis, the Fourier plan wave representation of HMIMO channels has been extended to the scenario including multiple user equipment (UEs) with each equipped by an HMIMO surface, based on which they investigated the system capacity using the maximum ratio transmission (MRT) and zero-forcing (ZF) precoding schemes, respectively. The study revealed that large spectral efficiency can be achieved by packing more antenna elements on HMIMO transceivers. Moreover, as spaces among antennas are reduced, strong mutual coupling deteriorates the spectral efficiency under a fixed number of antenna elements.

EM Wave Sampling

EM waves modulated by HMIMO are continuous in space. So, they have to be sampled and discretized for digital processing, which is related to the sampling of spatial EM waves aiming at retaining the maximum EM information with the minimum samples. The sampling has been investigated at the Nyquist rate, and allows to fully capture DOF of EM wave with a minimum number of samples. Specifically, the study has demonstrated using the redundancy of conventional half a wavelength interval sampling approach. They extended the proposed spatial domain Nyquist sampling to the non-isotropic scattering environment and made a preliminary design of the Nyquist sampling matrix for the complex environment to derive sampling efficiency. The results revealed that a reduction of 13% samples per square meter is realized compared to half a wavelength sampling for isotropic propagation with elongated hexagonal sampling structure, and more sample reduction is expected for non-isotropic propagation.

EM Information Theory

The communication channel, treated by Shannon's information theory, is merely the medium used to transmit the signal between transmitter and receiver, which is mathematically described as a conditional probability distribution. However, Shannon's theory neglects the physical effects of actual signal transmissions, which was emphasized by Dennis Gabor [1]. Metamaterials, metasurfaces, and HMIMO, and others need to use the EM information theory, which is an interdisciplinary framework to evaluate the fundamental limits of wireless communications with a fusion of EM theory and information theory. Beyond the conventional analysis and design framework built based upon Shannon's information theory, EM wave theory and circuit theory are two extra frameworks expected to be incorporated into the EM information theory. These frameworks are becoming more and more important and effective in analyzing newly emerged wireless systems, such as HMIMO. Table 24.2 highlights the metrics of HMIMO and Massive MIMO.

Artificial Intelligence and HMIMO

AI has been used to address the Terahertz band (0.1–10 THz) HMIMO communications problems. We have described earlier that there are so many complex technical issues of HMIMO. Recently, multihop RISs are used for realizing smart radio propagation environment where AI has been applied. A high-level modeling configuration of deep reinforcement learning (DRL)-based multihop RIS radio communications is depicted in Figure 24.6 [2].

Particularly, multiple passives and controllable RISs are proposed to be deployed to assist the transmissions between the BS and multiple single-antenna users. The joint design of digital beamforming matrix at the BS and analog beamforming

TABLE 24.2
HMIMO and Massive MIMO Communications Models Comparison

Metrics	HMIMO	Massive MIMO	Remarks
Aperture	Nearly continuous aperture	Discrete aperture	HMIMO is poised for near-field surface communications
Antenna element spacing	Much smaller than half a wavelength	Half a wavelength	HMIMO has much more capacity along with low latency
Mutual coupling	Ultra-high	Low (negligible)	More information is being transferred by HMIMO because of mutual coupling
Antenna array gain	High (super-directivity)	Low	HMIMO is capable of 360° virtual reality communications
Aperture area	Extremely large	Small or moderately large	Massive MIMO carries limited information compared to that of the HMIMO
Beam modes	Polarization and OAM modes	Mainly polarization modes	
Number of beam modes	Infinite modes (theoretically) (OAM modes)	Three modes (linear/circular/ellipse polarization)	EM modeling of HMIMO allows to carry more information, but not by others
Communication model	EM level model (Maxwell's equations, Helmholtz equation, tensor Green's function, Fresnel Kirchhoff diffraction). Circuit model (Ohm's law, Kirchhoff's current law, Kirchhoff's voltage law)	Mathematically abstracted model (Rayleigh scattering)	
Signal processing domain	EM-domain and hybrid EM-digital domain	Digital domain	EM domain signal processing in its infancy
Multiplexing space	Nearly infinite and continuous	Limited and discontinuous	New methods are being under development for HMIMO multiplexing
Multiplexing resolution	High (follow diffraction limit)	Low (limited by bandwidth and beam width)	
Sampling domain	Spatial sampling	Nyquist time/frequency sampling	HMIMO spatial sampling rate is under development
Mathematical tools	EM information theory, Kolmogorov's information theory, functional analysis, random process and probability theory	Shannon's information theory, random process and probability theory	Multidimensional EM information theory for HMIMO needs to be developed

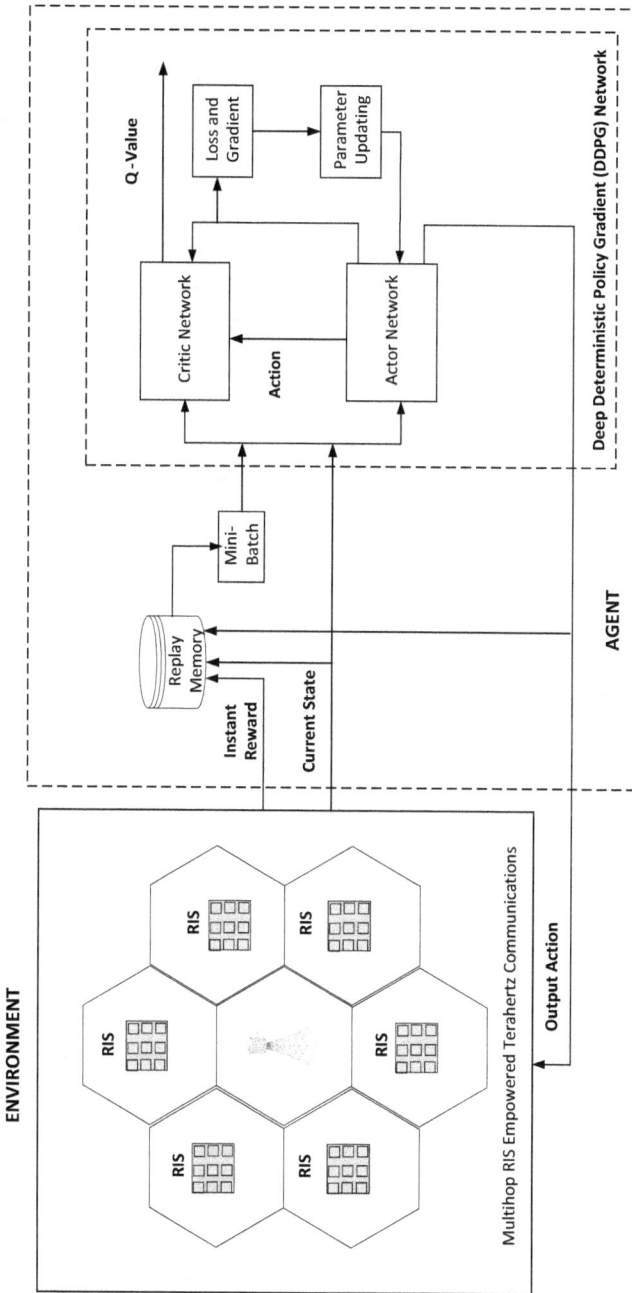

FIGURE 24.6   DRL-based multi-hop RIS communications model.

matrices at the RISs are used by leveraging the recent advances in DRL to combat the propagation loss. To improve the convergence of the proposed DRL-based algorithm, two algorithms are then designed to initialize the digital beamforming and the analog beamforming matrices utilizing the alternating optimization technique. Simulation results show that the proposed scheme can improve 50% more coverage range of THz communications compared with the benchmarks. Furthermore, it is also shown that our proposed DRL-based method is a state-of-the-art method to solve the NP-hard beamforming problem, especially when the signals at RIS-assisted THz communication networks experience multiple hops.

In another paper [3], the problem of associating (RISs to offer high-data rates and reliable low latency seamless communications with virtual reality (VR) experience to users is studied for a wireless VR network. This problem is considered within a cellular network that employs terahertz (THz) operated RISs acting as base stations. To provide a seamless VR experience, high data rates and reliable low latency need to be continuously guaranteed. To address these challenges, a novel risk-based framework based on the entropic value-at-risk is proposed for rate optimization and reliability performance. Furthermore, a Lyapunov optimization technique is used to reformulate the problem as a linear weighted function, while ensuring that higher order statistics of the queue length are maintained under a threshold. To address this problem, given the stochastic nature of the channel, a policy-based RL algorithm is proposed. Since the state space is extremely large, the policy is learned through a deep-RL algorithm. A recurrent neural network (RNN) RL framework is proposed to capture the dynamic channel behavior and improve the speed of conventional RL policy search algorithms. Simulation results demonstrate that the maximal queue length resulting from the proposed approach is only within 1% of the optimal solution. The results show a high accuracy and fast convergence for the RNN with a validation accuracy of 91.92%.

SUMMARY

In this chapter, we have presented a comprehensive overview of the key features and recent advances of HMIMO wireless communications. We first presented a multitude of holographic applications and listed representative holographic technology roadmaps for future HMIMO communications. We then emphasized upon three main components of HMIMO communications, namely physical aspects of HMIMO surfaces and their theoretical foundations, as well as enabling technologies of HMIMO communications. In the first component, we embraced the physical aspects of HMIMO surfaces in terms of their hardware structures, holographic design methodologies, tuning mechanisms, aperture shapes, functionalities, as well as representative state-of-the-art prototypes. In the second component, we presented theoretical foundations of HMIMO communications with respect to channel modeling, DOF, and capacity limits, and overviewed the HMIMO surfaces capability for EM-field sampling, as well as the resulting emerging research area of EM information theory. In the last component communications, we presented recent advances on physical layer HMIMO-enabling technologies, and on HMIMO channel estimation and HMIMO beamforming/beam focusing. We also compared HMIMO communications with existing technologies, especially mMIMO communications, and discussed a variety of extensions of HMIMO. We finally presented a comprehensive list of technical challenges and open research directions that we believe will drive unprecedented research promotions in the future.

REFERENCES

1. Gong, T. et al., "Holographic MIMO Communications: Theoretical Foundations, Enabling Technologies, and Future Directions," arXiv:2212.01257v3 [eess.SP] 28 Aug 2023.
2. Huang, C. et al., "Multi-Hop RIS-Empowered Terahertz Communications: A DRL-Based Hybrid Beamforming Design," *IEEE Journal on Selected Areas in Communications* June 2021, 39(6).
3. Chaccour, C. et al., "Risk-Based Optimization of Virtual Reality over Terahertz Reconfigurable Intelligent Surfaces," November 2023. https://ieeexplore.ieee.org/stamp/stamp.jsp?tp=&arnumber=9149411

ULTRA-DENSE NETWORK

OVERVIEW

Since the beginning of mobile industry, cell splitting and densification have been the most effective means to deliver ever-increasing capacity and improve user experience. In recent years, UDN has emerged as a prominent solution to meet the challenges of fulfilling IMT-2020 (5G) extremely high-capacity density requirements of up to 10 Mbps/m^2. Qualitatively, UDN is a network with a much higher density of radio resources than that in current networks, i.e., much denser small cell network in terms of either relative density or absolute density of the BSs. Quantitatively, the definition of UDN varies among the literature. UDN is defined as a network where the BS (or AP) density potentially reaches or even exceeds the user density, which is appropriate to characterize the scenario when the traffic per user increases while the number of users does not. An UDN is characterized as a network where the inter-site distance is only a few meters and a network reaching the point where its capacity grows sublinearly, due to the growing impact of interference, as the BS density increases [1].

CHALLENGES OF UDNs

Interference in an UDN becomes more severe, with higher volatility, and there may be a large number of strong interferers but nondominant. This leads to interference statistics different from those of an existing network with one (or a small number of) dominant interferers [1]. Under the assumption of heavy and uniform traffic load, all the BSs are always active in conventional cellular networks. In these networks with sparsely deployed BSs, the density of users often exceeds the density of BSs, at least during peak time. For such sparse networks, universal frequency reuse has long been believed as optimal to maximize the capacity, and the assumption that every BS has at least one user to serve (and hence all BSs should be activated) is reasonable. In a universal frequency reuse sparse network with time division multiple access, the average spectrum efficiency (SE) of the network increase with BS density linearly. When the network becomes dense where some BSs have no user to serve but are still activated, the SE first increases slowly and then decreases with BS density, and hence the density of BSs can be optimized [1]. The assumption of a constant path loss exponent in these papers might mask the UDN effect; nonetheless they showed that interference should be handled differently in an UDN. More interesting behaviors of the UDN network has been studied with variable path loss exponents.

Due to the traffic load fluctuation, turning off the BSs in the cells with low or no traffic load is an essential way for UDNs in improving energy efficiency (EE) as well as reducing interference. In practice, the network traffic fluctuates over different times and locations due to user behavior and mobility, which is especially true for UDNs and naturally calls for BS sleeping. In a universal frequency reuse sparse network with BS sleeping, where BS density is less than user density, the average SE still increases linearly with BS density as in the network without BS sleeping. In a universal frequency reuse UDN with BS sleeping where BS density is larger than user density, the SE only logarithmically increases with BS density.

Utilizing the massive amount of radio resources optimally in an UDN becomes increasingly complex. Misallocation of increased radio resources can cause higher interference, unbalanced load distributions, and higher power consumption. Furthermore, due to interference, local radio resource allocation may have a global impact to a UDN. In other words, "locality" does not really exist in the UDN, and radio resource allocation must be done based on a bigger picture of the UDN by taking into account of the tight coupling across the network. Sufficient bandwidth over wired connectivity to directly backhaul every BS in an UDN may be practically infeasible. Wireless self-backhauling has been proposed, which consumes valuable radio resources, generates additional interference, and leads to extra latency.

RETHINK UNIVERSAL FREQUENCY REUSE

Universal frequency reuse has been popular since the 3G era. It is also the common practice considered in UDN. Both the SE and EE are impacted by interference in such scenario. Complicated interference coordination in an UDN is undesirable due to the network scale and expensive backhaul. To manage the interference with less information sharing among the BSs, various semi-dynamic interference avoidance methods such as soft frequency reuse have been proposed. As the network becomes denser, i.e., as the ratio of BS density to user density increases, BS sleeping very effectively reduces the interference in the network. To further improve the average SE and EE of an UDN with BS sleeping, partial frequency reuse, i.e., with reuse factor greater than 1, has been investigated. When BS sleeping is allowed for the cells without active users, the frequency reuse factor that maximizes the SE or EE upper bound of the network with given ratio of BS density to user density was found, and the SE and EE gains of universal frequency reuse over partial frequency reuse in UDNs were quantified. It's found that universal frequency reuse is SE-optimal for the networks with arbitrary BS/user density ratios but is EE-optimal only when the ratio exceeds a threshold. This threshold is highly dependent on the total bandwidth of the network and the number of antennas at each BS. Both the normalized SE and EE gains of universal frequency reuse increase with the BS/user density ratio and slowly approach a constant that is dependent on the reuse factor.

INTEGRATED RESOURCE ALLOCATION, INTERFERENCE MANAGEMENT, AND TRAFFIC STEERING

There are complex interactions among resource allocation, interference management, and traffic steering in an UDN. Joint considerations of all three to reshape interference and steer traffic load as desired bring forth techniques such as variable and flexible resource reuse patterns, load aggregation/balancing, enhanced UE-BS association, carrier selection and BS on/off, etc., in semi-static time scales (e.g., hundreds of milliseconds or longer). Optimal resource allocation, taking into account of traffic distribution, interference, and performance requirements such as total latency, has shown considerable improvement in network resource utilization efficiency, which in turn causes less interference and leads to higher SE. Nonlocalized impact of interference in an UDN requires a large-scale optimization problem to be efficiently solved. Scalable algorithms have been pursued, including transforming a nonconvex programming into a sequence of convex programming, and distributed decision-making with network-wide iterations. Moreover, (sub-) optimal solution for an UDN may be pursued via optimal solutions for clusters of BSs by ignoring faraway interferers and considering cluster boundary constraints. Such solution for an UDN cluster of about 100 BSs and 1,000 UEs can be obtained within seconds on a regular PC, which is applicable in a practical network.

STANDARDS CONSIDERATION

The above approach also leads to a fast-adapting network, in terms of its BS and carrier on/off status and traffic load redistribution. The cellular standards may consider supporting fast accessible carriers, including fast carrier on/off, fast carrier selection and switching, reduction or removal of always-on signals, streamlined measurement procedures, simplified connection establishment mechanisms, and others. These features facilitate fast load balancing/shifting across BSs and carriers, as well as fast interference coordination and avoidance across BSs and carriers.

To address wireless in-band backhauling, an optimization study targeting the best E2E (i.e., multihop) performance and accounting for the split between backhaul links and access links has been conducted. Performance benefits have been demonstrated by numerical results were encouraging, while extending the in-band backhauling without radio resource split or multihop will be more attractive. Finally, in practice, an UDN is likely to be just part of a nonhomogeneous network, or part of a hierarchical network. Since the RAN architecture in 5G is more revolutionary having introduced in the centralized unit-distributed unit (CU-DU) structure, the realization of UDN may benefit from similar approaches of C-RAN.

SUMMARY

Even without a uniquely agreed quantitative definition, UDN is expected to be an essential element of 5G networks for various deployment scenarios. In addition to implementation-related technologies, standard-related designs are necessary to realize its full potential. Compared to the fast development of mm-Wave or massive-MIMO technologies, the progress of UDN-specific work requires more attention and effort. All emerging technologies described in this book will meet UDN requirements.

REFERENCE

1. Liu, J. et al., "Ultra-Dense Networks (UDN) for 5G," *IEEE 5G Tech Focus* March 2017, 1(1). https://futurenetworks.ieee.org/tech-focus/march-2017/ultra-dense-networks-udns-for-5g

25 Quantum Communications

OVERVIEW

Quantum communication is a field of applied quantum physics closely related to quantum information processing and quantum teleportation. Quantum information science (QIS) [1] has been evolving rapidly in recent years in terms of quantum communications and quantum computing (QC). Its most interesting application is protecting information channels against eavesdropping by means of quantum cryptography. It is envisioned that QIS will enable and boost future 6G systems from both communication and computing perspectives. For example, secure quantum communications such as quantum key distribution (QKD) can be leveraged to improve 6G security. QIS mainly consists of four branches: quantum mechanics, QC, quantum communications, and quantum sensing (QS) and metrology. However, Quantum mechanics provides the theoretical foundations and building blocks for quantum communications, QC, and QS and metrology, as depicted in Figure 25.1.

For example, entanglement can be leveraged for not only quantum communications but also QC and QS. In addition, quantum communications and QC can leverage each other and can be integrated together to revolutionize even to build quantum internet using quantum repeater nodes for amplifying quantum signals in the future. For example, quantum teleportation as one of basic quantum communication protocols can be used to facilitate teleportation of quantum states between quantum computers.

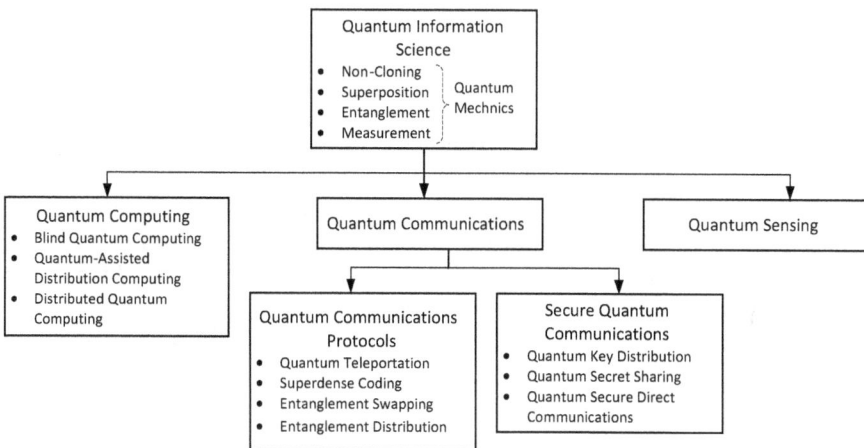

FIGURE 25.1 Quantum information science.

DOI: 10.1201/9781003499480-25

The basic element of QC is the quantum bit (qubit), which has two basic states as follows:

$$|0\rangle \text{ and } |1\rangle$$

A qubit $|\Psi\rangle$ can be seen as a generalization of the classical bit, which allows the superposition of $|0\rangle$ and $|1\rangle$ in a state as follows:

$$|\Psi\rangle = \alpha|0\rangle + \beta|1\rangle$$

where α and β are complex coefficients. The measurement of a qubit in superposition state involves that it will collapse to one of its basic states, although it is not possible to determine which one before measuring it. However, the probability of having $|0\rangle$ or $|1\rangle$ as the result of the measure is known, being $\|\alpha\|^2$ and $\|\beta\|^2$, respectively. Therefore, we have:

$$|\alpha|^2 + |\beta|^2 = 1$$

Operations with qubits are carried out by unitary transformations U; when U is applied to a superposition state, the result is another superposition state, obtained as the superposition of the corresponding basis vectors. This is an appealing characteristic of unitary transformations, which is called quantum parallelism because it can be employed to evaluate the different values of a function $f(x)$ for a given input x at the same time, although this parallelism may not be immediately useful [1] since the direct measurement on the output generally gives only $f(x)$ for one value of x.

Let $|y\rangle$ be in the superposition state $|y\rangle = \alpha|0\rangle + \beta|1\rangle$. The unitary transformation U_y may be defined as:

$$U_y : |y,0\rangle \rightarrow |y, f(y)\rangle$$

where $|y, 0\rangle$ is the joint state with the first qubit in $|y\rangle$ and the second qubit in $|0\rangle$ and $|y, f(y)\rangle$ stands for the corresponding joint output state. Therefore:

$$U_y : |y,0\rangle \rightarrow \alpha|0, f(0)\rangle + \beta|1, f(1)\rangle$$

which contains simultaneous information of $f(0)$ and $f(1)$, that is, two different values of $f(x)$. This process, known as oracle or quantum black box, can process quantum superposition states with an exponential speed-up compared to classical inputs [1]. The idea can be extended to an n-qubit system:

$$|\phi\rangle = |\Psi\rangle_1 \otimes |\Psi\rangle_{12} \otimes \ldots \otimes |\Psi\rangle_n$$

where $\otimes$ is the tensor product. The system shown in this equation can simultaneously process 2^n states but only one of them could be accessible by means of a direct measurement.

QUANTUM PHYSICS/MECHANICS

Quantum physics/mechanics, have principles such as non-cloning theorem, super-position, and entanglement, which do not have counterparts in classical systems. On the one hand, quantum mechanics describes unique constraints on quantum states of atoms and subatomic particles such as photons; on the other hand, quantum mechanics, if leveraged appropriately, can create many new quantum procedures and applications (e.g., quantum communications, QC, and QS) and enable new paradigms of QIS. These quantum mechanical principles are described in more detail as follows:

No Cloning and No Deleting

Suppose we have a cloning machine, which should perform the following information:

$$|\psi\rangle|0\rangle \rightarrow |\psi\rangle|\psi\rangle$$

for any qubit state $|\psi\rangle$. According to the laws of quantum mechanics, the transformation should be described by a unitary U. In particular, U should clone the standard basis states:

$$U|00\rangle = |00\rangle \text{ and } U|10\rangle = |11\rangle$$

But the action on a basis fixes the action on an arbitrary qubit state, due to the linearity of U. Thus, for $|\psi\rangle = a|0\rangle + b|1\rangle$, we find:

$$U|\psi\rangle|0\rangle = aU|00\rangle + bU|10\rangle = a|00\rangle + b|11\rangle$$

But what we wanted was:

$$|\psi\rangle|\psi\rangle = \left(a|0\rangle + b|1\rangle\right)(a|0\rangle + b|1\rangle) = a^2|00\rangle + ab|01\rangle + ba|10\rangle + b^2|11\rangle$$

which is not the same. Thus, $U|\psi\rangle|0\rangle \neq |\psi\rangle|\psi\rangle$ for arbitrary qubit states. Note that U does clone the basis properly, but by the linearity of quantum mechanics, it can therefore not clone arbitrary states. Instead, the cloning machine extends the superposition over two systems, producing an entangled state. No cloning theorem states that it is impossible to duplicate an existing quantum state. In other words, the state of a quantum particle or a qubit cannot be copied or reproduced (in opposition of classical bits that can be easily copied. This provides the foundation for information-theoretical security. For example, if an eavesdropper tries to intercept a photon in the middle of the path from a sender to a receiver, it will be automatically detected by the receiver due to the non-cloning theorem; this phenomenon is leveraged in QKD protocols.

No cloning proposition says that there exists no unitary operator U_{AB} on $\mathcal{H}_A \otimes \mathcal{H}_B$, such that, for fixed $|\varphi\rangle_B$ and all $|\psi\rangle_A$:

$$U_{AB}|\psi\rangle_A \otimes |\varphi\rangle_B = |\psi\rangle_A \otimes |\psi\rangle_B$$

A similar argument shows that it is also not possible to delete arbitrary quantum states, that is, turn two copies into one. Here we need three systems to include the state of the deleting apparatus.

No deleting proposition says that there exists no unitary operator U_{ABC} on $\mathcal{H}_A \otimes \mathcal{H}_B \otimes \mathcal{H}_C$, such that:

$$U_{ABC} |\psi\rangle_A \otimes |\varphi\rangle_B \otimes |\eta\rangle_C = |\psi\rangle_A \otimes |\varphi\rangle_B \otimes |\eta'\rangle_C$$

for fixed $|\varphi\rangle_B$, $|\eta\rangle_C$, $|\eta'\rangle_C$ and all $|\psi\rangle_A$ because we have proof of it. Let us consider two different input states, $|\psi\rangle$ and $|\psi'\rangle$, to the deleting machine. The overlap between the initial states is simply $\langle\psi|\psi'\rangle 2$, but at the output it is $\langle\psi|\psi'\rangle$, since the other states $|\psi\rangle$, $|\eta\rangle$, and $|\eta'\rangle$ are fixed. A unitary operation would however leave the overlap invariant.

SUPERPOSITION

Superposition is another unique quantum mechanical phenomenon describing quantum states. In classical systems, the state or the value of a classical bit is deterministic (i.e., 0 or 1). However, a quantum state is nondeterministic and can be a superposition of two (or more) basic states as explained earlier.

ENTANGLEMENT

Entanglement is an even stranger but extremely useful phenomenon, where quantum states of two (or more) qubits are maximally entangled. In other words, the quantum state of one entangled qubit is fully dependent on the state of any other entangled qubit, no matter how far apart the entangled qubits are physically located. If the quantum state of an entangled qubit A is known, the quantum state of the other entangled qubit will also be deterministic. We can explain entangled quantum states as four Bell states: $|\phi^+\rangle$, $|\phi^-\rangle$, $|\psi^+\rangle$, $|\psi^-\rangle$. Each of these states defines an entangled quantum state of two qubits as follows:

$$|\Phi^+\rangle = \frac{1}{\sqrt{2}}\left(|0\rangle_A \otimes |0\rangle_B + |1\rangle_A \otimes |1\rangle_B\right)$$

or

$$\frac{1}{\sqrt{2}}\left(|0_A 0_B\rangle + |1_A 1_B\rangle\right)$$

indicates that two qubits, A and B, are entangled with the identical state. Please note that Bell states can be generated easily using two quantum logic gates (e.g., Hadamard) with two basics $|0\rangle$ qubits as input. Entangled qubits are the most crucial resources in QIS systems, and they can enable new quantum procedures and applications such as quantum teleportation and quantum superdense coding.

Teleportation

Now imagine Alice and Bob are in the opposite situation: Instead of Alice wanting to send 2 classical bits and having only a quantum channel (plus pre-shared entanglement), she wants to send a qubit, but only has access to a classical channel. Can she somehow send the state to Bob using only a classical channel? If that is all the resources they share, the answer is no. Alice could try to measure the qubit in some way, for instance to learn the values of the coefficients a and b in the expression $|\psi\rangle = a|0\rangle + b|1\rangle$ by building up statistics (since $\Pr(0) = |a|^2$, and never mind, she also needs the relative phase between a and b), but she only has 1 copy of $|\psi\rangle$.

On the other hand, if Alice and Bob already share an entangled state, then it is possible to transfer $|\psi\rangle$ to Bob, and it only requires 2 bits! The "2 bits" are reminiscent of the four entangled states $|\Phi_j\rangle$ used in superdense coding, and they play the same role as measurement in teleportation. The protocol is very simple. Alice has a qubit prepared $|\psi\rangle_{A'}$ as well as half of a maximally entangled state $|\Phi\rangle_{AB}$. She then measures her two systems in the Bell basis, producing a two-bit outcome. What happens when the outcome corresponds to $|\Phi\rangle$?

$$_{A'A}\langle\Phi\,|\,\psi\rangle_{A'}\,|\,\Phi\rangle_{AB} = {}_{A'A}|\,\Phi\rangle_{AB}\frac{1}{\sqrt{2}}\left(a\,|\,000\rangle\rangle + a\,|\,011\rangle + b\,|\,100\rangle + b\,|\,111\rangle\right)_{A'AB}$$

$$= \frac{1}{2}(|00\rangle + |11\rangle)_{A'A}(a|000\rangle + a|011\rangle + b\,|\,100\rangle + b\,|\,111\rangle)_{A'AB}$$

$$= \frac{1}{2}(a|0\rangle + b|1\rangle)_B$$

$$= \frac{1}{2}|\,\psi\rangle_B$$

The state has been transferred to Bob! The squared norm of the output tells us the probability, so the chance that Alice obtains result $|\psi\rangle$ is $\frac{1}{4}$. And what about the other results?

Since $|\Phi\rangle_{AA'} = (\sigma_x)_{A'}|\Phi\rangle_{AA'}$, it follows that:

$$_{A'A}\langle\Phi|\psi\rangle_{A'}|\Phi\rangle_{AB} = {}_{A'A}|\Phi\rangle_{AB}\left(\sigma_x\right)_{A'}|\psi\rangle_{A'}|\Phi\rangle_{AB} = \frac{1}{2}\left(\sigma_x\right)_B|\psi\rangle_B$$

by replacing the above argument with $|\psi\rangle$ replaced with $(\sigma_x)_B|\psi\rangle_B$.

This works similarly for the other two outcomes. Thus, if Alice communicates the result of the Bell basis measurement to Bob, he can apply the corresponding Pauli operator to obtain the input state $|\psi\rangle$. Alice needs 2 bits to describe which outcome occurred, and since each projected state has the same weight, the probability of every outcome is $\frac{1}{4}$. The fact that the probability distribution does not depend on the input state is important; otherwise information about the state would essentially leak into other degrees of freedom, and the state could not be properly reconstructed by Bob.

Quantum Measurement and Disturbance

Even if a generic qubit is not definitely in one of the states $|0\rangle$ or $|1\rangle$, what happens after a measurement? Surely if we repeat the measurement, we should get the same result (provided nothing much has happened in the meantime). Indeed, this is the case in quantum mechanics. Starting from $|\psi\rangle = a|0\rangle + b|1\rangle$ and making the $|0\rangle/|1\rangle$ measurement leaves the system in state $|0\rangle$ with probability $|a|^2$ or the state $|1\rangle$ with probability $|b|^2$, so that a subsequent identical measurement yields the same result as the first.

We can measure in other bases as well. For instance, consider the basis $|\pm\rangle = \frac{1}{\sqrt{2}}(|0\rangle \pm |1\rangle)$. Now the probabilities for the two outcomes are:

$$\text{prob}(+) = |\langle +|\psi\rangle|^2 = \frac{1}{2}|a+b|^2$$

$$\text{prob}(-) = |\langle -|\psi\rangle|^2 = \frac{1}{2}|a-b|^2$$

Thus, if $|\psi\rangle = |0\rangle$, then $p_\pm = \frac{1}{2}$. That is, the measurement outcome is completely random. And after the measurement, the state is either $|+\rangle$ or $|-\rangle$. In this way, measurement disturbs the system by changing its state.

This phenomenon makes QKD possible. Very roughly, a potential eavesdropper attempting to listen in on a quantum transmission by measuring the signals will unavoidably disturb the signals, and this disturbance can be detected by the sender and receiver. Quantum measurement is the process of transforming a quantum state to a classical state. One quantum measurement principle is deferred measurement, which means that any quantum measurement can be moved to the end of a quantum circuit. In other words, to leverage QIS, the last step is often to store and measure quantum states or physical qubits (e.g., photons, trapped ions) in the target quantum system and to generate measurement results in classic bits for future use. This process is a type of quantum capability, referred to as quantum measurement or readout. Quantum measurement is required for quantum communications, QC, and QS. For example, in typical QKD protocols, the quantum receiver receives physical qubits (e.g., polarized photons) from the quantum sender and measures them to generate classical bits, the quantum sender via a classical channel as a part of QKD protocols. In QC systems based on trapped ions, ions as physical qubits pass through quantum logic gates and then are read out to generate classical bits as the solution to a computation problem.

QUANTUM COMMUNICATIONS ARCHITECTURE MODEL AND PROTOCOLS

Leveraging quantum mechanics, some fundamental quantum communication procedures become possible, such as quantum teleportation, superdense coding, and entanglement distribution. The physical and logical layer of quantum communications are shown in Figure 25.2 following the classical OSI protocol architecture model.

The first part of the physical layer deals with the lattice of physical q-bits that are processed by the physical quantum processor. The second part of the physical layer provides encoding of logical q-bits for quantum error correction. Quantum communication

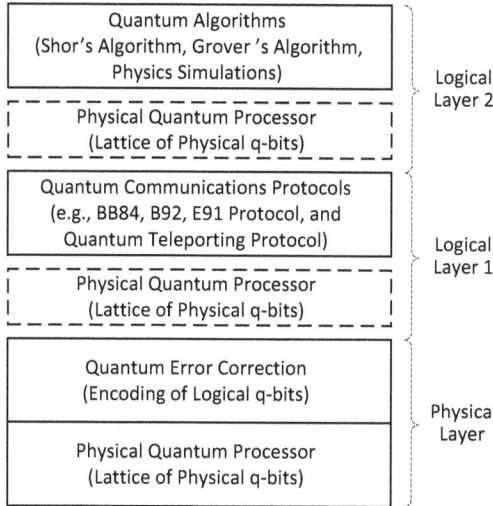

FIGURE 25.2   Physical and logical layer architecture model for quantum communications.

protocol layer belongs to logical layer 1, but it also needs physical quantum processor for processing of the quantum communication protocols. Examples of quantum communication protocols are BB84, B92, E91, and quantum teleporting protocol.

In the BB84 protocol, Alice can transmit a random secret key to Bob by sending a string of photons with the private key encoded in their polarization. The no-cloning theorem guarantees that Eve cannot measure these photons and transmit them to Bob without disturbing the photon's state in a detectable way.

B92 protocol is a modified version of the BB84 protocol with the key difference between the two being that while BB84 protocol uses four different polarization states of photon, the B92 protocol uses two (one from the rectilinear basis, conventionally H-polarization state, and one from the diagonal basis, conventionally $+45°$-polarization state).

The E91 protocol is based on the principle of quantum entanglement and a clever use of the identities prescribed by Bell's test to entangled qubits. *Quantum teleportation protocol* is designed to transmit all possible (unknown) input states; thus, a measure of averaging over all inputs is used.

The logical layer 2 provides services of the quantum algorithms/applications (e.g., Shor's algorithm, Grover's algorithm, and physics simulation). This logical layer 2 also needs to use a physical quantum processor dealing with q-bits.

QUANTUM TELEPORTATION PROTOCOL

The quantum state of a particle can be transmitted from one location to another without physically moving the particle. This is known as teleportation in quantum mechanics. Science fiction depicts teleportation as the instantaneous transport of objects, but in reality, it does not work that way. Entanglement plays a crucial role in quantum teleportation. Whenever two or more particles become entangled, their states cannot be

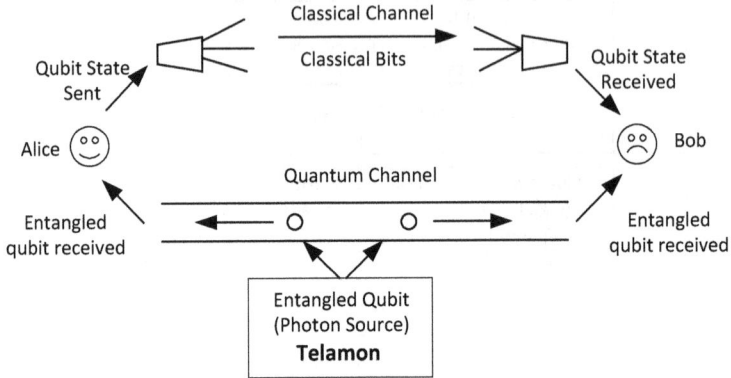

FIGURE 25.3 Quantum teleportation.

described independently of one another. No matter how far apart the particles are, the state of one particle affects the state of the other instantly. In teleportation, quantum is crucial because it allows quantum information to be transmitted without violating quantum laws. In the real world, QC, secure communication, and cryptography all benefit from this. As we move further into the decade, developing quantum technologies will depend heavily on quantum teleportation, a fundamental concept in QIS.

Quantum teleportation is the transfer of quantum states (in qubit) to another quantum state. It's not about transferring the information physically, but instead on transferring the state of the information. To transfer a quantum bit, Alice and Bob must use a third party (Telamon) to send them an entangled qubit pai (Figure 25.3). Alice then performs some operations on their qubit and sends the results to Bob over a classical communication channel. Bob then performs some operations on their end to receive Alice's qubit (Figure 25.3).

This allows transmitting the state of a data qubit $|\varphi\rangle$ from one quantum node A (Alice) to another quantum node B (Bob) without physically transmitting the qubit A, but relying on:

- Consuming a pair of entangled qubits
- Leveraging Bell measurement
- Transmitting two classical bits

Both quantum nodes A and B basically perform local operations and classical communication (LOCC).

- First, a shared entanglement has been established between A and B (i.e., there are two entangled qubits: $|q_A\rangle$ at A and $|q_B\rangle$ at B.
- Second, A generates the data qubit $|\varphi\rangle$.
- Third, A performs a Bell measurement of the entangled qubit $|q_A\rangle$ and the data qubit $|\varphi\rangle$.
- Fourth, the result from this Bell measurement will be encoded in two classical bits, which will be physically transmitted via a classical channel from A to B.

- Fifth, based on the received two classical bits, B modifies the state of the entangled qubit $|q_B\rangle$ in a way to generate a new data qubit, which state is identical to the data qubit $|\varphi\rangle$ at A.

To generate a data qubit with the same state of $|\varphi\rangle$, a node (B or other nodes) must have received two classical bits and possess $|q_B\rangle$. Essentially, quantum teleportation uses two classical bits to transmit the state of one qubit. As a basic quantum communication protocol, quantum teleportation protocol can be used to enable more advanced quantum procedures and applications. For example, in distributed quantum computing (DQC), data qubits from one noisy intermediate-scale quantum (NISQ) computer to another NISQ computer are very sensitive and cannot be lost. For this purpose, quantum teleportation can be leveraged to teleport sensitive data qubits from one quantum computer A to another quantum computer B. Note that measurement-based DQC may not need quantum teleportation. Entanglement swapping protocols are also based on quantum teleportation.

SUPERDENSE CODING

This can be regarded as the opposite process of quantum teleportation. Using super-dense coding, quantum node A achieves the goal of sending two classical bits to quantum node B without physically transmitting these two classical bits but relying on:

- Transmitting one qubit
- Consuming a pair of entangled qubits
- Leveraging Bell measurement

Like quantum teleportation, both quantum nodes A and B in superdense coding perform LOCC.

- First, a shared entanglement has been established between A and B (i.e., there are two entangled qubits: $|q_A\rangle$ at A and $|q_B\rangle$ at B).
- Second, A has two classical bits to be sent to B.
- Third, according to these two classical bits (i.e., four possibilities), A performs a quantum gate to qubit $|q_A\rangle$ to change the state of $|q_A\rangle$ (i.e., encode two classical bits on $|q_A\rangle$).
- Fourth, A transmits $|q_A\rangle$ to B; fifth, B receives $|q_A\rangle$.
- Last, B performs quantum gate on $|q_A\rangle$ and $|q_B\rangle$ to decode these two classical bits.

To decode the two classical bits, a node (B or other nodes) must possess $|q_A\rangle$ sent by A and $|q_B\rangle$ hosted by B. Essentially, superdense coding uses one qubit to transmit two classical bits from A to B, which can be regarded as increased quantum channel capacity.

ENTANGLEMENT DISTRIBUTION

We have described the entanglement distribution procedure and related communication protocols that can distribute entangled qubits to quantum nodes in remote

$|q_C>$

Node C

$|p_A>$ QM   QM

Alice   Bob

Node A   Node B

$|q_A>$ $|q_B>$ $|p_B>$

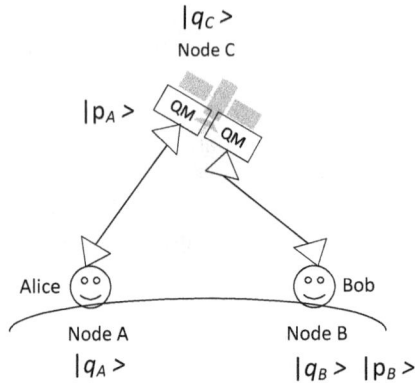

FIGURE 25.4 Entanglement distribution between Alice (Node A) and Bob (Node B) via satellite (Node C).

locations. Quantum networks facilitate numerous applications such as secure communication and distributed quantum computation by performing entanglement distribution. Multiuser quantum applications where quantum information is shared between multiple users require access to a shared multipartite state between the users. As described above, entanglement is required in both quantum teleportation and superdense coding; it can also enable more advanced quantum applications such as QKD and DQC. However, the two quantum nodes A and B (Figure 25.4) that need to use entanglement may be too distant to directly share entangled bits.

As such, a third node, C, in the middle can act as an entanglement relay to establish and distribute entanglement between A and B.

For example,

- A satellite as node C can first establish a pair of entangled qubits (e.g., $|q_C\rangle$ at C and $|q_A\rangle$ at A) between itself and a ground station A, and a pair of entangled qubits (e.g., $|p_A\rangle$ at C and $|p_B\rangle$ at B) between itself and a ground station B.
- Leveraging quantum teleportation and consuming $|p_C\rangle$ and $|p_B\rangle$, the satellite C can teleport the state of $|q_C\rangle$ to the ground station B (referred to as $|q_B\rangle$).
- Eventually, both ground stations A and B now share a pair of entangled qubits (i.e., $|q_A\rangle$ at A and and $|q_B\rangle$ at B).

This process is referred to as entanglement distribution or entanglement swapping. Pease note that there could be multiple nodes in the middle (similar to the satellite C) to perform entanglement swapping, which can be referred to as Multihop entanglement distribution. The basic quantum communication procedures described above can be extended to create even more advanced quantum communication protocols such as QKD, quantum secure direct communications (QSDC), and quantum secret sharing (QSS).

QUANTUM KEY DISTRIBUTION

QKD leverages quantum mechanics to establish a security key between quantum node A and quantum node B, which is information-theoretically secure. QKD protocols typically consist of two stages: the key-sharing stage and post-processing stage. In the first key-exchange stage, A and B employ prepare-and-measure protocols (e.g., BB84) or entanglement-based protocols (e.g., E91) to share a set of classical bits, based on private quantum measurement with random measurement basis at each side. In prepare-and-measure protocols, A prepares qubits and transmits them via quantum channels such as optical fibers to B; then B measures received qubits on a random measurement basis (e.g., vertical, or horizontal polarizations for photons) to generate classical bits; A and B exchange their measurement basis (not the classical bits from measurement) via classical channels; A and B take the classical bits from the same measurement basis as the shared ones.

In the second post-processing stage, A and B aim to generate the final security key from the shared classical bits that have been produced in the first stage; basically, A and B perform information reconciliation and privacy amplification to correct errors and reduce privacy leakage of A and B. The post-processing stage is usually done via an authenticated classical channel. A QKD protocol could be discrete-variable QKD (DV-QKD) or continuous-variable QKD (CV-QKD): DV-QKD may need dedicate quantum channels (e.g., optical fibers), while CV-QKD in general can easily work over deployed optical fibers in telecommunication infrastructure.

QUANTUM SECURE DIRECT COMMUNICATIONS

QSDC provides direct secure communications between nodes A and B based on quantum mechanics without relying on security key distribution and management. In contrast, QKD only provides secure key establishment, and the follow-up communications are still based on classical communications only encrypted using the established key. Using QSDC, node A can send secret information directly and securely to node B via quantum channels without using an established security key. The first QSDC protocol, as described in [1], leverages entanglement (e.g., Bell states) to achieve secure direct communications. Basically, node A as the sender prepares a list of N Bell states according to the message bit sequence to be sent to node B; note that each Bell state stands for a pair of two entangled qubits. Node A splits N Bell states to two sequences: Seq1 and Seq2. Seq1 (or Seq2) contains N qubits, and each qubit in Seq1 is entangled with a qubit in Seq2. Then node A sends Seq2 to node B; node B receives Seq2 and measures some of the qubits in Seq2 to work with node A to detect eavesdropping, similar to eavesdropping detection in QKD.

Last, node A sends Seq1 to node B; node B performs Bell measurement on each entangled qubit pair from Seq1 and Seq2 to recover the encoded classical bits; node B also works with node A by announcing the measurement basis to node A to estimate error rate and detect any eavesdropping for the second time. Like quantum teleportation, QSDC realizes secure transmission of classical bits using qubits, but quantum teleportation does not have an eavesdropping check.

Quantum Secret Sharing

QSDC aims to distribute a secret key among multiple nodes by letting each node know a share or a part of the secret key. However, no single node can know the entire secret key. The secret key can only be reconstructed via collaboration from enough nodes. QSS typically refers to the scenario: The secret key to be shared is based on quantum states instead of classical bits. QSS enables splitting and sharing such quantum states among multiple nodes.

Quantum Computing

Examples of QC include blind quantum computing (BQC), also referred to as secure quantum computing with privacy preservation. Other examples are quantum machine learning (QML) and DQC. These enable new types of computing services and applications for quantum-enabled 6G. BQC provides a way for a client to delegate a computation task to one or more remote quantum computers without disclosing the source data to be computed over. Basically, a client node with source data delegates the computation of the source data to a remote computation node (i.e., a server); furthermore, the client node does not want to disclose any source data to the remote computation node and thus preserves the source data privacy. There is no assumption or guarantee that the remote computation node is a trusted entity from the source data privacy perspective. As a new client/server computation model, BQC generally enables the following:

- The client delegates a computation function to the server.
- The client does not send original qubits to the server but sends transformed qubits to the server.
- The computation function is performed at the server on the transformed qubits to generate temporary result qubits, which could be quantum-circuit-based computation or measurement-based quantum computation. The server sends the temporary result qubits to the client.
- The client receives the temporary result qubits and transforms them to the final result qubits.

During this process, the server cannot figure out the original qubits from the transformed qubits. Also, it will not take too much effort on the client side to transform the original qubits to the transformed qubits or transform the temporary result qubits to the final result qubits. One of the very first BQC protocols follows this process, although the client needs some basic quantum features such as quantum memory, qubit preparation and measurement, and qubit transmission.

DQC refers to distributing quantum computing tasks to multiple quantum computers. This is especially useful when it is impossible or highly inefficient to perform the task on any single quantum computer. For example, an NISQ computer may only support less than 100 physical qubits and thus cannot execute complex QC tasks. However, multiple NISQ computers can be connected via classical and/or quantum channels to gain higher computation power before full-fledged quantum computers become available. As a result, it becomes possible for them to perform the complex QC task jointly and efficiently. In a broader context, DQC may refer to quantum-aided classical distributed computing, where the solution is to leverage quantum mechanics to solve classical distributed computing problems with better performance. For instance, entanglement and entangled qubits can be exploited to improve leader election in classical distributed computing.

Quantum Sensing

QS is an advanced sensor technology that vastly improves the accuracy of how we measure, navigate, study, explore, see, and interact with the world around us by sensing changes in motion, and electric and magnetic fields. The analyzed data is collected at the atomic level. A conceptual system for quantum sensor using diamond is shown in Figure 25.5. Quantum sensors have applications in a wide variety of fields including microscopy, positioning systems, communication technology, electric and magnetic field sensors, as well as geophysical areas of research such as mineral prospecting and seismology. Classical inertial sensors can provide the bandwidth and range, while the quantum sensors provide extreme accuracy without error or noise (Figure 25.5). For quantum sensors, the bandwidth that sensing takes place is typically 1 Hz (once per second) (Figure 25.5).

QS can measure electric and magnetic fields very accurately across many frequencies, physical quantities against atomic properties, so there is no drift or need to calibrate, and use quantum entanglement to improve sensitivity or precision. Quantum sensor technology may still be in its infancy; however, the research and development carried out by Advanced Navigation demonstrates that it can be put into practical use. By using a fusion of the Advanced Navigation Boreas digital fiber-optic gyroscope (DFOG) as the classical sensor with quantum sensor hardware, the first inertial navigation systems of this type are being prepared for practical use.

A common challenge in QS is the conflict between isolating the sensitive quantum states from external disturbances, while at the same time being able to manipulate the quantum states and expose them to the physical quantity that is to be measured. In theory, quantum sensors have the advantage of far more accurate measurements than

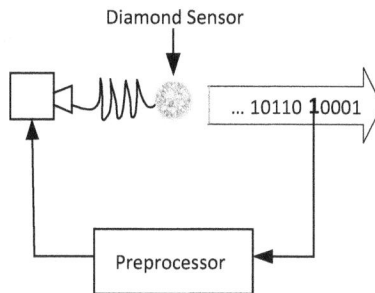

FIGURE 25.5   Quantum sensor.

conventional sensors. But to become commercially attractive, the advantage must be realized in practice at a reasonable price.

QUANTUM INFORMATION SCIENCE FOR 6G NETWORKS

The progress is QIS has demonstrated that the next-generation 6G networks can be quantum enabled. Quantum communications is expected to play a big role in quantum-assisted radio access network, quantum non-terrestrial satellite networks, quantum-assisted edge networks, quantum-assisted data center, quantum-assisted blockchain, quantum-assisted wireless artificial intelligence (AI), quantum management and control, and quantum capabilities-as-a service in the future. Figure 25.6 depicts those conceptualized quantum-enabled networks and services.

Quantum technologies are progressing fast. Even if significant developments are still needed, in light of the potential opportunities generated by the expected industrial impacts of quantum technologies. Investments are made to improve quantum technologies throughout the world. Feasibility demonstrations and performance testing are very important and required: this thread could be facilitated by the development of platforms where innovators (e.g., research institutions, software developers, hardware industry, internet service industry, and security professionals) meet to share open innovation capable of boosting the creation and development of new ecosystems on quantum.

Coordinated standardization efforts are required: for example, the topics concerning the integration of quantum nodes and systems in current infrastructures are still limitedly covered by the standardization activities. Integration and interoperability aspects are fundamental to planning the exploitation of such disruptive technologies and services. In this respect, the physical layer of hybrid classical/QKD systems remains unaddressed. What system parameters should be monitored and how this information should be processed to infer the impact on the system performance are also important matters of standardization.

Moreover, the current lack of commercial or open-source modeling tools for the simulation of classical/QKD systems does not help in this aspect, and it is an obstacle that should be removed. These tools would indeed allow us to validate any specification using an agreed simulation framework and without making use of expensive experimental setups. Another main challenge might be the development of human resources with appropriate skill: the know-how should involve understanding telecommunications, computer science, as well as engineering of quantum systems.

QUANTUM RADIO ACCESS NETWORK

Higher-performance systems now remain impractical largely only because their algorithms are extremely computationally demanding in the radio access network. For optimal performance, an amount of computation that increases at an exponential rate both with the number of users and with the data rate of each user is often required. The base station's computational capacity is thus becoming one of the key limiting factors on wireless capacity. Both QC and quantum communications can be leveraged to improve RAN efficiency and security. Some existing work [1] has well

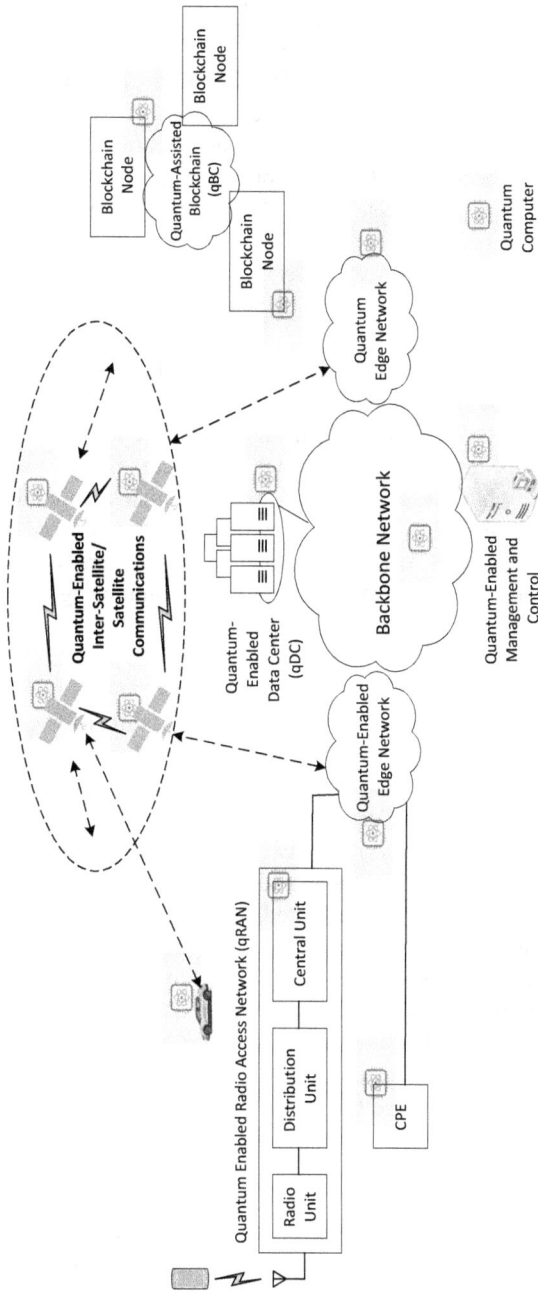

FIGURE 25.6 Quantum-enabled network and services.

demonstrated the optimal wireless resource allocation using quantum annealing over D-Wave quantum computers. In general, with powerful QC, it becomes possible to find the optimal solution for radio resource allocation and cell planning, which leads to higher energy efficiency and spectrum efficiency; for instance, leveraging quantum search algorithms [1]. Similarly, all other components of Open RAN (O-RAN), O-RAN distribution unit (O-DU), and O-RAN central unit (O-CU) can be made quantum enabled as well as QKD quantum secured.

MIMO channel traces, showing that 10^{-6} second of compute time on the 2,000Q, can enable 48 user, 48 access point antenna binary phase shift keying (BPSK) communication at 20 dB signal-to-noise ratio (SNR) with a bit error rate of 10^{-6} and a 1,500-byte frame error rate of 10^{-4}.

Quantum Edge Network

It is expected that an increasing number of vehicles in the edge network will be deployed in 6G networks. Challenges in such pervasive edge computing environments include security, task offloading, and edge resource allocation. Secure quantum communications can be leveraged to guarantee secure communications to, from, and between edge nodes, while QC can help to find the optimal solution for task offloading and edge resource allocation. It is predicted that room temperature photonic-based QC may be widely available. At that time, quantum computers may be deployed in edge networks and provide power QC, referred to as quantum edge computing.

Quantum Satellite Networks

Geostationary, medium Earth orbit (MEO), low Earth orbit (LEO), and micro/pico satellites will continue to be deployed in different space orbits and form inter-satellite mesh communications satellite networks (CSNs). Satellite nodes are usually connected via free-space optic links, and a powerful satellite node can host a quantum computer. Quantum CSNs (qCSNs) have twofold advantages. First, quantum communications between two satellite nodes and between a satellite node and ground stations have been practically demonstrated, using free-space optics as quantum channels. If satellite nodes aim to leverage entanglement for quantum communications, entanglement distribution for distant satellite nodes needs to be carefully designed. Moreover, powerful satellite nodes with a quantum computer can provide QC as a service for other satellite nodes and ground stations. Second, a satellite node can be used as a trust node or a quantum repeater to aid and improve quantum communications such as satellite-based QKD.

Quantum Data Center

Optical links such as optical fiber and free-space optics are used in modern data centers for inter-rack communications to increase data rates [1]. These optical links can be leveraged as quantum channels simultaneously for quantum data center (QDC). As a result, quantum communication and quantum cryptography can be realized to improve inter-rack security using a hybrid approach that jointly leveraged quantum random number generation, QKD, and post-quantum cryptography (PQC) to a multi-level cryptographic solution with quantum-safe data communications between server racks. QC can also be leveraged for solving computation-intensive problems in data centers such as optimal data flow and energy consumption management [1]. Even an architecture combining quantum random access memory (QRAM) and quantum networks has been proposed [Liu] – a unified concept referring to some specific quantum hardware that could efficiently deal with the quantum data and would provide an efficient interface between classical data and quantum processors. The key component of the proposed QDC is a QRAM, which is a device that allows a user to access multiple different elements in superposition from a database (which can be either classical or quantum). At minimum, a QDC consists of a QRAM coupled to a quantum network.

QUANTUM BLOCKCHAIN

Blockchain or distributed ledger technology elegantly combines a set of mechanisms (e.g., distributed consensus protocols, distributed database, cryptography, and hashing) to realize a decentralized system with multiple advantages such as transparency and immutability. As a 6G enabler, blockchain technology may be used in many applications such as decentralized authentication and distributed wireless resource sharing among parties that may not trust each other. However, blockchain technology also inherits some potential issues such as security attacks from malicious nodes, low transaction speed due to consensus protocols, and privacy breach due to transparent data contained in blocks. These issues may be solved or mitigated using QIT, referred to as quantum-assisted blockchain (qChain). For example, quantum communications can be used to improve communication security among blockchain nodes; non-separability of entangled qubits can be used to simulate the link relationship between blocks; entanglement can also be used for designing new consensus protocols without introducing high communications overhead and in turn increase transaction speed. Reference [1] provides more examples on quantum-assisted blockchain protocols.

QUANTUM ARTIFICIAL INTELLIGENCE

Quantum artificial intelligence (QAI) combines QC with AI. It seeks to use the unique properties of quantum computers which leverage quantum mechanical effects, such as, superposition, entanglement, and other quantum properties to enhance the capabilities of AI systems. 6G system will be more intelligent and autonomous with the availability of massive data from ubiquitous devices and network nodes, and the application of AI algorithms such as deep learning (DL), deep reinforcement learning, federated learning, and transfer learning. QIS can benefit wireless AI in several ways. First, decentralized AI such as federated learning introduces model exchange and communications among participants, which can be quantum-safe secured using quantum cryptography. Second, both the AI training process and inference process are computation intensive, which can be expedited using QC. Third, BQC could be leveraged to realize privacy-preserving AI training. Finally, QML and quantum deep learning (QDL) may introduce novel and more efficient wireless AI algorithms.

Quantum Internet Management and Control

The concept of quantum internet is a network of quantum computers that will send, compute, and receive information encoded in quantum states. The quantum internet will not replace the modern or "classical" internet; instead, it will provide new functionalities such as quantum cryptography and quantum cloud computing. Today's quantum networking experiments rely on a set of devices with limited functionality and performance. However, it can be inferred from classical networks that to create wide area, operational quantum networks, more capable devices with additional functionality are needed. These new devices must satisfy suitable requirements for reliability, scalability, and maintenance.

Quantum Network Devices

Considering the classical network, quantum network requires some essential devices like quantum memory with efficient optical interface and satellite-to-fiber connections, high-speed, low-loss quantum switches, multiplexing technologies, and transducers for quantum sources and improved sources themselves, as well as transduction from optical and telecommunications regimes to quantum computer-relevant domains, including microwaves. In addition, the following quantum components are needed:

- Quantum-limited detectors, ultra-low loss optical channels, space-to-ground connections, and classical networking and cybersecurity protocols.
- Entanglement and hyper-entangled state sources and transmission, control, and measurement of quantum states.
- Transducers for quantum sources and signals from optical and telecommunications (telecom) regimes to quantum computer-relevant domains, including microwaves.
- Development of quantum memory buffers and small-scale quantum computers that are compatible with photon-based quantum bits in the optical or telecom wavelengths.
- Long-range entanglement distribution (terrestrial and space-based) using quantum repeaters, allowing entanglement-based protocols between small- and large-scale quantum processors.

Generally, key quantum network components available today remain at a laboratory-level readiness and have yet to run operationally in a full network configuration. Moving forward will require overcoming critical challenges aimed to achieve cascaded operation and connectivity, among them [2]:

- Integrating existing components by unifying their operational properties (bandwidth, wavelength, duty-cycle) using systems-level engineering
- Achieving high-rate (GHz) quantum entanglement sources, quantum memory buffers, and detectors to compensate for cascading operation losses (Figure 25.7)
- Further development of key quantum network components, such as high-speed, low-loss quantum switches and multiplexing technologies

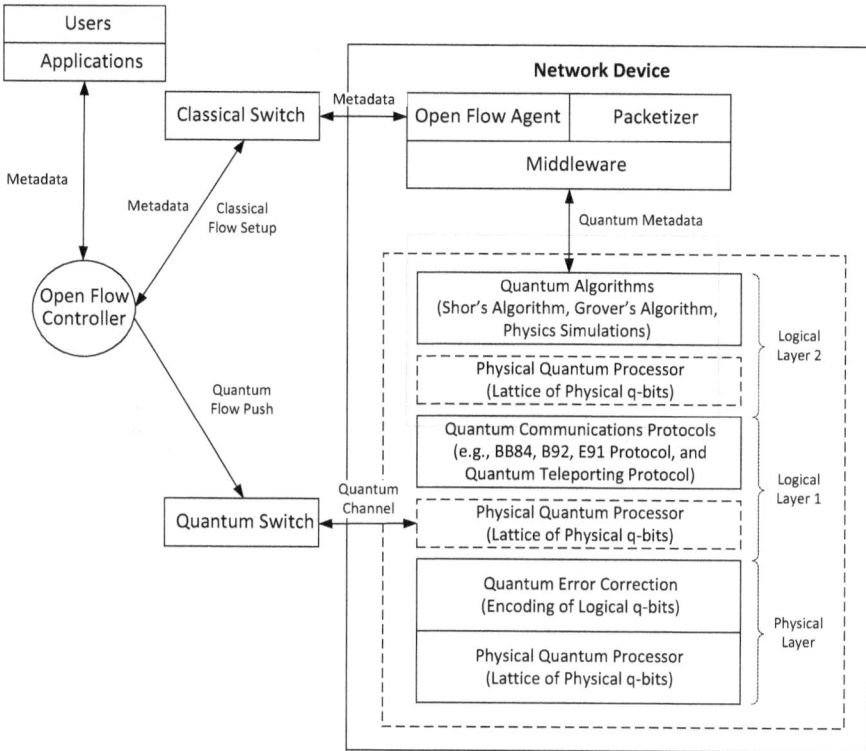

FIGURE 25.7   Concept of classical quantum network device.

REPEATER, SWITCH, AND ROUTER FOR QUANTUM ENTANGLEMENT

Multihop networks require a means of strengthening and repeating signals along with selecting paths through the network. While physical and software solutions are used in classical networks, an equivalent has not been found for quantum networks. Challenges include different forms of quantum entanglement generation, swapping, and purification protocols over multiple users, as well as coordination and integration of traditional networks with quantum network technologies for optimal control and operations.

Routing is a fundamental network function. Multihop networks require a means of selecting paths through the network. First prototypes of the entanglement-based quantum network will not use quantum memory and repeaters. Instead, they will employ software-defined network (SDN) technology to perform traditional wavelength routing and assignment in optical networks to establish quantum paths among quantum nodes or between quantum nodes and entangled photon sources (EPS). Optical switches will be dynamically programmed to establish multiple quantum paths. These nontrivial topologies will allow for establishing entanglement distribution and quantum teleportation protocols over multiple users. Coexistence with traditional networks in the same optical fiber transmission systems will be critical for sharing the same dense wavelength division multiplexing (DWDM) network component.

To move from simple forms of entanglement distribution between a fixed pair of destinations (limited by the attenuation of single-photon telecom fibers to distances on the order of ~100 km or less), it is necessary to extend the entanglement pair distribution range using quantum repeaters. Unlike the operation of a classical repeater, the quantum repeater does not amplify photons in an entangled state while they are transiting. Instead, the quantum repeater can "hop" the entanglement property across an additional distance interval by consuming the resource of a second entangled pair. The innovation needed to fulfill this is the quantum process of entanglement swapping.

The potential for improved distance transmission with a quantum repeater is realized with the addition of working quantum memories, which can buffer a pair of photons after a successful transmission without requiring the other pair's survival at the same time. A prototype of a one-hop quantum repeater requires four heralded quantum memories. The probability of the whole entanglement swapping procedure succeeding then requires two successful pair transmissions and storage occurring within the memory's storage time, resulting in a much higher order than coincident success in a single test interval and allowing for higher distribution rates of entanglement covering longer distances. Other approaches to quantum repeaters rely on all-optical quantum processing, as well as those based on atomic ensembles.

QUANTUM ERROR CORRECTION

A fundamental difference for quantum networks arises from the fact that entanglement, whose long-distance generation is an essential network function, is inherently present at the network's physical layer. This differs from classical networking, where shared states typically are established only at higher layers. In this context, solutions must be found to guarantee network device fidelity levels capable of supporting entanglement distribution and deterministic teleportation, as well as quantum repeater schemes that can compensate for loss and allow for operation error correction.

The last step in the evolution of quantum communication networks will be reached when devices capable of error correction and concatenated quantum repeaters are available. This quantum network stage would be able to support all protocols required for DQC and quantum sensor applications, including distributed memory many-computer architectures and their applications. Realizing such a quantum network will require several advances in quantum network technology, including quantum links with high repetition rates, fidelity capable of supporting entanglement distribution and deterministic teleportation, and quantum repeater schemes that allow for both loss and operation error correction and can support the necessary operation depth at each node (which have to be performed fault tolerantly). The ability to generate multipartite entanglement also is especially relevant for sensing applications.

A common requirement for all these applications, especially those requiring a higher level of network functionality, is the need for developing an architecture and for defining a quantum network stack. The architecture should allow the quantum protocols to connect to the underlying hardware implementation transparently and provide the necessary functionality to counteract the effects of limited qubit and gate lifetimes.

QKD Implementation

When network operators deploy the QKD network to secure data in their communication network, they need to consider how the QKD network will be incorporated into their classical communication network for QKD-derived keys' delivery to secure application entities in classical communication network. Secure application entities can reside in various network domains within classical communication network. While QKD network domain and secure application entities' network domain can be managed and configured independently via its own SDN controller, a network operator can introduce multidomain SDN orchestrator for a single and integrated management for both network domains. Therefore, an SDN controller is deployed for a given network domain while the whole network system is orchestrated by an SDN orchestrator.

For the use case of QKD-derived keys' delivery to secure application entities in optical transport network (OTN), with the rich fiber environment in telecom companies and because of steep performance degradation of QKD over optical fiber length compared with the performance of OTN, network operators can choose the deployment of a QKD network with dark fiber to be separated from classical OTN to operate and manage each network without the performance degradation of each network.

With this separated deployment in the network configuration and each network's operation under the network operator's integrated network management, QKD network domain and OTN network domain need to be interconnected between two nodes which, respectively, belong to each network domain for QKD-derived keys' delivery to secure application entities in OTN.

Usually, this kind of interconnection plays the role of providing QKD-derived key delivery API between the key supplier, that is, QKD node and the key receiver, that is, secure application entity. Under this configuration, QKD does not guarantee that QKD-derived key will be used securely in secure application entity because its use in secure application entity is beyond the responsibility of QKD. However, in this case, as QKD node and secure application entity in OTN node belong to the same network operator's telecom network, the network operator can control both networks and guarantee the secure use of QKD-derived keys with telecom operators' security. For the security from key generation to the use of the key in telecom network, the address matching between QKD node and OTN node in two network domains needs to be resolved before key delivery under the network operator's management.

Therefore, the network operator needs to coordinate both QKD and OTN network domains and a multidomain SDN orchestrator is required for this reason. In this use case, SDN orchestrator can play the coordinating role with the information received from the SDN controller of QKD network and the information from the SDN controller of OTN, respectively. With this configuration, the network operator can ensure the secure end-to-end QKD service provisioning between QKD network and OTN.

For the SDN orchestrator to play the coordinating role between QKD and OTN network domains, the interface between an SDN orchestrator and an SDN controller of QKD network needs to be defined. This interface describes the flow of information between the SDN controller performing as a server and the SDN orchestrator operating as a client. Through this interface, SDN orchestrator can orchestrate QKD network in terms of discovery of QKD network topology, monitoring of QKD network status and resource inventory, end-to-end QKD service provisioning with path calculation in QKD network, management policy, performance management, as well as the address matching, as described above.

With the unique quantum features such as superposition, entanglement, teleportation, and other quantum properties, quantum communications and computing would enable more secure web applications. For example, QKD allows to secure distribution of security keys between a client and an HTTP server, and ultimately, with such QKD-enabled security keys, the HTTP client and HTTP server can establish more secure internet connections. In the meantime, many recent QKD protocols rely on entanglement, which is also required in quantum teleportation and quantum superdense coding.

Essentially, it is both critical and challenging to manage and control quantum networks efficiently due to the transience of quantum capabilities of a quantum node. Potential research issues include quantum capability management and quantum link services. Quantum capability refers to quantum resources (e.g., quantum memory, quantum channel, and entangled qubits) and quantum operation abilities (e.g., entanglement creation, entanglement swapping, and entanglement distillation) at a quantum node such as a quantum repeater. To enable a quantum network, the first demand is to efficiently manage quantum capabilities of all quantum nodes within the quantum network. A quantum network may consist of many quantum nodes, which are connected by quantum channels and classic channels. In addition, quantum capability could be dynamically changing, for example, due to qubit memory time and entanglement coherence time. This poses challenges for efficiently managing quantum capabilities of a quantum network. For example, available entangled qubits with required fidelity at each quantum node change rapidly (at the second or even millisecond level), an immediate research issue is how to efficiently create entangled qubits (on demand or proactively) to meet application requirements and how to accurately know the up-to-date list of available entangled qubits at every quantum node to schedule precise entanglement distribution.

The quantum link layer should provide flexible and complete services to a higher layer. Ideally, link layer services should be designed in the way that the higher-layer quantum applications can easily access and manage the quantum link layer in an efficient way considering unique characteristics and constraints of qubits, quantum channel, entanglement, and so on. For example, the higher layer may need to request the creation of entangled qubits in advance, cancel a pending request for creating entangled qubits, query the fidelity of any created entangled qubits, and/or request to consume entangled qubits (e.g., for entanglement swapping).

Quantum Cloud Computing

The concept of quantum-as-a-service (QaaS) is a cloud service that provides customers with access to QC platforms over the internet. QaaS uses the software-as-a-service (SaaS) delivery model. QaaS is still in its infancy. In the context of 6G, such quantum measurement as a service could be hosted within the 6G core network (CN), by central units of the 6G RAN, and/or even by powerful 6G devices such as quantum-capable drones and vehicles.

Systems with Entangle Distribution

Entanglement distribution is a crucial phase for the construction of the entangled CN structure of the quantum internet. In a quantum internet scenario, quantum entanglement is a preliminary condition of quantum networking protocols. As described earlier, a quantum network is to establish two (or more) entangled qubits between any two (or more) quantum nodes so that such entangled qubits can be consumed to enable quantum applications such as QKD, quantum teleportation, and quantum superdense coding, and eventually support distributed quantum applications. In the quantum network, the existence of entangled qubits which are distributed and maintained among multiple quantum nodes are referred to as quantum connection. As such, to establish a quantum connection means to distribute entangled qubits to two or more quantum nodes, referred to as entanglement distribution. Entanglement distribution heavily relies on entanglement swapping techniques to extend quantum connection to distant quantum nodes across multiple hops.

However, current quantum physics and devices set limitations on entanglement generate rate (not high) and entanglement coherence time (short). This makes it challenging to efficiently establish quantum connections. As a result, quantum connection management is even harder, but it is one of the most essential components to enable new distributed quantum applications. Most existing entanglement distribution solutions are designed for generic network topologies. It remains a question if they are applicable or efficient for 6G systems. There are a few specific scenarios in 6G systems where entanglement distribution is needed and may require redesign.

6G systems will include satellite communications network (SCN). An SCN will likely be a multihop mesh network consisting of thousands of satellites with free space optics as both classical channels and quantum channels. Entanglement distribution is required to establish quantum connection between two or more distant satellites. The issue is how to efficiently generate entangled qubits and distribute them (e.g., using entanglement swapping) to any pair of two distant satellites considering multiple factors such as entanglement coherence time, entanglement fidelity, quantum memory size, and entanglement consumption rate. SCNs in 6G systems will not be an island but will be interconnected with ground networks to form space air ground integrated networks (SAGINs). Entanglement distribution will not only consider satellites in space but also quantum nodes on the ground such as ground stations and vehicles. This introduces new variables and new complexity, which need new entanglement distribution solutions. In addition, with the introduction of O-RAN, 6G systems will consist of multiple hops or segments (i.e., fronthaul, middle haul, and backhaul). Nodes in each hop will have different quantum capabilities and different quantum channels. How to efficiently distribute entangled qubits among them remains unsolved.

Hybrid Classical-Quantum Network Evolution

We have described earlier that it would be a combination of both classical and quantum network evolution of hybrid classical-quantum networks. QIS will not fully replace classical ICT including the classical internet. Instead, classical ICT and QIS will complement each other and evolve together to enable hybrid classical-quantum systems with better performance. For example, QKD and PQC are likely to coexist as cryptography options satisfying different security requirements. Furthermore, a hybrid PQC-QKD will have hierarchical and better security solution for data centers.

In fact, even QKD itself is a hybrid protocol in the sense that it requires both quantum channels and classical channels; CV-QKD and DV-QKD can also be jointed used as a hybrid approach with better performance. When deploying QKD protocols in ICT backbone networks, a hybrid solution is to use existing optical fibers to support both classical transmissions and QKD quantum photon transmissions, for example, over different wavelengths. In another example, satellite QKD and ground QKD may be jointed leveraged as a hybrid solution to potentially achieve better performance in terms of end-to-end distance and key rate.

From the 6G network computing perspective, classical computers will coexist with quantum computers as quantum computers are specialized for certain types of problems. In addition, quantum computers usually need classical computers or classical circuit for controlling the quantum unit. Hence, it is expected that classical computers and quantum computers will be combined to realize hybrid classical- QC algorithms to maximize the advantages of each technology. The quantum internet [2] will drive the classical internet to evolve toward a hybrid internet in a systematic evolution, which demands substantial improvements and adaptation from both technologies. For example, small quantum networks could be deployed in the cloud or toward edge networks to provide quantum resource and services, while classical internet will utilize these quantum resources and services to strengthen traditional applications or enable novel applications; on the other hand, classical internet also provides traditional communication capabilities, on which a control and management plane for these quantum networks can be based. A completely new resource in such a hybrid Internet is entanglement (or entangled qubits), which will empower many unprecedented applications in future hybrid internet in the future. In addition, gateway proxies could be designed between small quantum networks and classical internet.

SUMMARY

QIS can benefit and empower 6G, the next-generation wireless cellular system. We have described the basic principles of QC and communications: qubits, no cloning and deleting, superposition, entanglement, teleportation, quantum measurement and disturbance, quantum communications (teleportation protocol, superdense coding, quantum entanglement and key distribution, QSDC, and quantum secret sharing), QC, and QS. Finally, we have described QIS for 6G Networks. 6G networks need to be more efficient, secure, intelligent, and autonomous, to support emerging applications and services such as holographic and haptic communications. QIS itself has been envisioned as one of the critical enabling technologies to enable efficient, secure, and intelligent 6G networks. We have described how QC can be leveraged to solve challenging wireless resource optimization problems in 6G systems. QIS can also benefit and be jointly used with other 6G enablers such as AI and space air ground integrated networks. We expect QIS will play an important role in numerous 6G areas, including but not limited to quantum-assisted radio access networks, quantum non-terrestrial networks, quantum-assisted edge network, quantum- assisted data center, quantum-assisted blockchain, and quantum-assisted wireless AI. However, QIS is still evolving, and quantum-enabled 6G networks face many challenges.

REFERENCES

1. Wang, C. and Rahman, A. "Quantum-Enabled 6G Wireless Networks: Opportunities and Challenges," *IEEE Wireless Communications*, February 2022, 29, 58–69.
2. Quantum Internet, "From Long-distance Entanglement to Building a Nationwide Quantum Internet," DOE Report, February 2020.

26 6G Cybersecurity

OVERVIEW

Data processing, threat detection, traffic analysis, and data encryption are considered the most critical issues in sixth-generation (6G) networks. The security issues due to massive traffic processing can be solved using decentralized security systems, in which the traffic can be handled dynamically and locally. Multiple novel schemes have been considered to manage different kinds of threats to improve data privacy and communication security for 6G services, especially threats specific to computing. Secure channel coding, channel-based adaptation, artificial interfering signals, and secret sequence extraction are the principal physical layer techniques based on classical information theory. Based on the no-cloning theorem and entanglement nature in quantum physics, information can be securely transmitted with classical bits and channels using quantum key distribution and quantum teleportation. According to the trends of computing centric services and 6C convergence, cyberattacks will present another critical problem for 6G security. For detecting malicious activities, artificial intelligence/machine learning/deep learning (AI/ML/DL)-based attack prediction methods are promising potential solutions for these constantly changing environments.

DIGITAL SERVICES

Contrasting with the scenario centric services based on 5G and Beyond 5G (B5G) evolutions, the fifth 6G core service is focused on providing secure wireless computing in the digital world. With the development of data mining, cloud-edge computing, mobile caching, and AI/ML/DL, general computing has become an important service objective of wireless communications. The utilization of network function visualization and network slicing means that wireless systems must be scheduled more flexibly. Therefore, wireless computing is also essential for networks to self-optimize transceiver schemes globally. This kind of service applies to computing data nonlocally and transmitting results, without specific requirements to achieve concrete experiences in actual events and scenarios [1]. Thus, data rate, latency, and connectivity are not the basic requirements of computing and must be considered as a new service class with corresponding performance requirements in the 6G system.

Cyber and physical layer security issues within wireless networks is widespread in daily life. As a result, for wireless computing among individuals and communities, privacy leakage is a predominant concern caused by perpetual data uploading, caching, and transmitting. Providing wireless computing with trusted communications is a crucial objective. Security should therefore be considered as a basic performance requirement of wireless computing in 6G and is referred to as secure wireless computing for private data (SWCPD). By combining the evolutionary and revolutionary services detailed above, a comprehensive vision of 6G core services can be

DOI: 10.1201/9781003499480-26

identified. The aim of these services is not only to provide comfortable experiences to human and machines but also to guarantee security for general computing.

To intelligently support comfortable and secure 6G services, the following KPIs need to be supported: (a) common KPI, capacity, spectrum efficiency (SE), energy efficiency (EE), and intelligence, and (b) distinctive KPI, security, connectivity, latency, and data rate. Note that connectivity, latency, and data rate have the basic KPIs for traditional networks. However, 6G is required to support additional KPIs on the top of the traditional KPIs because of its AI/ML/DL-based intelligence services to have the best quality-of-experience (QoE). Assigning more bandwidth is a direct method to improve data rate. As the spectrum of present frequency band is increasingly scarce, developing new frequency bands higher than the extremely high frequency (EHF) band, such as submm-Wave, terahertz, and visible light, is a critical process.

As indicated above, KPIs include both distinctive KPIs for specific services and common KPIs for general services. Specifically, the exact units of KPIs are bit/s for data rate, milliseconds for latency, m^3 for connectivity range, per/m^3 for connectivity density, bit/s for capacity, bit/s per hertz (bit/s/Hz) for spectrum efficiency (SE), and bps per watt (bit/s/W) for EE. Multiple aspects need to be considered when analyzing security, including physical layer security, network information security, and DL-related security. Taking security capacity as an example, its unit is bit/s. For intelligence, the most used measurement is accuracies; thus, the unit is percentage.

CORE SERVICE INTEGRATION

Both computing and scenario centric services are needed for 6G services interactions. Scenario-centric services produce large amounts of data for wireless computing and analyzing, while computing-centric services feedback computed results to improve the performance of the scenario-centric services. AI is a common technology to improve the performance of both scenario and computing centric services. Therefore, scenario and computing centric services will also be considered as integrations for globally improving system performance and user experience.

LAYER INTEGRATION

It is necessary to improve individual service KPIs globally in addition to multiple-service integration. Based on the communication models with multiple layers, for example open system interconnection (OSI) transport protocol and internet engineering task force (IETF) transmission control protocol/internet protocol (TCP/IP) reference models, traditional schemes are always considered for optimizing KPIs on a single layer. Previous research into communications have also encountered this issue. As a result, it is necessary to integrate the layers and jointly optimize the KPIs for each service.

CHALLENGES AND POSSIBLE SOLUTIONS

The new proposals also present a series of additional problems. In this section, foreseeable challenges for 6G implementation are discussed and possible solutions are presented. Peak data rate: The 6G radio must be designed to utilize EHF to increase the peak data rate. It is also critical to build EHF mobile channel models and consider flexible spectrum management methods according to the various 6G scenarios. Another more serious concern is how to guarantee accurate beam arrivals in high-mobility scenarios. Signals on EHF bands experience large path losses, and the communication range is very short. Thus, a massive number of access points (APs) must be deployed, leading to extremely dense wireless networks (eDWNs). With the deployment of eDWNs, frequent handoffs will be another problem in mobile scenarios, along with the economic drawbacks of continuously deploying eDWNs over entire service areas.

Potential solutions to these issues are introduced as follows. First, AI/ML/DL can be employed as a powerful tool for designing hardware and software. The benefits of AI/ML/DL for managing resources intelligently have been witnessed during the communication trends of digitalization and visualization. Outdoor positioning technology with centimeter-level accuracy is also proposed for achieving accurate beam arrivals. In addition, macro-base station (macro-BS), micro-BS, and multi-AP-based cooperative transmissions are possible solutions to reduce the frequency of handoffs.

LATENCY AND RELIABILITY

Although cooperative processing can reduce computing latency, it raises the cost of communications for link negotiating and data sharing, increasing communication overheads and transmission latency. Similarly, current AI/ML/DL-based predictions rely on big data, which also introduces more overheads for data collection. Most DL schemes must train deep neural networks (DNNs) offline. Thus, if the channel state, network topology, and user policy change, the DNN must be retrained, leading to latency for online applications. Hierarchical execution of tasks is an excellent solution for reducing the overheads and latency caused by cooperation. Cross-layer optimization and end-to-end DL schemes are helpful to optimize the transceiver and propagation environment as a whole and control it with meta-surfaces for improving communication reliability.

CONNECTIVITY

There are numerous challenges involved in integrating networks with different media to increase the quality of connectivity, including hardware issues of device compatibility and software issues of system generality. Digitalization, visualization, and DL are possible answers. To increase the density of connectivity, the main challenge is to improve positioning accuracy, particularly in outdoor and mobile situations, as nearly 100 devices should be connected in the space of $1 \ m^3$. Advanced positioning technology with centimeter-level accuracy should thus be utilized for managing densely deployed devices.

System capacity: The challenges of enlarging capacity lie in optimizing link quality for A2G networks and taking advantage of interference power. Both DRL and intelligent meta-surface are proposed to address these issues as technologies based on intelligent policy and environment optimizations, respectively. Global power, subchannel, and beam optimization are also crucial, and can jointly control interference in 6G systems.

SPECTRAL EFFICIENCY

Dense reuse of the same frequency to enhance SE will introduce additional interference. In addition, super-Massive MIMO (sm-MIMO) has high demands for beam optimization, channel modeling, and transceiver design. Challenges for orbital angular momentum (OAM) include the limited number of available OAM modes, joint OAM-mode, and frequency/time partition and channel estimation for different OAM modes. A common and critical challenge for frequency reusing among users and systems lies in effective interference utilization. For the interference-related challenge, deep reinforcement learning (DRL), intelligent surface, and global resource optimization are also potential solutions.

ENERGY EFFICIENCY

The main challenges when applying the technology to raising energy efficiency (EE) lie in the hardware design. It is first necessary to improve harvesting efficiencies of wireless power transferring (WPT) and simultaneous WPT (SWIPT). With the utilization of EHF, new hardware must then be developed for energy harvesting. It is also difficult to design feasible antenna architectures for complicated environments, leading to a waste of transmitting power. To deal with the hardware designing difficulties, AI/ML/DL, function digitalization, and reconfigurable meta-materials have recently been proposed as solutions for significantly improving EE.

NETWORK INTELLIGENCE

The current DNNs are almost black box and heavily rely on training datasets. Developed DNN models for wireless communication are also limited and are mainly borrowed from the fields of computer vision. As a result, AI/ML/DL-based wireless communication networks only focus on simple tasks with a single objective. Most schemes also require offline training with a large dataset and lack of generalization ability, limiting the capabilities of instant deployment and dynamic self-adaption. The existing intelligent communication research works have only aimed at optimization of the performance of 5G and B5G schemes. For 6G, the systems and schemes must be originally produced by cognition, instead of optimization.

Additionally, existing AI/ML/DL and DRL-based methods rely heavily on formulated datasets and fixed spaces of states and policies and cannot operate in unpredictable situations. Current DL methods also only teach agents the best policies under different determined conditions; thus, it is difficult to carry out trade-offs among various performance optimizations in dynamic and complicated 6G environments. To solve the explainability and generalization problems, models based on AI/ML/DL schemes, such as deep unfolding neural networks (NNs), are potential choices, and should be developed based on strict mathematical deductions. Model-driven DNNs, such as deep unfolding NNs, also provide better explainability for utilization, while transfer learning can improve the generalization capability. In addition, DL cooperating with analysis of game theory could possibly achieve trade-off policies among various performance requirements for multiple agents.

NEW APPROACHES TO SECURING USER DATA

Numerous issues remain in classical physical layer security, AI/ML/DL-related cyber security, and quantum-based communication security. The necessity of jointly considering the physical layer and cyber security with quantum methods for 6G services is emphasized here, however, particularly for those scenarios related to computing and AI/ML/DL. With the popularity and necessity of data collecting for learning in computing centric services, designing a joint secure framework will be the main challenge for promoting total security of the 6G system. Federated learning is a possible solution based on distributed learning and transferring learning, meaning that all user data is saved locally, and only abstracted models will be shared. Endogenous security has also been put forward to assist systems to establish protection strategies proactively and automatically through joint consideration of physical layer-cyber securities and classical-quantum channels.

6G cybersecurity systems do away with once-ubiquitous security tactics. Passwords might be eliminated for 6G security. The cyber environment will be certificate-based and encrypted. This means that 6G cybersecurity systems will verify whether users are authorized to access a given software, for example. 6G will also benefit from novel existing security approaches. With micro-segmentation, for example, systems will be able to isolate communications and create a virtual 'bubble' to increase security. These new approaches to cybersecurity solutions will enable the robust zero-trust network architecture required for the planned 6G cybersecurity infrastructure.

Zero-trust architecture assumes that no entity is implicitly safe and that the network cannot trust anyone unless they have the appropriate credentials. This architecture promises to make communication and data access points far more difficult to breach 6G is not only protecting just the service, but also the users' data or a business' critical trade secrets. With zero-trust, micro-segmentation, and tight security controls combined, you can really protect your data.

Data goes far beyond the daily web browsing and financial data that is most thought of in the context of cybersecurity and privacy. The past five years have seen an explosion in the number of connected devices across verticals and sectors where cybersecurity plays a literal life-and-death role. The internet pf medical things (IoMT), for example, helps care teams to deliver more tailored care and is improving medical care and patient outcomes. This data is equally valuable to criminals who would hold this sensitive data hostage and release it for a price.

The deployment of 5G has improved cybersecurity but the exponential increase in devices and data that 6G promises will require robust real-time cybersecurity responses. Researchers at Keysight are investigating new cybersecurity testing techniques that leverage digital twin technology to flag potential threats and take corrective measures in real-time. This process takes weeks under today's 5G network, but 6G is anticipated to cut this to a few hours, resulting in a more resilient cybersecurity infrastructure for device users and manufacturers.

CYBERSECURITY COMPLEXITY WITH AI/ML/DL

AI, ML, and DL are critical components of 6G, and vital to training cybersecurity systems and algorithms. They also offer additional layers of complexity that create more robust cybersecurity systems. However, as AI becomes more pervasive, so too do the number of bad actors with the skills and incentive to exploit the technology for nefarious purposes. Conquering the vulnerabilities within AI and ML training algorithms is critical to building a resilient, scalable, and secure 6G-powered future.

Cyberattacks can manipulate ML models during training or testing times and will undermine the AI's predictions. Just as worrying, these attacks can reverse engineer the algorithm to extract proprietary information. A second threat that researchers must account for is the creation of models built for nefarious purposes, either to commit cybercrimes or with military or law enforcement context. An AI algorithm's quality is determined by its reliability, accuracy, and consistency. Tactics that undermine any of these have cascading repercussions for any future developments that rely on AI. Thus, accelerating innovation across the spectrum of emerging technologies is dependent on ensuring that AI algorithms are trained to identify and block adversarial events.

For 6G cybersecurity, the long-term goal is an autonomous, self-preserving network that can respond independently to potential threats without causing disruption to normal use. While these cyber-resilient networks are still on the horizon, researchers are investigating adversarial ML approaches to train models to identify probable threats and determine an appropriate correlating response. This is a core responsibility of AI in the planned 6G cybersecurity architecture. New technologies always bring new threats to light which must be addressed alongside existing threats that accompany current technologies. The research focused on preventing cybercrimes particularly emphasize the development of 6G cybersecurity solutions that will scale and combat threats inherent in the growing multivendor marketplace. Already, 6G cybersecurity research is creating a paradigm shift in how we think about securing digital data today.

BLOCKCHAIN

It is evident that 6G will emerge as highly software-based and open networks allowing the participation of multiple stakeholders. This undoubtedly will make 6G more flexible, agile, autonomous, intelligent, and cost-efficient networks. However, the programmability and openness will make 6G networks more prone to issues like security, privacy, traceability, interoperability, auditability, resource manageability, spectrum efficiency, and 3D mobility. To address these issues, a deep integration of blockchain technology with 6G networks is foreseen. We have not addressed the BC technology in this book for the sake of brevity.

FUTURE RESEARCH AREAS

According to the enabling technologies and possible solutions, AI will significantly improve the system performance of 6G. Blockchain is another potentially useful method to manage the 6G systems flexibly and securely, while quantum computing will also provide increased computing efficiency. Furthermore, future developments in mathematics, biology, psychology, sociology, and energy and material sciences will also provide possible inspirations for progress in 6G. Future research trends such as applying meta-learning, lightweight NN, graph NN, capsule NN, and NN architecture searching to intelligent 6G research is also of significance. Based on blockchain, all available resources including spectrum and data can then be stored and shared through distributed blockchain transactions and protected by consensus protocols and cryptographic security, with no need to entrust any central party for ledger maintenance. Quantum computing can also be used to accelerate the speed of information processing, leading to more optimal solutions for 6G communications with shorter time. Furthermore, future research will not only rely on optimization algorithms, compressive sensing, and game theory for digital processing, but will also require biology, energy, and material science technologies to develop friendly devices for humans and the environment, aiming at health protection, convenient operation, service life extension, pollution prevention, and energy consumption reduction. Additionally, psychology and sociology are essential for improved understanding of individual user and societal demands within the 6G scenarios.

AUTONOMOUS VEHICLES

6G-based communications have many applications and are emerging as a new system to utilize existing vehicles and communication devices in autonomous vehicles (AVs). Electric vehicles and AVs not supporting the integration of intelligent cybersecurity will become vulnerable, and their internal functions, features, and devices providing services will be damaged [2]. Here we present an intelligent cybersecurity model integrating intelligent features according to the emerging 6G-based technology based on evolving cyberattacks. The model's novel design was developed using the necessary algorithms to provide quick and proactive decisions with intelligent cybersecurity based on 6G (IC6G) policies when AVs face cyberattacks. In this model, network security algorithms incorporating intelligent techniques are developed using applied cryptography. Money transaction handling services implemented in an AV are considered an example to determine the security and intelligence level depending on the IC6G policies. Intelligence, complexity, and EE are assessed. Finally, we conclude that the model results are effective for intelligently detecting and preventing cyberattacks on AVs.

PROBLEM STATEMENT

AVs have one of many compulsory services involved with money transactions for automatic charges when using AVs. Many users have reported the loss of millions of dollars after paying charges for bogus services while traveling. Hackers act as authorized persons and steal users' money, and, unfortunately, banks are unable to directly stop those transactions, as they still operate under the assumption of protecting deposited money from thieves, hackers, and physical violators. The problems this creates are many, and the services established by the service providers and the providers' policies create even more cybersecurity problems, as they inadvertently support hackers. Thus, these policies should be handled intelligently and according to the situation, location, time, and other major relevant factors.

In cyberattacks such as phishing, solutions with a 6G-based intelligent cybersecurity model can solve these problems intelligently and proactively. Scientists have developed many cybersecurity solutions for many illegal activities, but it is the policies that block personal interests and encourage hackers to get involved in illegal activities when they see the ease with which these transactions can be attacked. In AVs, the following policies are executed proactively when the system works intelligently. When these policies are handled intelligently and with political support, each transaction can be secured. The policies enacted should protect both users and service providers from the vulnerabilities created by the communication devices used in AVs. Further, these policies should encourage service providers to make the necessary decisions proactively.

Intelligent sensors placed peripherally around the AV are in direct contact with the AV's electronic devices, including the communicating transmitters and receivers. An intelligent cybersecurity model detects the vulnerabilities of these devices when they face cyberattacks and threats. The total energy consumption E_d is a function of several transceiver variables, with the most important variable being distance d and is summarized as:

$$E_d = E_{sd} + \eta \omega d^n$$

In in the above equation, E_{sd} is the distance-independent term that accounts for the overhead of radio electronics and digital processing. $\eta w d^n$ is the distance-dependent term, where η stands for the amplifier inefficiency factor, ω is the free-space path loss, d is the distance, and n is the environmental factor. n can be set as a number between 2 and 4 depending on the condition of the environment and the vulnerability of devices and communication channels; η determines the inefficiency of the transmitter when producing maximum power ωd^n at the antenna. Energy (E) is equal to the multiplication of power (P) and time (t). $E_0 = E_i - E_d$.

$$E_e = (E_0 / E_i) 100\%$$

E_i is the input energy, E_o is the output energy, and E_e is the ratio of output and input energy in percentage. The sum capacity E_C is proportional to E_{sd}; we can also assume $E_{sd} = E_C$ because the energy during secure and insecure communication is different due to many factors and influences, and the total capacity of 6G system is provided by:

$$E_C = \sum_{k=1} \sum_{i=1} B_{ki} \log_2 \left[1 + P_{ki} / (N_{ki} + I_{ki}) \right]$$

E_C is the total capacity of 6G technologies, k is the kth network coverage, B is the bandwidth, N is the losses due to noise, I losses due to interference, i is the i^{th}channel, and P power. Intelligent cybersecurity depends on these parameters. However, intelligence related to security policies, conditions of operations, and types of attacks also requires additional processing power. That is, overhead/processing power increases with the increase in intelligence. When accounting for all vulnerabilities, the overhead increases with the level of intelligence, which is dependent on the operation of services, which in turn influences the policies that service providers set. The parameters considered for determining vulnerabilities are proportionally equal to energy consumption, as given in E_d. The parameters given in the above equation are dependent on the policies of technical and operational limits which affect the sum capacity (E_C) and energy consumption (E_{sd}) of devices used in AVs.

PROPOSED MODEL

The experimental setup and actual parameters for each AV should be considered in each result. Generally, security limits (high, medium, and low) should be set either by the experts or the intelligent approach of the systems designed by the experts. In other words, the service providers advised by these experts must provide the necessary security solutions that would allow us to update the IC6G approach considered in the AVs. In this research, an autonomous service is considered an example of a feature integrated within the proposed model. The proposed method used the model developed in this study, as shown in Figure 26.1.

In this method, intelligent cybersecurity is considered, using intelligent features and the intelligent cybersecurity based on 6G (IC6G) policies. The proactive AV features and IC6G-based policies considered in the proposed model were implemented in the novel design of this method. Intelligence-based policies are created from available or collected data related to intelligence dependent services. In this study, we collected data from service users who were influenced by cyberattacks. In the 6G-based intelligent cybersecurity solution, network security algorithms incorporating intelligent techniques developed from applied cryptography were used.

All cybersecurity policies that allow service providers to secure their services is considered here, where the results will be focused on the reflection of those policies. 6G-based intelligent cybersecurity solutions (Figure 26.1) depend on the policy and conditions of the parameters used in expression of E_c. To secure a user's identity or personal information, a remote procedure call (RPC) can be used to secure remote

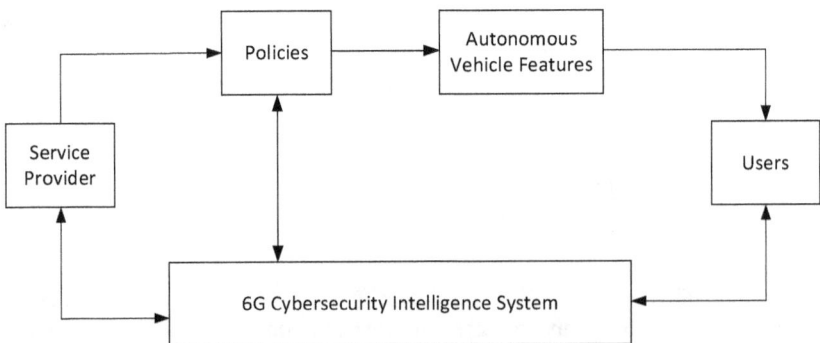

FIGURE 26.1 Proposed model for intelligent cybersecurity for autonomous vehicle.

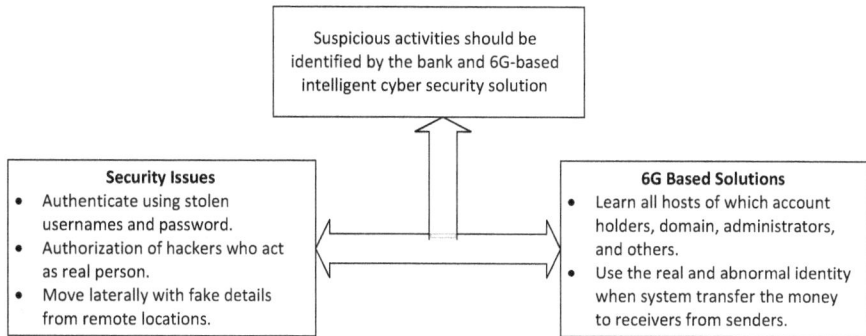

FIGURE 26.2 Security issues and 6G-based intelligent cybersecurity solutions.

procedures with an authentication technique. The host and the user who is requesting a service are both authenticated through the Diffie-Hellman authentication technique. Data encryption standard (DES) encryption is used by that authentication mechanism (Figure 26.2).

Here is a scenario: Travelers can use AVs for short visits or other such journeys. After a long day, the user or the traveler is tired and sleeps during the journey. When they finally arrive at home, they receive a call from a visa office regarding identity verification of a visa they had applied for. Tired, they take the call, not realizing that it is not genuine, and answer "Yes" to their questions, after which they go back to sleep. This was, in fact, a call by a hacker. The next morning, they wake up to messages from the bank, and upon checking their bank account, find that their money has been stolen by the hacker. According to the messages from the bank, 18 transactions happened during that night from that single "Yes." In this situation, what are the bank's and account holders' responsibilities?

The bank should have contacted the client personally and verified the situation. If their phone was switched off or if the bank was unable to contact the person during the night, the bank should have stopped all transactions; what happens instead is that the blame is directed solely at the account holder for having said "Yes." In this situation, the user/traveler/account holder could not have done anything because they were unaware and asleep.

Many hackers find opportunities to attack when users or passengers of AVs transfer or pay money from their accounts to real senders or vendors. Intelligent and automated networks supported by 6G-based communication technologies enhance the cybersecurity solutions during transactions established between the two authorized nodes (sender and receiver). Here, 6G-based intelligent cybersecurity solutions depend on the following questions, which simplify the transactions within AVs:

What type of AI-based cybersecurity algorithms does the proposed model use? How many AI-based cybersecurity algorithms does your 6G-based intelligent cybersecurity model have? How frequently do service providers (banks) update security policies, such as transactions limits? How long until AI-based cybersecurity algorithms can trigger detections in each 6G-based transaction? How many 6G-based intelligent algorithms require a learning period for normal and abnormal transactions? How does your transaction prioritize critical and high-risk hosts that require immediate attention from the service provider or bank? What is the complexity reduction that the proposed model provides for security analysts?

RESULTS

The experimental setup and actual parameters for each AV should be considered in each result. Generally, security limits (high, medium, and low) should be set either by the experts or the intelligent approach of the systems designed by the experts. In other words, the service providers advised by these experts must provide the necessary security solutions that would allow us to update the IC6G approach considered in the AVs. In this experiment, we collected data from 100 random users attacked by hackers from different banks. Table 26.1 lists the structure of the data used in this experiment. However, we have elaborated on the details of the data sizes, columns, and rows considered in this table.

Moreover, 70% of bank users are attacked a few times (less than 3% of the users within a fixed time) by hackers when the security limit is set to the low bank balance of the users. In addition, 20% of bank users are attacked several times (less than 17% of the users within a fixed time) by hackers when the security limit is set to the medium bank balance of the users. Finally, 10% of bank users are attacked more times (less than 50% of the users within a fixed time) by hackers when the security limit is set to the high bank balance of the users. To improve the results, six random places where international banks are located were chosen when AV is moving. The average percentage of all three security limits when hackers' activities are involved is recorded in Table 26.1.

The different security limits are sometimes set according to a user's earnings and preference and are set by the users. In many places, it is set up by the banks or systems authorized by expert service providers. Within the current system of bank transactions for paying expenses and services, clues were left that indicated they were hacked. In these studies, people who kept their withdrawal limit low never lost their money but were still attacked in multiple ways. The people with a medium limit had mixed attacks (2% lost the money, 15% were attacked, but did not lose money) in public locations, where they were most probably targeted by expert hackers who were sacked from public organizations. People with high limits were also attacked by hackers; in those cases, a high limit was set by the service providers without the users' official authorization.

TABLE 26.1
Hackers' Activities Against Autonomous Vehicle Users Who were Attacked

	User 1 (%)	User 2 (%)	User 3 (%)	User 4 (%)	User 5 (%)	User 6 (%)
Low (70 users)	2	2.5	2.4	1.7	2	1.2
Medium (20 users)	15	10	11	17	9	14
High (10 users)	31	43	49	27	34	42

Vulnerabilities based on Energies (E_d) of Cyberattacks against different cyberattacks based on security limits shows the different security limits when AVs face cyberattacks, or threats as follows.

1. **High Limit**: The threats encountered by the high limit tend to damage the configurations of the communication services, which include services such as transferring cash for users' expenses. This specific feature, integrated as AV onboard diagnostics (OBD), sends a warning when a high limit is set. The limits may be set by the bank or users or autonomous system, but they must be set intelligently and recorded with maximum evidence or verifications and/or mutual understanding of users. These recorded verifications must be kept for at least a few weeks for minimizing illegal transactions. When we use the IC6G approach in AVs, users get the correct information on verification procedures through the OBD.
2. **Medium Limit**: These threats weaken and slow down the communication services of AVs. In all communication services, both users and service providers should be alert during the number of continuous transactions.
3. **Low Limit**: Selected threats, such as cyberbullying, may be extracted from the profiles of users because the transaction is set to a low limit. It is the users' responsibility. The policies of the devices used in an AV will change the vulnerabilities and secrecy rate of the services, respectively. Using IC6G, the overall security facilities of an AV can be better maintained dynamically and proactively.

All policies set for improving cybersecurity solutions need to be reviewed according to the users' financial circumstances. The service providers' responsibilities should be to support all depositors who expect protection and security above other facilities. The parameters considered for determining vulnerabilities are proportionally equal to energy consumption, as given in E_d. The parameters given in E_C are dependent on the policies of technical and operational limits which affect the sum capacity (E_C) and energy consumption (E_{sd}) of devices used in AVs.

The results of this research depend on the policies written by experts and expert systems intelligently. The management of financial transactions by AVs is seen as an illustration of an intelligent cybersecurity solution based on 6G. The proposed model's cybersecurity solutions rely on the intelligence levels which would in turn influence policies, such as, five different services: (1) banking, (2) ticketing, (3) school fees, (4) hospital charges, and (5) parking payment.

It is assumed that all services are policy-dependent, and that these policies support the levels of intelligence considered in the solutions of intelligent cybersecurity integrated with AVs. Intelligence, security, complexity, energy efficiency, trustworthiness, scalability, and privacy were used in this study. The following explanations are provided below:

- **Intelligence**: Although the behavior of the same user is acceptable, intelligence can be noted from policies or keywords entered in the field of the service. Furthermore, intelligence analyzed against policies or keywords depends on the previous behaviors of users when the service is being used.

- **Security**: Strong policies increase the security of all services when cyber-attacks occur during mobile transactions. The automation of these policy generations will improve the security of services considered in AVs with some delays, which is the trade-off between policy and security.
- **Complexity**: The complexity increases when users expect maximum security because there is a trade-off between the cost of energy and security.
- **Energy efficiency**: Analyzing the enhancement of EE with the complexity and intelligence levels and the strength of the policies is a common technique for enhancing security.
- **Trustworthiness**: The reputation of the packet and its trustworthiness are evaluated based on one or more of the four verifications: data quality, location of service users, time of accessing services, and travel direction of the AV.
- **Scalability**: The use of sensors with intelligent cybersecurity increases when more service users and AVs are involved.
- **Privacy**: Policies will also enhance intelligent cybersecurity because some of the data used in automated and connected vehicles are personal and sensitive.

DISCUSSION AND ANALYSIS

Although appropriate cybersecurity solutions are assessed in this study, the following points are noted as having a substantial impact on the outcome results, as they provide zero or minimum cybercrime, which can result in loss of control of critical equipment used in AVs. Furthermore, cybercrime attacks the warning systems responsible for services integrated into AVs. In addition, they can cause damage to human health and the environment resulting from catastrophic spills, waste discharges, and air emissions. All results we obtained in this research depend on the limits set by the intelligent experts who provide the intelligent cybersecurity solutions to many sectors such as business.

Within the business sectors, the banking system is considered as an example or scenario in these results. Although many sectors and systems (medical, business, etc.) use the secure services through the intelligent cybersecurity, we have considered some selected services in this result. Intelligent cybersecurity solutions vary with the EE affected by the security limits and vulnerabilities Although five services are considered in the results, a specific service is to provide the necessary discussion and analysis. In some international banks and their services, the transferring procedures of the policies used in the system need to be investigated, as they are the real problem. A hacker can fool people and transfer millions of dollars ($) or Saudi riyals (SR) within a minute if the transferring policy in some international banks is not secured. For example, a hacker can act as a legal officer and ask for verification from a person who has paid visa fees from their account to an official account. The average person trusts third parties in many situations and circumstances to enact such payments. Intelligent experts and systems should have some procedures which depend on the policies, steps, and evidence collected from banks. To design and develop the intelligent procedure, the following evidence is collected from the bank:

1. The receiver's account details were not properly checked; the receiver can open the account and delete the account without references.

2. The senders' confirmation must be verified personally for securing the transactions.
3. The bank must have the proper verifications before sending the one-time password.
4. Account holders must trust the banks, but banks must not trust the receivers without proper verifications.
5. The bank should make sure that the receiver's account number is active for at least the last three months and valid for at least the next three months after the transactions.

Among the many services used in AVs, communication services are deployed for users who would like to communicate or exchange online transactions when they pay for their expenses during a journey. Users should be able to use the services (banking, ticketing, schooling (e.g., tuition and other fees for academic services) and others) comfortably and securely. In this discussion, five different services are considered, as previously mentioned: services 1, 2, 3, 4, and 5, banking, ticketing, school fees, hospital charge, and parking payment, respectively. When we deploy the IC6G approach in our proposed model, all five services are improved because policies are set up intelligently according to users' financial situation and transaction history. Whatever the situation, one of all three security limits should be selected and issued intelligently, instantly, and dynamically by the service provider. If the account holder's phone is switched off and the bank has allowed the hackers to transfer money, then the bank has allowed a fraud to occur (in fact, the bank should have waited until verbal confirmation from the account holder).

In this discussion, the evidence mentioned above should be considered carefully to improve security when transferring or withdrawing money from an account. In addition, we proposed a model with solutions using the IC6G-based policies to prevent cyberattacks and cybercrimes. The bogus services during movement, unintelligent behaviors, and the interruption of the handling services attacked by hackers are the problems discussed in this study. Intelligence levels were obtained from the policies concluded by the previous behaviors of the users of the services. We solved the research problem by analyzing intelligence levels with these policies.

CHALLENGES AND LIMITATIONS

The AVs with an IC6G will have many challenges which affect the users' daily life. The architecture we proposed in this research will present new opportunities for many potential systems and future applications:

- AVs' basic and luxury features will influence 6G-based gadgets.
- An introduction of cybersecurity solutions in 6G networks and related platforms used in AVs.
- An increase in intelligent features and proactive cybersecurity solutions.

The above points will spur research that support improving transportation policies. Brain-controlled vehicles (BCVs) may be introduced for simplifying the operations of the devices used in autonomous systems, including AVs. Further, the functions of 6G networks will make BCVs possible and will support IC6G in improving the intelligent features of the AVs. Regarding the cost of energy and intelligent cybersecurity, the most challenging aspect of cost and EE is determining the trade-off between five aspects:

- Evolution of AV technology
- Access to AV technology by stakeholders (communication service providers, road operators, automakers, AV consumers, repairers, and the general public)
- Limiting hackers' access to AV technology
- Widespread dynamic strategy for avoiding hacker amplification
- Efficient usage of AV operating logfiles

Tons of CO_2 emissions and millions of hours of driving every year will be saved with AVs, creating vulnerabilities in the communication devices used in those AVs. To solve these challenges, we need tough security policies that need to be applied intelligently. Our research model and approach provide a basic idea: intelligent cybersecurity with ML and AI algorithms should be considered to solve these problems. Intelligent cybersecurity with UAVs may offer some unique security challenges to 6G networks, especially regarding AVs used on land; it is possible that advances in UAV will lead to AVs getting low-cost energy and security.

The strength of this work lies in the IC6G policies, which should be the best for improving cybersecurity solutions because these policies are generated from the users' behaviors noted in each previous handling of the services. For instance, the limits and changes in bank transactions in banking services are noted to generate policies. On the other hand, the weakness and limitations of the research lie in the collection of previous behaviors for the last seven days to three months, which will increase the time complexity and storage, creating unnecessary delays when services are being used during the transactions. In addition, there are several other limitations regarding the cost of energy and intelligent cybersecurity for users and others: the collection of confidential data and generated policies depends on the behavior of the previous history of the services allocated in the Avs and regarding the importance of licensed details of the final official 6G release in relation to the IC6G-based policies.

CONCLUSION AND FUTURE WORK

This study presented the results from the proposed model that might be effective for intelligently detecting and thwarting cyberattacks on AVs and intelligent cybersecurity solutions that maintain secure services from all vulnerabilities created by attackers, faulty devices, or fake messages. Policies developed for AVs should enhance the protection of all users and communication devices integrated within the Avs. When securing service policies are maintained by intelligent experts, both users and service providers can secure services using a proactive approach. As the strength of the policies increases, the intelligence level also provides more intelligent cybersecurity solutions.

Therefore, the security limits discussed in the results should be set and fit by service providers based on the situation and important security factors, such as authentication. The main contribution of the proposed approach is intelligent cybersecurity solutions that provide the necessary security to all services used in AVs when cyberattacks occur. Furthermore, cyberattacks affect the electronic functions of AVs, which damage the AVs' operations and maneuvering of vehicle movements. The influence of intelligent cybersecurity not only solves the AVs safety issues of electronic control systems but also provides secure services to passengers using the AV.

Insights from this study are provided through the proposed model, which includes 6G-based cybersecurity solutions and policies. Intelligent cybersecurity is considered to maximize security and minimize energy costs for all passengers using autonomous and mobile services while traveling. The proposed solutions use IC6G-based policies to prevent cyberattacks and cybercrimes and intelligently enhance the effectiveness of cybersecurity solutions. Furthering the work of the proposed model, we can add more features and services to keep up with the emerging security technology if it is suitable for the situation and environmental conditions. Securing future services with intelligent cybersecurity in AVs will depend on emerging security technology (7G) and the strength of policies at the time. Once 7G fixes all its weaknesses, the capacity range and hand-off issues will no longer be a concern. Then the user will only be concerned with the cost of mobile phone calls, internet, and other services at that time. Furthermore, these features and services depend on energy-efficient algorithms and emerging technologies considered at the time. This research will continue to develop AVs with intelligent vision and 'human-like' thinking capabilities.

SUMMARY

Future 6G security issues were explored here, including core services, use cases, KPIs, enabling technologies, architectures, typical scenarios, existing challenges, possible solutions, opportunities, and research directions. Through opening new horizons for integration of comfort, security, and intelligence, the content of the chapter aims to provide inspiration for future research into 6G [1]. In addition, we have provided detail security consideration for AVs over 6G networks.

REFERENCES

1. Gui, G. et al., "6G: Opening New Horizons for Integration of Comfort, Security, and Intelligence," *IEEE Wireless Communications* October 2020, 5, 126–132.
2. Algarni, A. M. and Thayananthn, V. "Autonomous Vehicles With a 6G-Based Intelligent Cybersecurity Model," *IEEE Access* February 2023, 11, 15284–15296, Machine learning, deep learning and neural networks.

Index

Pages in *italics* refer to figures and pages in **bold** refer to tables.

For Product Safety Concerns and Information please contact our EU
representative GPSR@taylorandfrancis.com
Taylor & Francis Verlag GmbH, Kaufingerstraße 24, 80331 München, Germany

* 9 7 8 1 0 3 2 8 1 3 7 1 4 *